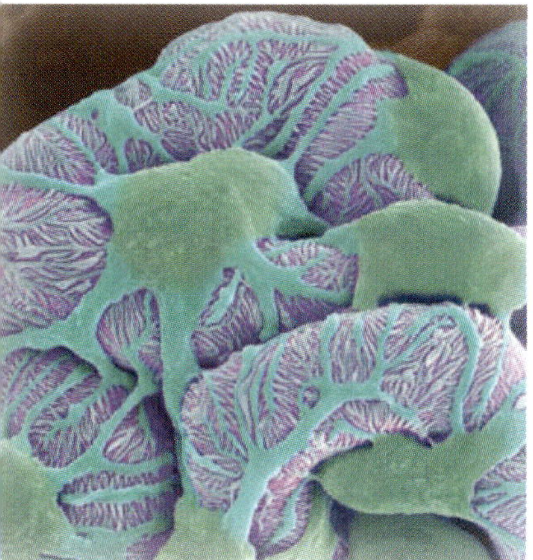

2020
The University of Akron
Jeff Spencer

Human Anatomy & Physiology

Laboratory 3100:203

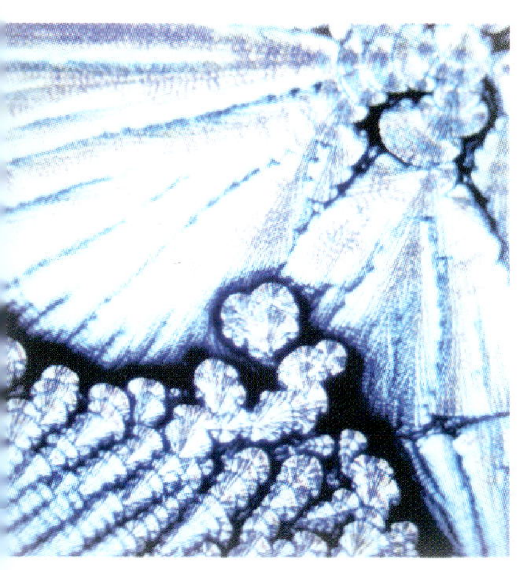

Human Anatomy & Physiology Laboratory

2020
The University of Akron
3100:203
Jeff Spencer

Photography of anatomical models: Ramsey Langford and Jeff Spencer.
Class activities developed by Luke Stetzik and Jeff Spencer.

Printed in the United States of America
10 9 8 7 6 5 4 3 2 1
ISBN: 978-1-61740-692-8

Van-Griner Learning
Cincinnati, Ohio
www.van-griner.com

President: Dreis Van Landuyt
Project Manager: Maria Walterbusch
Customer Care Lead: Lauren Houseworth

Spencer 692-8 F19
316239
Copyright © 2021

Table of Contents

Human Anatomy & Physiology Laboratory

Syllabus v

Lab Assignments viii

Chapter 1 — Endocrine *Hormone Production, Secretion, and Effects* 1-1

Chapter 2 — Special Senses *Gustation, Olfaction, Sight, Equilibrium, and Sound* 2-1

Chapter 3 — Perception *Touch, Smell, Hearing, Dynamic Equilibrium, and Vision* 3-1

Chapter 4 — Blood *Plasma and the Formed Elements* 4-1

Chapter 5 — The Heart *Cardiac Structures and Autorythmic Electrical Activity* 5-1

Chapter 6 — Blood Vessels *Transport and Exchange* 6-1

Chapter 7 — Lymphatic *Fluid Recovery, Lipid Absorption, and the Immune System* 7-1

Chapter 8 — Respiration *Airways and Gas Exchange* 8-1

Chapter 9 — Urinary *Kidneys, Renal Corpuscles, and Nephrons* 9-1

Chapter 10 — Digestion *Gastrointestinal Tract and Accessory Structures* 10-1

Chapter 11 — Reproduction *Female and Male Hormones and Reproductive Systems* 11-1

Chapter 12 — Development *Fertilization, Embryonic, and Fetal Development* 12-1

Acknowledgments A-1

Notes N-1

Prelab *Assignments for Chapters 1–12* P-1

Class Activity *Assignments for Chapters 1–12* CA-1

Quiz *Quizzes 1–6* Q-1

Midterm Practical Exam E-1

Final Practical Exam E-3

Flashcards

Syllabus 

Teaching Assistant:

Office Hours:

Required Text: **Human Anatomy & Physiology Laboratory 3100:203 Manual** by **Jeff Spencer**

The Human Anatomy and Physiology Laboratory utilizes models of various human anatomical structures, histology images, physiological instruments and preserved specimens to enhance your awareness of human form and function.

Learning Outcomes: After studying all materials and resources presented in the course, the student will be able to:
1. describe the human body's major systems.
2. use anatomical terminology to identify and describe various structures of the body.
3. demonstrate an understanding of physiology from cellular to organ system level.
4. describe the anatomical structures of the human body relative to systems, location, size and planes of the body.
5. be able to apply this knowledge to correctly spell answers to questions based on form and function.

Attendance: You are expected to attend all lab sessions, to arrive on time, and to remain until the lab assignment is completed. Missing class may lower your performance on exams and you will not receive credit for prelabs or in class activities. You are responsible for any announcements made in class or posted on Springboard.

Disabilities, Test Anxiety, and Other Issues: If you have a disability or think you have a disability (physical, learning, hearing, vision, psychiatric) or chronic health problems which may require accommodation, please contact the Office of Accessibility located in Simmons Hall, Room 105, 330-972-7928 as early as possible. If you arrange to take your exams through the Office of Accessibility, you must schedule 1 week prior of the scheduled test. You must also schedule your exam on the SAME day the class is taking the exam, and your exam time MUST OVERLAP our class time.

Class Behavior: Free discussion, inquiry, and expression are encouraged in this class. Classroom behavior that interferes with either (a) the instructor's ability to conduct the class or (b) the ability of students to benefit from the instruction is not acceptable. Students are expected to treat one another with the utmost respect. In order to minimize distractions, I ask that all electronics be turned off and PUT AWAY while you are in class (this means that cell phones, MP3 players, PDAs, etc. should be in your bag and NOT on your desk or in your lap). The only exception to this is if you are using a personal laptop to view or take class notes. No form of tobacco (cigarettes, e-cigarettes, chewing etc.) is allowed in class.

Academic Honesty: No form of cheating will be tolerated. Cheating is defined in the Student Academic Integrity Policy (www.uakron.edu/sja) and should be reviewed by each student. The consequences of violating this policy may result in zero points for an assignment, examination, or quiz and in severe cases the assignment of a failing grade may be given for the course, at the discretion of the instructor.

Tutors: You are encouraged to take advantage of free tutoring resources offered at UA. The Tutorial Center is located in the basement of Bierce Library, 330-972-6552. http://www.uakron.edu/tutoring/

January 26th is the last day to drop classes without "WD" appearing on your transcript. March 1st is the last day to process course withdrawals for Spring Semester 2019 (11:59 pm).

Grading: 60% of the laboratory grade is obtained through two practical examinations, 30% is achieved through six quizzes, and 10% from prelab exercises and class activities.

Grades are posted online at: **https://springboard.uakron.edu/**

Cell phones and smart watches are not permitted during quizzes and practical exams. A cell phone or smart watch seen during a quiz or practical exam will result in a zero. Students will not be permitted to make up any missed quiz, prelab exercise, class activity, or practical exam without a University approved excuse in writing. If you have any questions in regard to the *Human Anatomy & Physiology Laboratory,* contact Mr. Jeff Spencer in Auburn Science B220, 972-7996 or spence6@uakron.edu.

You are responsible for your grade in this class. You are responsible for reading and following this syllabus, attending lab, keeping track of course deadlines and policies, keeping track of your grades, knowing and avoiding the penalties for nonperformance, taking advantage of resources, studying, getting help if you need it, understanding how grades are assigned, and completing course work satisfactorily. No extra work will be available for extra credit.

Prelab exercises points (5%) are earned by completing the week's prelab exercises prior to lab and submitting upon arrival to lab. You will not receive credit if you arrive more than 10 minutes after lab begins.

Class activities points (5%) are earned by completing an activity during lab. Activities vary week to week and from lab section to lab section.

There will be six quizzes during the semester (5% each quiz for a total of 30%). Full credit for each answer is dependent upon correct spelling. Anatomical quiz questions will be based on the same images in your manual and physiology quiz questions from the text. You will submit each quiz on the pages provided near the end of the manual.

The midterm practical exam (30%) and the final practical exam (30%) will be held during normal lab day and time period. Each lab section will be divided into two groups for exams. The first group will begin the exam at class start time and will not be permitted to exit the room until one hour has passed. The second group will begin the exam one hour after class start time. Unique practical examinations will be given for each lab section. Dissected specimens, models, and images from your manual will comprise the practical examinations. The midterm and final practical examinations will be comprised of 50 questions from approximately 20 stations. Exams will be submitted on the pages provided near the end of the manual. Full credit for each answer is dependent upon correct spelling. Punctuality is essential. You will not be permitted to take an exam once the door is closed (i.e., be on time or you will receive a zero for the exam).

Student Performance Evaluation

Assignment	Percent of Lab Grade		Letter Grade	Percent of Total Points in Course
Prelab Exercises	5%		A	93% – 100%
Class Activities	5%		A-	90% – 92.9%
Quizzes	30%		B+	87% – 89.9%
Midterm Practical Exam	30%		B	83% – 86.9%
Final Practical Exam	30%		B-	80% – 82.9%
			C+	77% – 79.9%
			C	73% – 76.9%
			C-	70% – 72.9%
			D+	67% – 69.9%
			D	63% – 66.9%
			D-	60% – 62.9%
			F	Below 60%

Grading Rubric

Grading Rubric for A&P Lab Exams:

All information is based upon 100%.

No Points Off
One vowel added, missing or wrong.
- Examples: dermal papillaea, simple squamus, neutraphil

−¼%
Two or more vowels added, missing or incorrect order. One consonant added or missing. Not answering left or right or identifying left or right incorrectly.
- Examples: alvealer, xipoid process, maxila

−½%
The beginning or ending of the term is wrong but the root word is there. Two or more consonants added or missing. Not answering artery or vein or incorrectly identifying artery or vein. Partial answers that are correct but incomplete of entire answer.
- Examples: intercalalation disk, areolator connective tissue

No Points Given
If none of the above apply, the answer is wrong!
If a misspelling causes the word to change into a different structure, it is wrong.
- Example: coracoid vs. coronoid

Assignment	Lab Grade
Prelab Exercises	5%
Class Activities	5%
Quizzes (6, each worth 5%)	30%
Midterm Practical Exam	30%
Final Practical Exam	30%
Final Grade	100%

Calculated as Weighted Grade
Formula for Calculation of Final Grade (your test grade in number, e.g. 90% = 0.9)

Final Grade = Midterm × 0.3 + Final × 0.3 + Quiz 1 × 0.05 + Quiz 2 x 0.05 + Quiz 3 × 0.05 + Quiz 4 × 0.05 + Quiz 5 × 0.05 + Quiz 6 × 0.05 + 18 Prelab Exercises × 0.05 + 10 Class Activities × 0.05

- Example: $72 \times 0.3 + 80 \times 0.3 + 67 \times 0.05 + 75 \times 0.05 + 80 \times 0.05 + 78 \times 0.05 + 83 \times 0.05 + 80 \times 0.05 + 100 \times 0.05 + 90 \times 0.05 = 21.6 + 24 + 3.35 + 3.75 + 4 + 3.9 + 4.15 + 4 + 5 + 4.5 = 78.25 = C+$

Lab Assignments

Week	Date	Chapter	Laboratory Topics
1	1/13–16	1 2	The Endocrine System Special Senses Prelab 1 Completed during Class
2	1/21–23	 3 4	*MLK Day—No Labs Monday ONLY* Perception Blood Prelabs 2–4 Due upon Arrival, Class Activity 1 Completed during Class
3	1/27–30	5	The Heart Prelab 5 Due upon Arrival, Class Activity 2 Completed during Class
4	2/3–6	6 **1 & 2**	Blood Vessels Prelab 6 Due upon Arrival, Class Activity 3 Completed during Class **Quiz 1**
5	2/10–13	 **3 & 4**	Class Activity 4 & 5 Completed during Class **Quiz 2**
6	2/17–20		*Presidents' Day—No Labs This Week*
7	2/24–27	 **5 & 6**	Review Class Activity 6 Completed during Class **Quiz 3**
8	3/2–5		**Midterm Practical Examination** *Last Day to Withdraw 3/1*
9	3/9–12	7 8	The Lymphatic System The Respiratory System Prelabs 7 & 8 Due upon Arrival, Class Activity 7 & 8 Completed during Class
10	3/16–19	9	The Urinary System Prelab 9 Due upon Arrival, Class Activity 9 Completed during Class
11	3/23–29		*Spring Break—No Labs This Week*
12	3/30–4/2	10 **7 & 8**	The Digestive System Prelab 10 Due upon Arrival, Class Activity 10 Completed during Class **Quiz 4**
13	4/6–9	11 **9 & 10**	Female and Male Reproductive Systems Prelab 11 Due upon Arrival, Class Activity 11 Completed during Class **Quiz 5**
14	4/13–16	12 **11 & 12**	Development Prelab 12 Due upon Arrival, Class Activity 12 Completed during Class **Quiz 6**
15	4/20–23		Review
16	4/27–30		**Final Practical Examination**

Endocrine

Hormone Production, Secretion, and Effects

- **Hypothalamus (all of these affect the anterior pituitary)**
  <u>Hormone</u>
  Corticotropin releasing hormone (CRH)
  Gonadotropin releasing hormone (GnRH)
  Growth hormone releasing hormone (GHRH)
  Prolactin inhibiting hormone (PIH)
  Somatostatin
  Thyrotropin releasing hormone (TRH)

- **Adenohypophysis (anterior pituitary)**

<u>Hormone</u>	<u>Target organ</u>
Adrenocorticotropic hormone (ACTH)	Adrenal gland
Follicle stimulating hormone (FSH)	Ovaries or testes
Growth hormone (GH)	All cells
Luteinizing hormone (LH)	Ovaries or testes
Prolactin (PRL)	Mammary glands
Thyroid stimulating hormone (TSH)	Thyroid gland

- **Neurohypophysis (posterior pituitary)**
  Produced in hypothalamus

<u>Hormone</u>	<u>Target organ</u>
Antidiuretic hormone (ADH)	Kidneys
Oxytocin (OT)	Uterus, mammary glands

- **Thyroid gland's follicular cells**

<u>Hormone</u>	<u>Effects</u>
T_3 (triiodothyroxine)	Raises metabolic rate
T_4 (thyroxine)	Raises metabolic rate

- **Thyroid gland's parafollicular cells**

<u>Hormone</u>	<u>Effects</u>
Calcitonin	Bone deposition

- **Pancreatic islets**

<u>Hormone</u>	<u>Effects</u>
Glucagon (alpha cells)	Raises blood sugar
Insulin (beta cells)	Lowers blood sugar

- **Ovaries**

<u>Hormone</u>	<u>Effects</u>
Estradiol	Prepare reproductive system
Progesterone	Reproductive regulation

- **Testes**

<u>Hormone</u>	<u>Effects</u>
Testosterone	Maleness, growth libido

- **Adrenal cortex**

	<u>Category</u>	<u>Hormone</u>	<u>Effects</u>
Zona glomerulosa	Mineralocorticoid	Aldosterone	Water and Na retention
Zona fasciculata	Glucocorticoid	Cortisol	Tissue repair, anti-inflammatory
Zona reticularis	Sex steroids	Androgens	Secondary sexual characteristics
		Estrogens	

- **Adrenal medulla**

	<u>Category</u>	<u>Hormone</u>	<u>Effects</u>
	Catecholamine	Epinephrine	Increase metabolism and heart rate
		Norepinephrine	

The endocrine system is a collection of glands that help maintain homeostasis. The chemical messengers of the endocrine system are hormones that are secreted directly into the bloodstream and travel to all cells of the body. Only cells that have receptors for a particular hormone will experience an effect. The nervous system is intertwined with the endocrine system and shares a few molecules that influence other cells. For example, norepinephrine is released in a synapse in the nervous system as a neurotransmitter that binds to receptors of the postsynaptic cell. The same chemical, norepinephrine, is released by the adrenal medulla in times of stress directly into the blood stream and affects those cells in the body that have receptors. Hormone receptors can be part of the plasma membrane or inside of the cell. Water soluble hormones bind to receptors on the plasma membrane, turning enzymes on or off, altering the cell's metabolism. A second messenger that activates protein kinases is a common pathway of water soluble hormones. All hormones except those listed below are water soluble. Hormones that are lipid soluble enter target cells freely and bind to receptors in the cytoplasm or in the nucleus. These types of hormone/receptor complexes usually initiate transcription of specific proteins. In skeletal muscle actin, myosin, and other proteins are produced in response to testosterone. Another example is the thyroid hormone triiodotyronine (T_3), which binds to nuclear receptors and causes transcription/translation of Na^+/K^+ pumps that help raise body temperature. Estrogen, progesterone cortisol, and aldosterone are other lipid soluble.

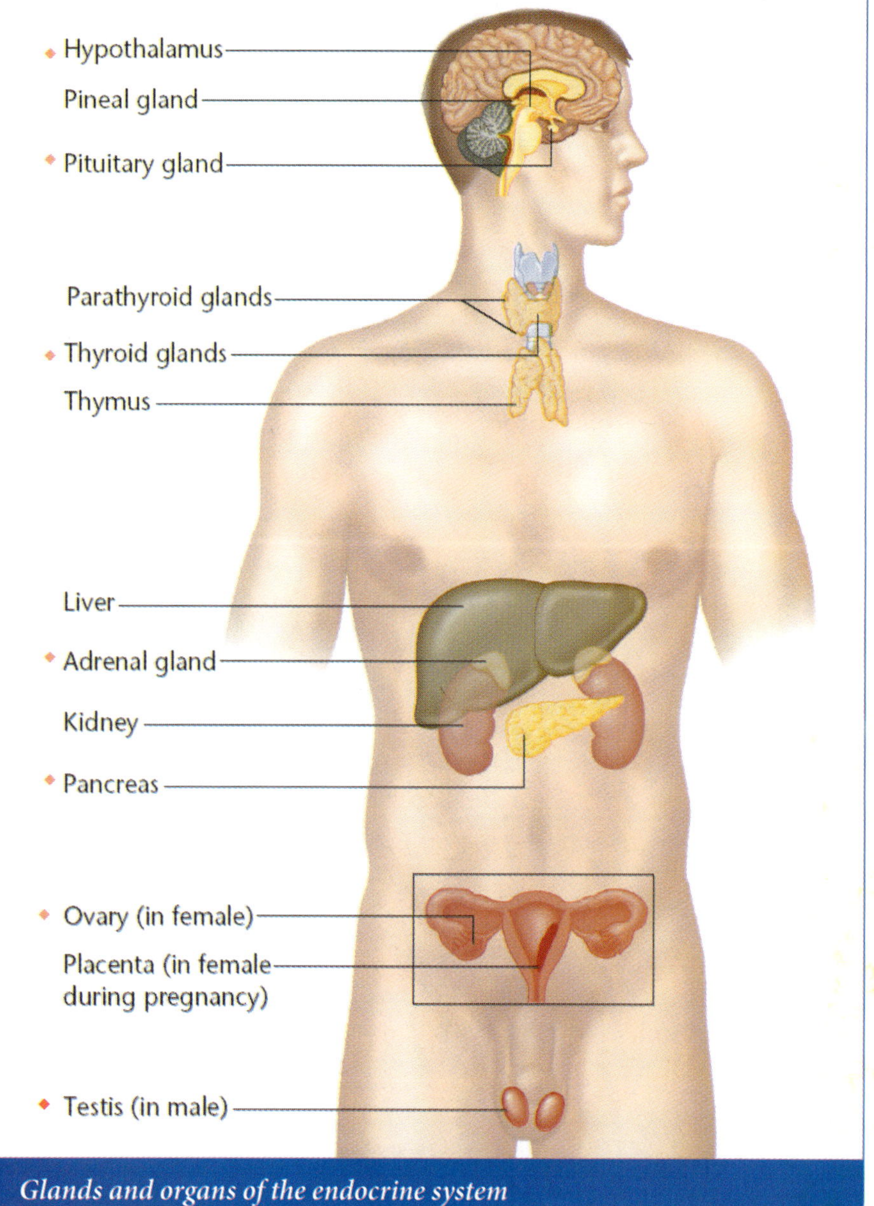

Glands and organs of the endocrine system

Most glands and organs of the endocrine system are under control of the hypothalamus and pituitary gland.

The nuclei (dense regions of neuron cell bodies) of the hypothalamus monitor many aspects of maintaining homeostasis. Just as a thermostat in your house keeps the temperature within a narrow range, the hypothalamus keeps body temperature, water and salt balance, sexual function, and growth within physiological limits. It does so by sending releasing hormones to the pituitary gland which stimulates the production of pituitary hormones that will affect their target gland. The target gland will secrete the corresponding hormone and as its levels rise in the blood stream, the hypothalamus is inhibited from producing and secreting its releasing hormone. This complicated series is an example of negative feedback, similar to the way a thermostat works in your house.

Hypothalamus (This is artificially highlighted to illustrate the region.)

Infundibulum (This is the stalk connecting hypothalamus to the pituitary. The hypophyseal portal system, a network of capillary beds in the hypothalamus and pituitary, is connected by venules traveling through the infundibulum.)

Pituitary gland (This model does not distinguish between the adenohypophysis and the neurohypophysis.)

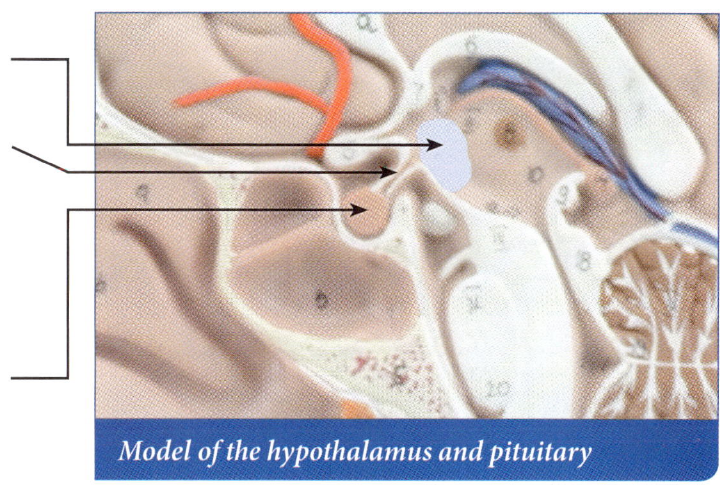

Model of the hypothalamus and pituitary

Hypothalamus

Hormone	Adenohypophysis (anterior pituitary) Hormone	Target organ
Corticotropin releasing hormone (CRH) →	Adrenocorticotropic hormone (ACTH) →	Adrenal gland
Gonadotropin releasing hormone (GnRH) →	Follicle stimulating hormone (FSH) →	Ovaries or testes
	Luteinizing hormone (LH) →	Ovaries or testes
Growth hormone releasing hormone (GHRH) →	Growth hormone (GH) →	All cells
Prolactin inhibiting hormone (PIH) →	Prolactin (PRL) →	Mammary glands
Thyrotropin releasing hormone (TRH) →	Thyroid stimulating hormone (TSH) →	Thyroid gland
Somatostatin →	Inhibits GH and TSH	

Neurohypophysis (posterior pituitary)

Hormone-stored and released here	Target organ
Antidiuretic hormone (ADH) →	Kidneys
Oxytocin (OT) →	Uterus, mammary glands

ADH and OT are produced in nuclei of the hypothalamus and are transported down their axons to the neurohypophysis for storage until released.

The pituitary gland is located in the center of the head. Although it is the smallest of the endocrine glands, it helps regulate all of the systems of the body. It is made of two distinct regions: the anterior (adenohypophysis) and the posterior (neurohypophysis). The anterior pituitary is epithelial tissue. It produces and secretes six types of hormones directly into the blood stream. The neurohypophysis is not glandular but rather nervous tissue. Both anterior and posterior are controlled by the hypothalamus, a region of nuclei (dense areas of neurons) above and connected to the pituitary via the infundibulum. The hypothalamus secretes releasing hormones that travel down the infundibulum (hypophyseal portal system) to the anterior pituitary, causing the secretion of one of six major types of hormones. The hypothalamus produces two hormones (OT & ADH) and transports them down axons to the neurohypophysis for storage until a signal for release occurs. The posterior pituitary does not make these hormones; it just stores and releases them. Diabetes insipidus is an uncommon disorder resulting in excessive thirst/drinking (polydipsia) and large urine output (polyuria). It is

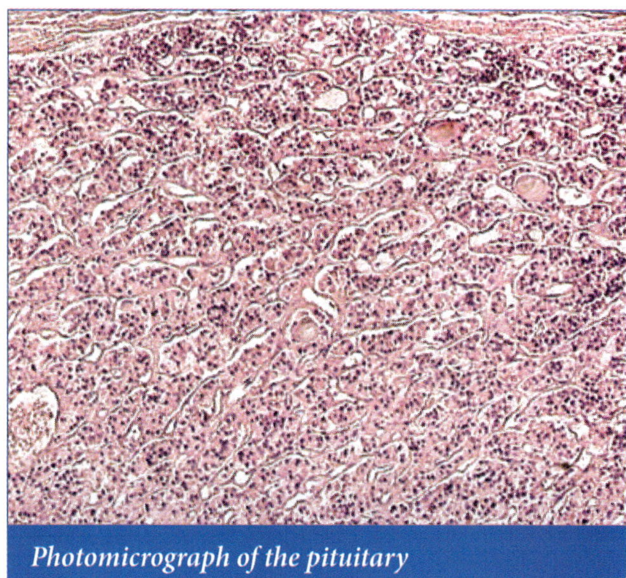

Photomicrograph of the pituitary

caused by inadaquate levels of ADH. This is not related to diabetes mellitus. The term diabetes refers to excessive urine output, polyuria. Notice all the white spaces (sinusoids) that allow hormones to easily enter the bloodstream in the photomicrograph above.

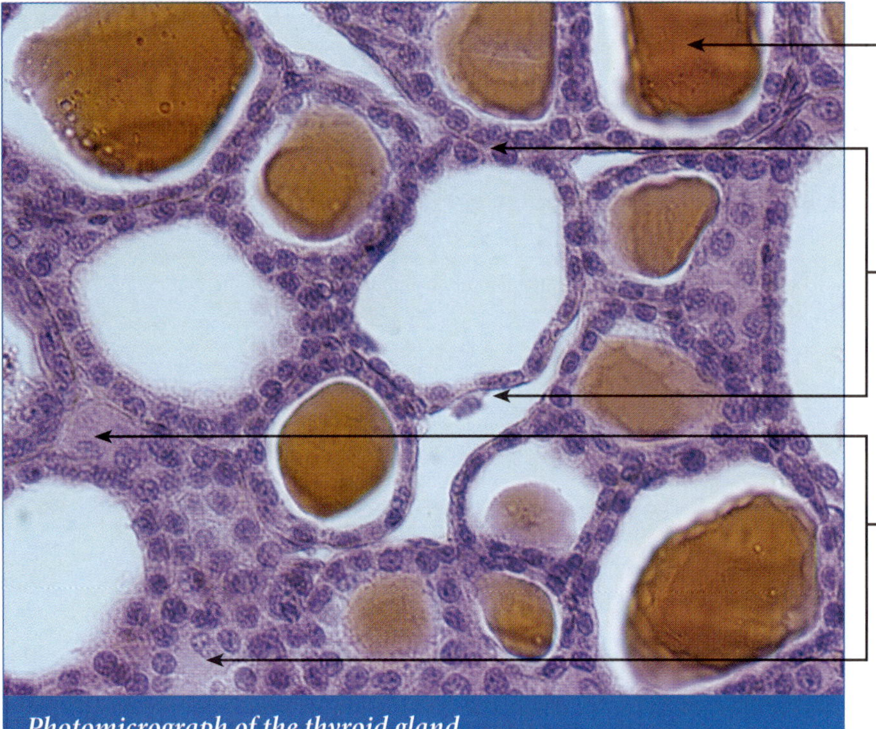

Photomicrograph of the thyroid gland

The thyroid gland is a bilobed gland that wraps around the anterior portion of the trachea and part of the thyroid cartilage of the larynx. Thyroid hormone is secreted in response to an increase in the body's metabolic demands (cold weather, for example). Our internal thermostat is located in the preoptic nucleus of the hypothalamus. When a drop in body temperature occurs, the hypothalamus secretes TRH that travels via the hypophyseal portal system to the pituitary, which secretes TSH. In response to thyroid stimulating hormone (TSH) from the anterior pituitary, the thyroid gland secretes T_3 (triiodothyroxine) and T_4 (thyroxine) which raises basal metabolic rate. Both T_3 and T_4 are formed by follicular cells and stored within follicles. They each contain iodine, and therefore it is important to have dietary sources (our table salt is iodized). T_4 is the form of over 95% of all circulating thyroid hormone, although it is converted into the active form, T_3, inside target cells. T_3 increases oxygen consumption, metabolism, and the formation of sodium/potassium pumps which increase body temperature. The thyroid gland's parafollicular cells produce calcitonin, which promotes calcium deposition (important in bone growth).

Thyroid gland's follicular cells

Hormone	Effects
T_3 (triiodothyroxine)	Raises metabolic rate
T_4 (thyroxine)	Raises metabolic rate

Thyroid gland's parafollicular cells

Hormone	Effects
Calcitonin	Bone deposition

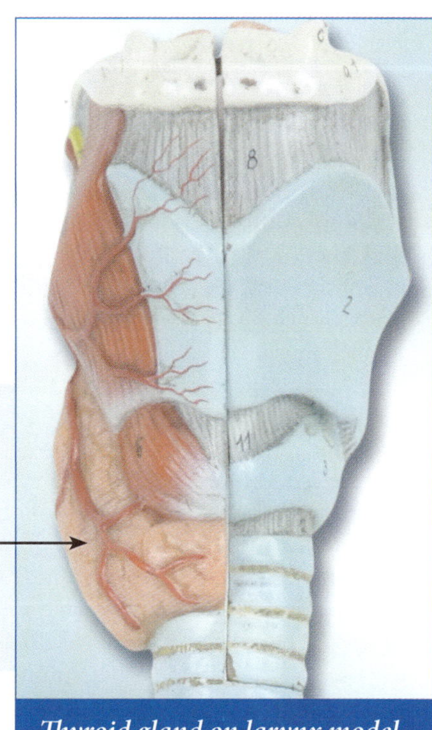

Thyroid gland

Thyroid gland on larynx model

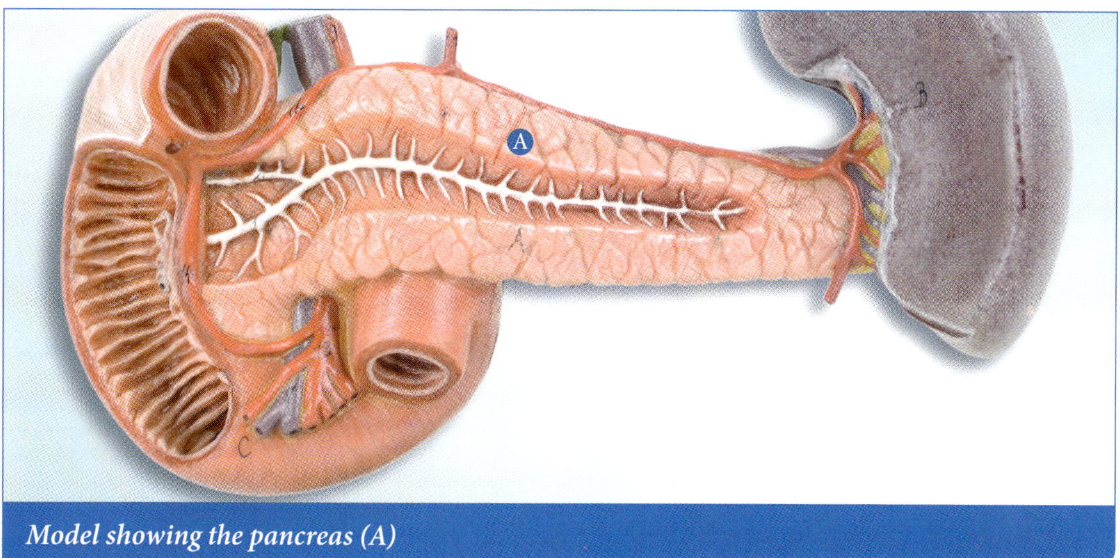

Model showing the pancreas (A)

The pancreas is a mixed organ of both exocrine (ducts) and endocrine (ductless) portions. About 98% of pancreatic cells produce digestive enzymes that empty via ducts into the small intestine, i.e., the exocrine portion. The remaining 2% are lightly staining (see below) groups of endocrine cells called pancreatic islets or islet of Langerhans. Most of these cells are alpha cells (produce glucagon that raises blood sugar by breaking down glycogen inside of cells and releasing glucose) and beta cells (produce insulin that lowers blood sugar by transporting glucose from bloodstream to inside cells; requires insulin receptor to be present on the cells). Diabetes mellitus, type I, occurs when beta cells fail to produce adequate concentrations of insulin. Diabetes mellitus, type II, accounts for about 90% of all diabetics and is not caused by insufficient levels of insulin, but rather a low density of insulin receptors, or a failure to bind insulin and transport glucose into the cell.

Pancreatic islets

Hormone	Effects
Glucagon (alpha cells)	Raises blood sugar by breaking down glycogen inside of cells, releasing glucose
Insulin (beta cells)	Lowers blood sugar by transporting glucose from bloodstream to inside cells

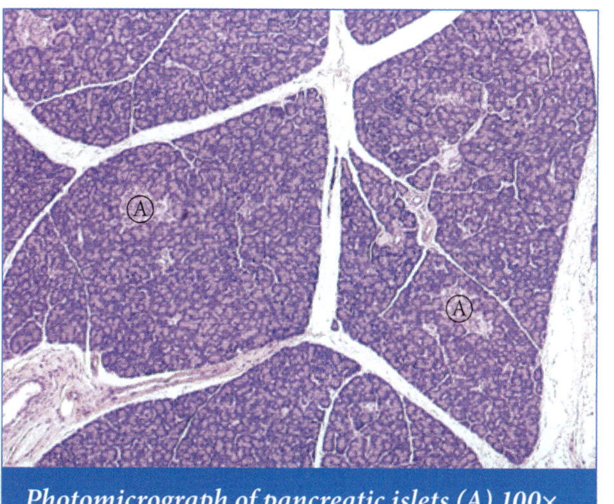

Photomicrograph of pancreatic islets (A) 100×

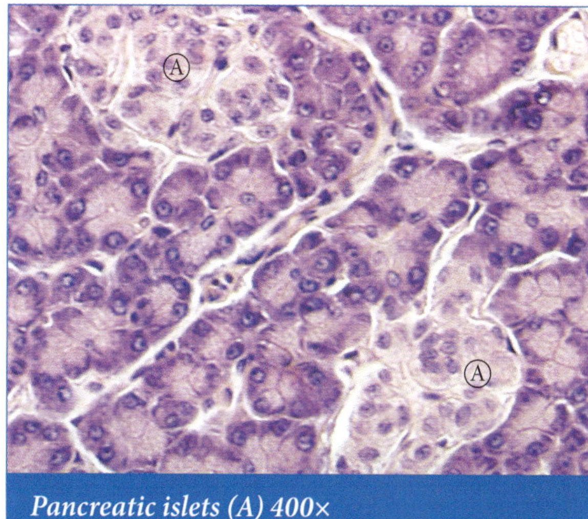

Pancreatic islets (A) 400×

Examine the two photomicrographs above. Notice the small, lightly staining patches of cells, the pancreatic islets. These are mostly alpha and beta cells that are responsible for controlling blood sugar.

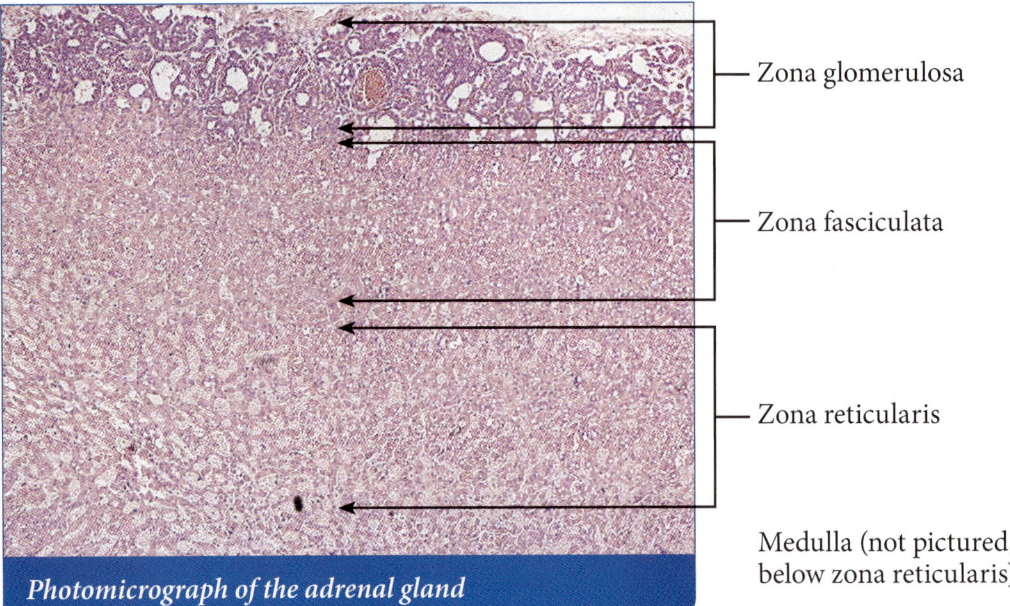

Photomicrograph of the adrenal gland

Adrenal cortex	Category	Hormone	Effects
Zona glomerulosa	Mineralocorticoid	Aldosterone	Water and Na^+ retention, K^+ excretion
Zona fasciculata	Glucocorticoid	Cortisol	Tissue repair, anti-inflammatory
Zona reticularis	Sex steroids	Androgens	Secondary sexual characteristics
		Estrogens	
Adrenal medulla	Category	Hormone	Effects
	Catecholamines	Epinephrine	Increase metabolism, heart rate, and breathing
		Norepinephrine	

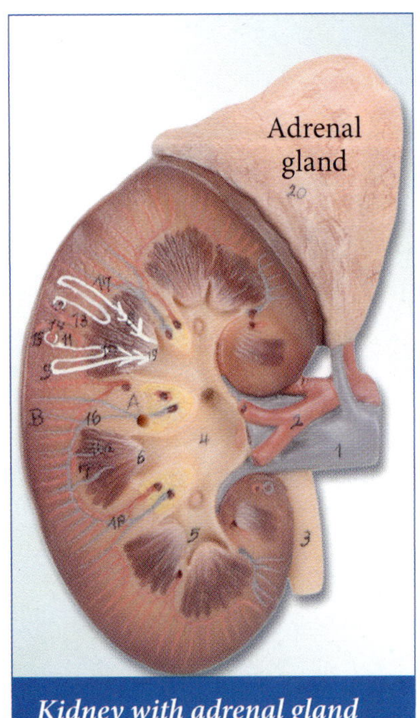

Kidney with adrenal gland

The adrenal gland rests on the superior portion of the kidney. It produces about 30 unique hormones, and we will look at six. There are two unique regions of different developmental backgrounds in the adrenal gland. The inner region, the medulla, is innervated by preganglionic sympathetic neurons that stimulate specialized postganglionic neurons (Chromaffin cells) to release the catacholamines epinephrine and norepinephrine. These catacholamines are used as neurotransmitters by most neurons, but the medulla's Chromaffin cells secrete them directly into the bloodstream, as all endocrine glands do. Since these neurons are under the control of the autonomic nervous system, the release and effects (increase heart rate, breathing, and overall metabolism) occur almost instantly during times of anxiety or excitement.

The outer region, and the majority of the adrenal gland, is the cortex. The cortex is divided into three layers. The outermost layer, the zona glomerulosa, produces mineralocorticoids which affect salt and water balance. Aldosterone is the chief hormone from this layer of the cortex and promotes water and Na^+ retention and K^+ excretion.

The zona fasciculata is the next and thickest layer. It produces glucocorticoids, mainly cortisol, in response to the pituitary hormone ACTH. Cortisol mobilizes glucose and fatty acids for repair and stress adaptation, and is anti-inflammatory (hydrocortisone creams).

The innermost layer of the cortex is the zona reticularis, which produces androgens and estrogens in both males and females.

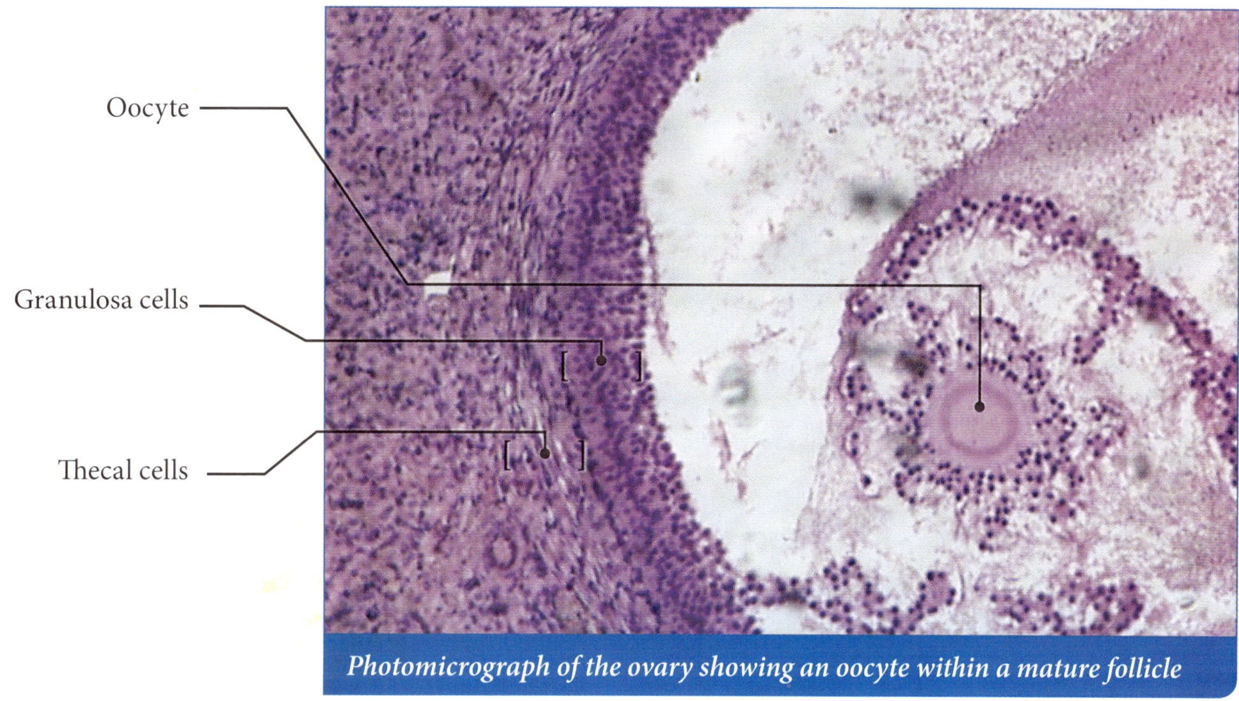

Photomicrograph of the ovary showing an oocyte within a mature follicle

In developing follicles within the ovary, granulosa and Thecal cells produce estradiol (an estrogen) in the preovulatory stage. After ovulation, the follicle transforms into the corpus luteum, which produces progesterone in the postovulatory stage and pregnancy. Both of these hormones contribute to the feminine physical and reproductive capacity.

Ovaries	Hormone	Effects
Granulosa cells	Estradiol	Prepare reproductive system
Thecal cells	Estradiol	Prepare reproductive system
Corpus luteum	Progesterone	Reproductive regulation

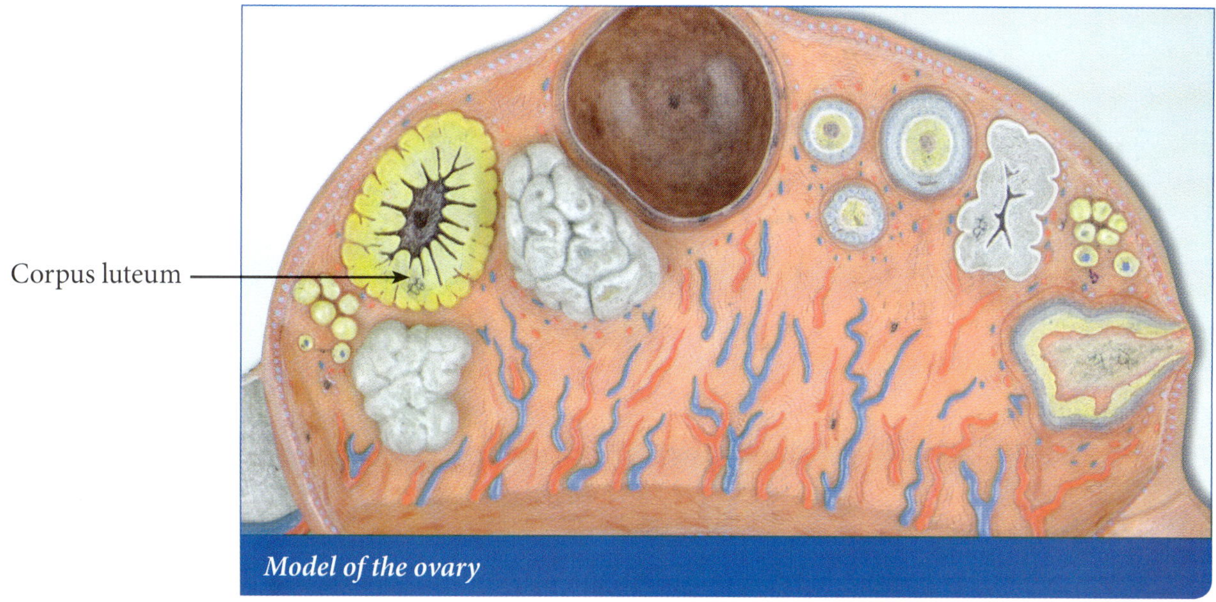

Model of the ovary

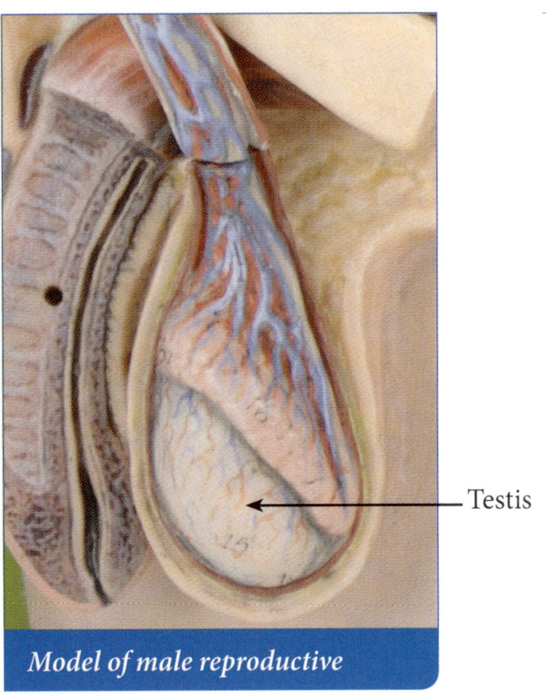

Testis

Model of male reproductive

The testis produces spermatozoa and androgens. This mixed exocrine/endocrine gland begins making sperm cells and testosterone at the onset of puberty (pituitary release of gonadotropins, FSH, and LH) and never stops. The majority of the testis consists of seminiferous tubules where spermatozoa are produced. Sperm production is stimulated by FSH. The hormone testosterone is produced by interstitial, or Leydig, cells. LH is the hormone that stimulates these cells to produce androgens, mainly testosterone. A dramatic example of the importance of hormone interaction with its receptor is the case of androgen insensitivity syndrome. This occurs in genetic male individuals with defective testosterone receptors, causing the individual to develop as female. As testosterone levels rise in the bloodstream, negative feedback in the hypothalamus reduces GnRH secretion which lowers FSH and LH, keeping testosterone levels in normal concentrations. Anabolic steroids used to gain muscle mass mimic testosterone in both muscle cells and the endocrine pathway. Anabolic steroids cause negative feedback inhibition of FSH and LH causing atrophy (shriveling) of the testis.

Interstitial cells of testis

Hormone	Effects
Testosterone	Maleness, growth, libido

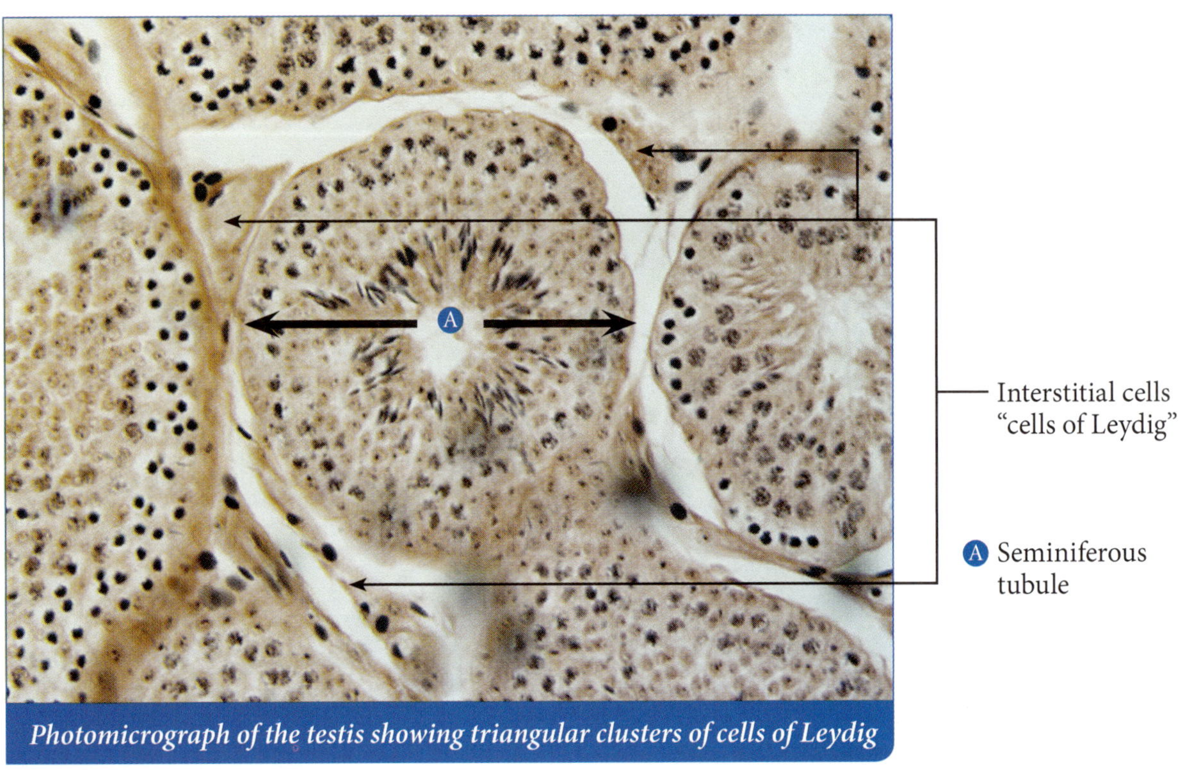

Photomicrograph of the testis showing triangular clusters of cells of Leydig

Interstitial cells "cells of Leydig"

Ⓐ Seminiferous tubule

Special Senses

Gustation, Olfaction, Sight, Equilibrium, and Sound

Our sensory receptors tell us everything we know about ourselves and the world around us. We examined general, or somatic, senses (touch, pressure, movement, heat, cold, and pain) last semester. The special senses (taste, smell, equilibrium, hearing, vision) provide us with details of our world and are composed of complex sense organs that are located in the head. All the information that makes you who you are enters your brain from the stimulation of sensory cells.

Gustation

- Taste buds – Contain gustatory (taste) cells
- Taste sensations
  - Bitter
  - Salty
  - Sour
  - Sweet
  - Umami
- Lingual papillae
- Taste pore

Olfaction

- Olfactory bulb (cranial nerve I)
- Olfactory receptor cells

Hearing

- Cochlea
- Vestibule
- Semicircular ducts
- Auditory ossicles
  - Stapes
  - Incus
  - Malleus
- Tympanic membrane
- Tensor tympani muscle
- Auditory tube
- Helix of the Auricle
- Lobule of the Auricle
- Auditory canal
- Cochlear nerve
- Vestibular nerve
- Scala vestibuli
- Cochlear duct
- Scala tympani
- Tectorial membrane
- Inner & outer hair cells
- Basilar membrane
- Vestibular membrane

Vision

- Superior rectus muscle
- Lateral rectus muscle
- Medial rectus muscle
- Superior oblique rectus muscle's tendon
- Sclera – Whites of the eyes
- Iris – Responsible for eye color
- Pupil – The opening that allows light to enter
- Choroid – Pigmented and vascular layer
- Cornea – Transparent, anterior layer
- Frontal sinus
- Vitreous body

- **Retina** – Light sensitive, transparent membrane. Made of the following layers and contains the following features:
  - Ganglion cells
  - Bipolar cells
  - Rod and Cone nuclei
  - Rod and Cone light sensitive outer segment
  - Choroid

- **Macula lutea** – A small, yellow patch on which an image is focused when looking at a specific area. The center of focus.

- **Fovea centralis** – A depression in the center of the macula lutea, where our highest visual acuity (resolution) occurs. Cone cells, which perceive color, are the only photoreceptors found here, as it is absent of rod cells (responsible for night vision).

- **Optic disk** – Optic nerve exits and blood vessels enter and exit.

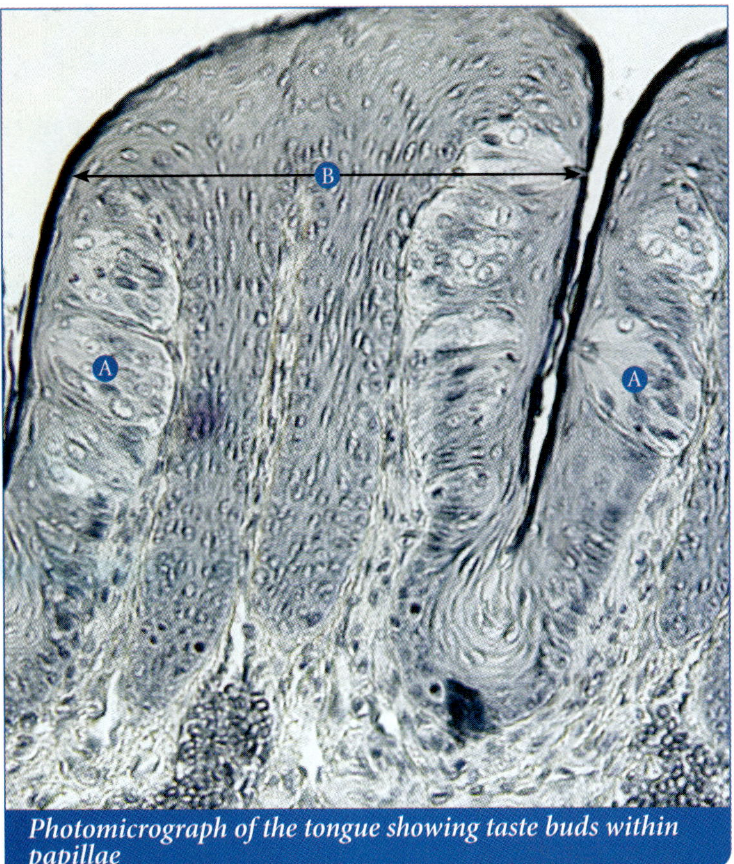

Photomicrograph of the tongue showing taste buds within papillae

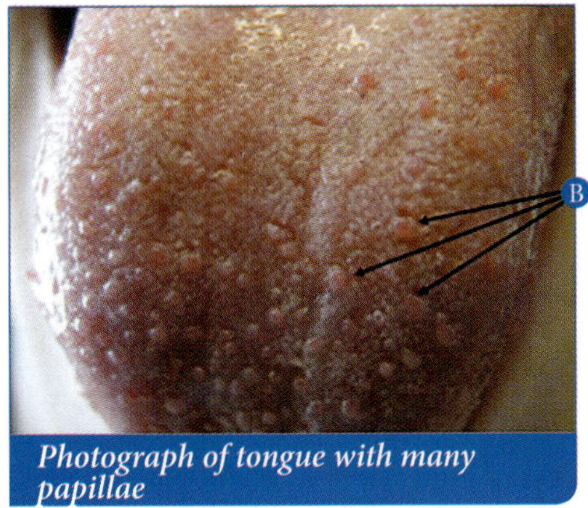

Photograph of tongue with many papillae

A   Taste bud

B   Lingual papillae

C   Taste pore

Both gustation and olfaction require molecules or ions to dissolve in saliva or mucus and bind to specific chemoreceptors to elicit a sensation.

Gustatory (taste) cells are sensitive to at least one of the five types of the following taste sensations.
- Bitter
- Salty
- Sour
- Sweet
- Umami

Taste buds are located throughout the tongue with regional specialization for a specific taste. An individual taste bud may be able to distinguish more than one type of compound, eliciting different sensations; i.e., the same cell can produce a sweet sensation and a bitter sensation, depending on the molecule that binds to the receptors of that cell. Our sensitivity to these taste molecules at different concentrations varies greatly. We are most sensitive to bitter compounds (0.000008 M), then sour (0.0009 M), umami (0.0007 M), sweet (0.01 M), and salty (0.01 M).

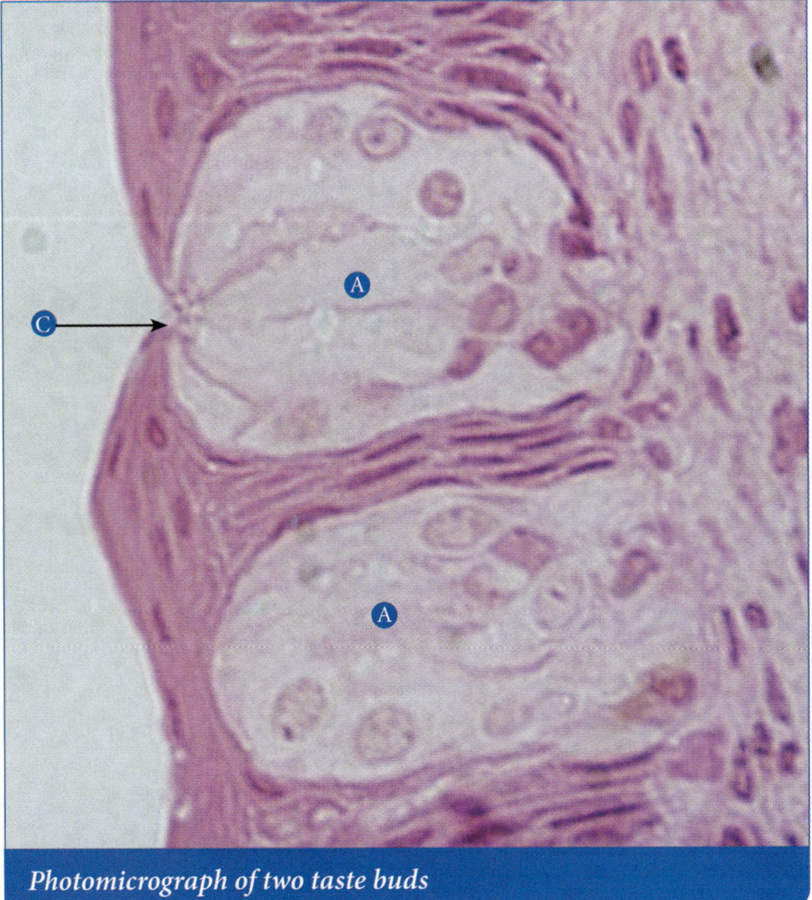

Photomicrograph of two taste buds

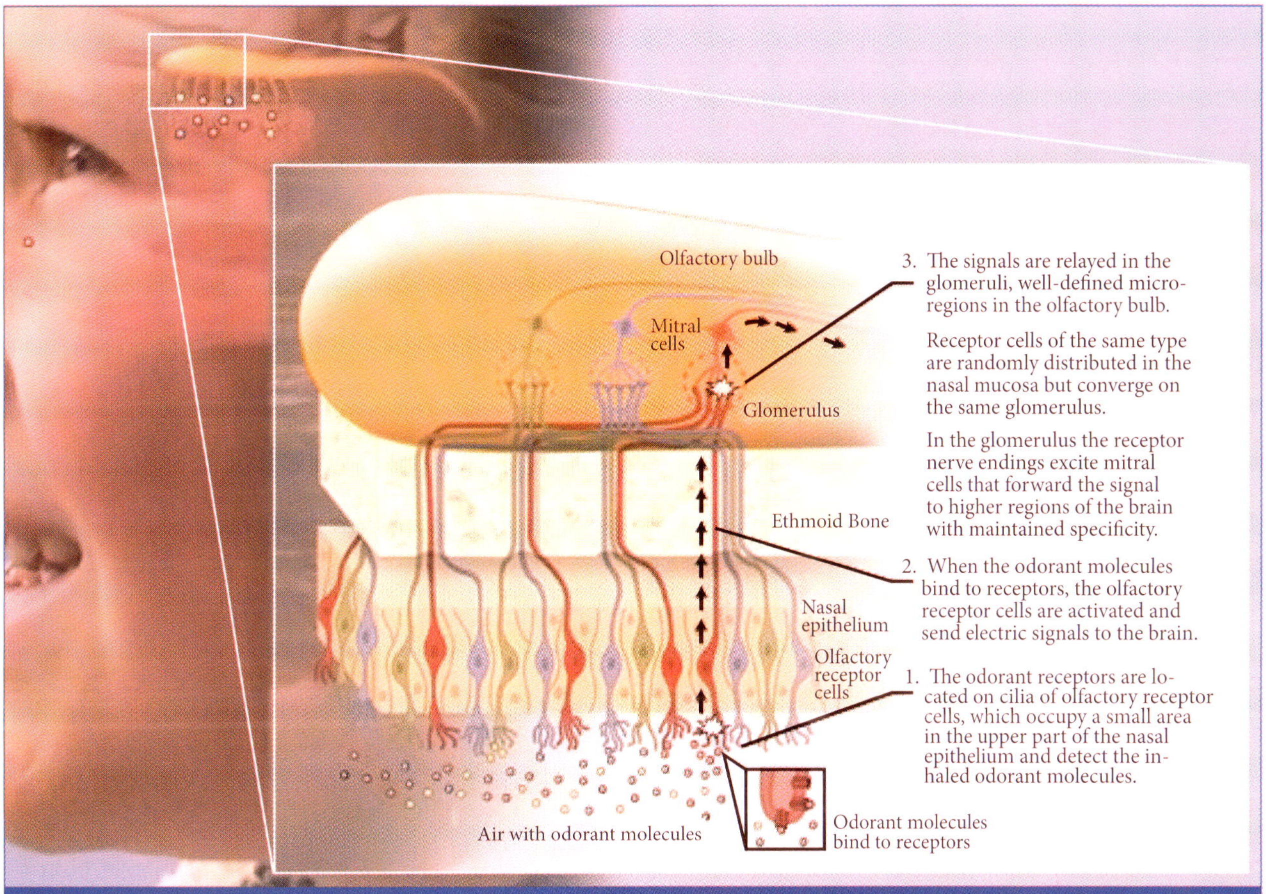

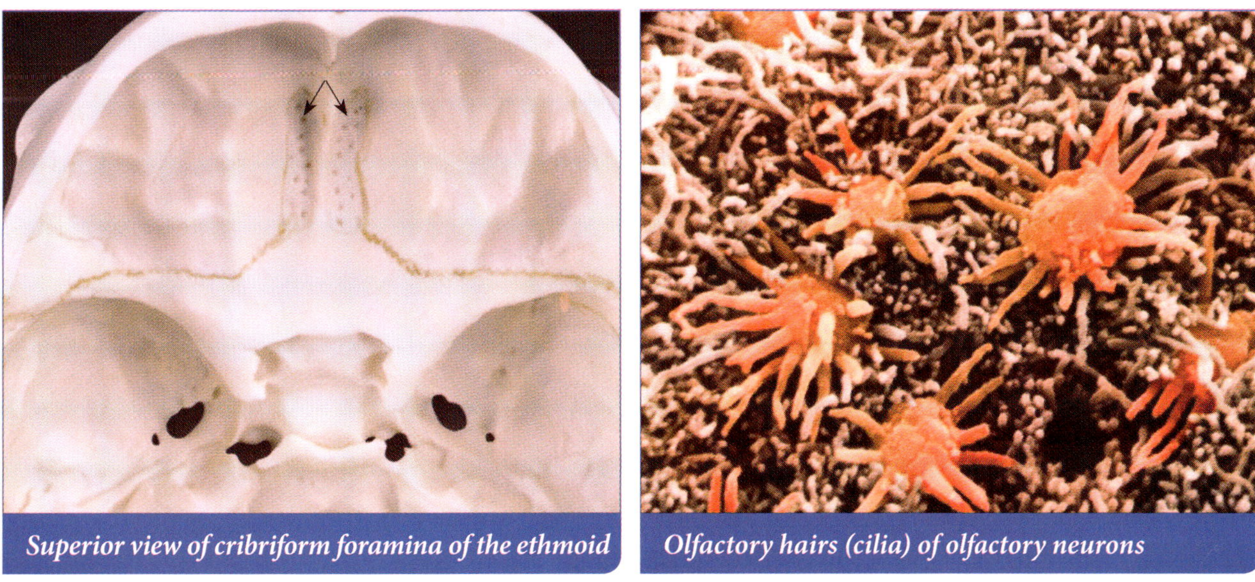

The process of olfaction from odor molecules to nervous impulse to the brain

Superior view of cribriform foramina of the ethmoid

Olfactory hairs (cilia) of olfactory neurons

Cilia of olfactory neurons (above right) bind specific odor molecules and transmit an action potential to the olfactory bulb through tiny passageways (cribriform foramina) through the ethmoid bone (above left). The approximate 400 unique receptor types are able to distinguish up to a trillion different odors. As we eat, we are more sensitive to the smell of our food than the taste. This is evident when nasal passageways are plugged from a cold or allergies and the loss of taste is noticeable.

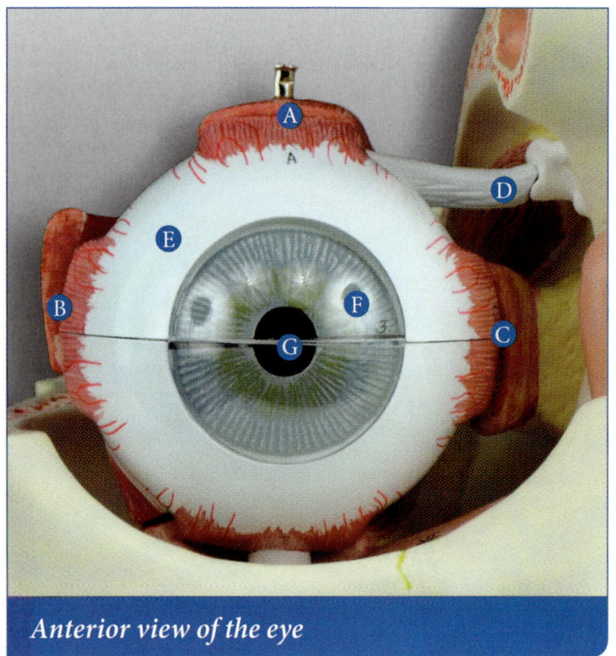

Anterior view of the eye

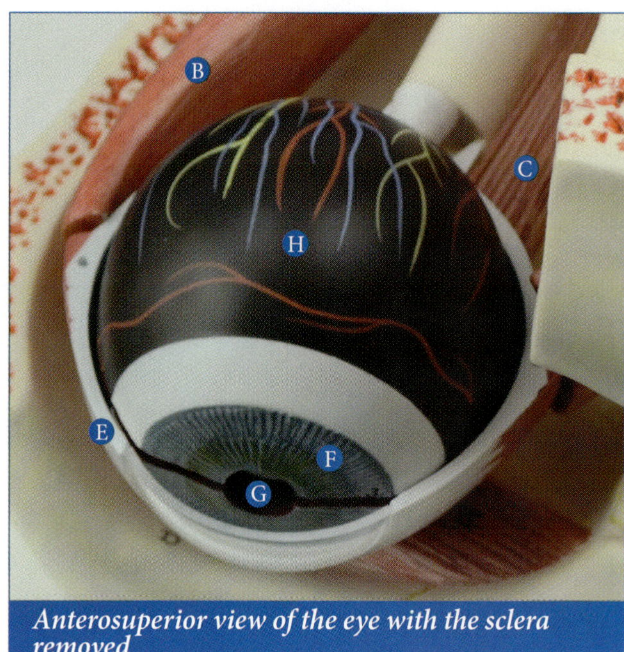

Anterosuperior view of the eye with the sclera removed

- Ⓐ Superior rectus muscle
- Ⓑ Lateral rectus muscle
- Ⓒ Medial rectus muscle
- Ⓓ Superior oblique rectus muscle's tendon
- Ⓔ Sclera
- Ⓕ Iris
- Ⓖ Pupil
- Ⓗ Choroid
- Ⓘ Cornea
- Ⓙ Frontal sinus

Extrinsic eye muscles, the medial rectus muscles in particular, are responsible for **convergence** (the medial movement of the eyes) when viewing objects that are approaching or close-up.

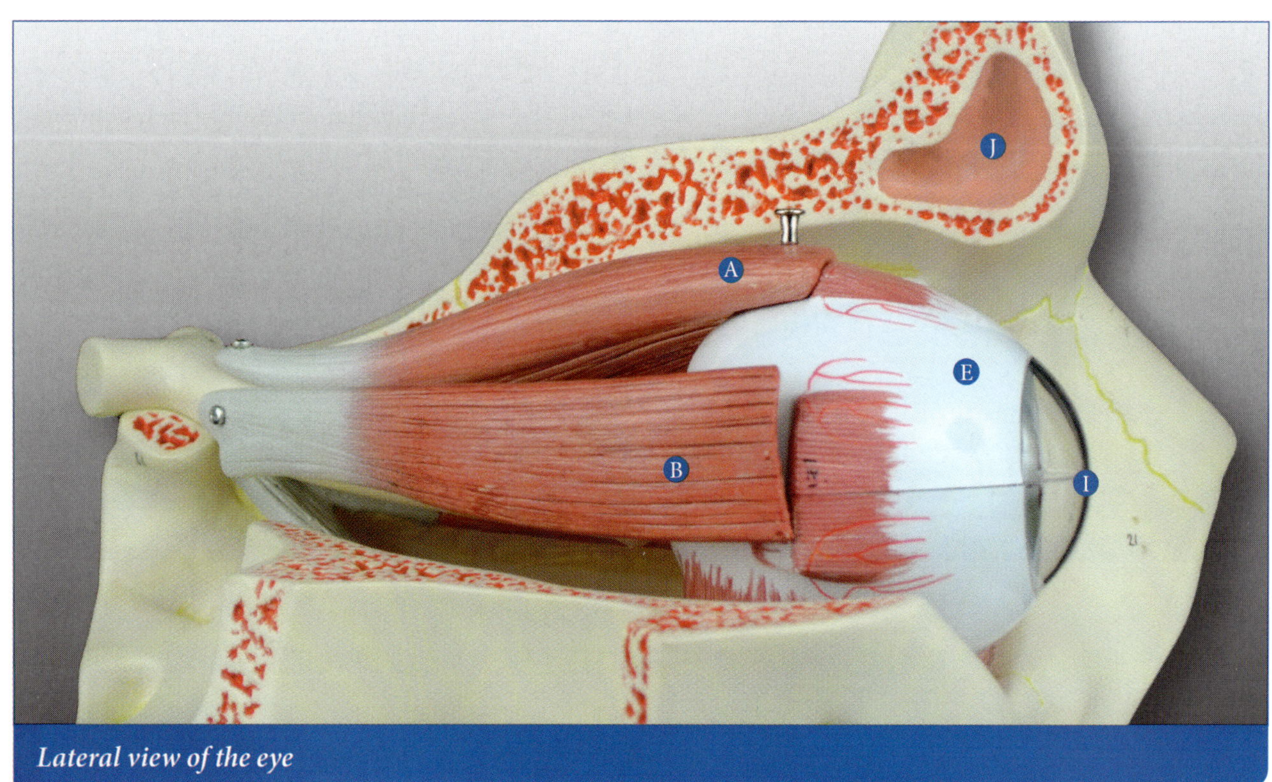

Lateral view of the eye

A Sclera

B Cornea

C Iris

D Choroid

E Ciliary muscles

F Suspensory ligaments

G Pupil

H Lens

I Retina

J Vitreous body in posterior cavity

As you are reading these words, three events are occurring in the eyes to ensure proper focus and sharpness: convergence, accommodation of the lens, and pupil constriction.

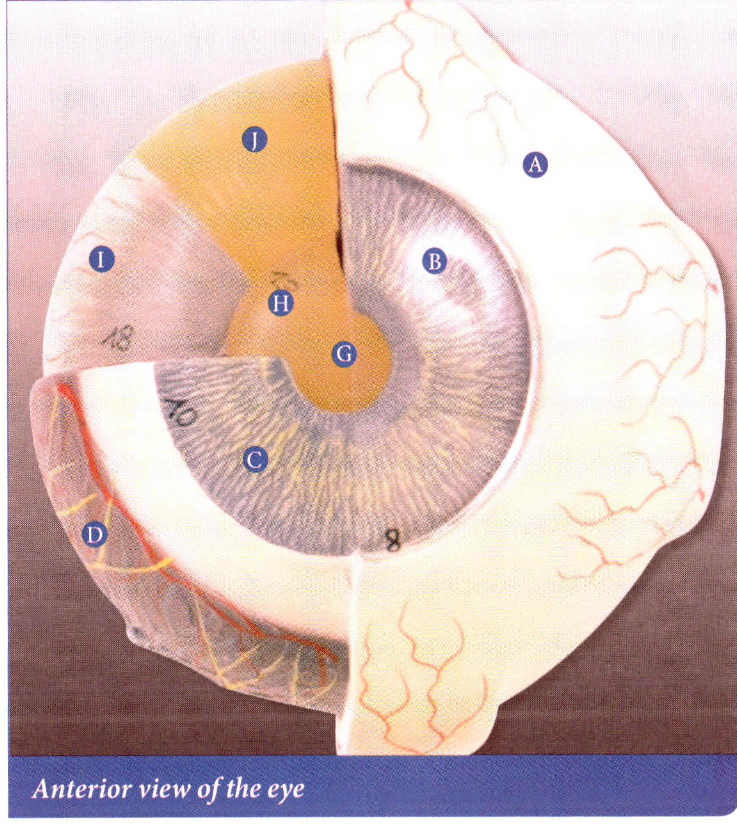

Anterior view of the eye

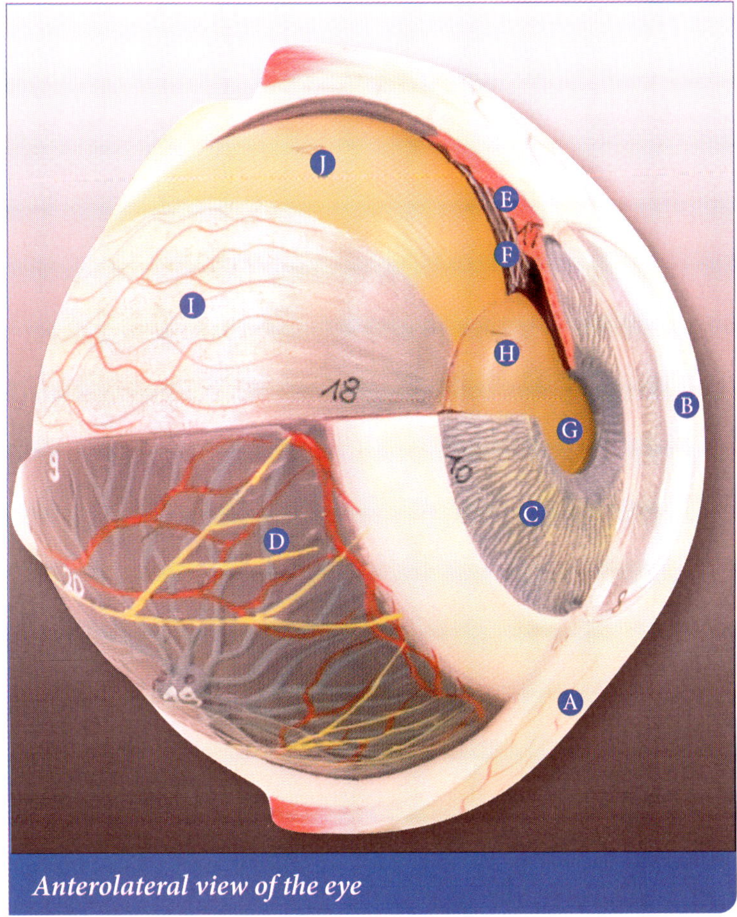

Anterolateral view of the eye

The cornea is the anterior, transparent portion of the superficial fibrous tunic. The posterior, white portion is the sclera. As light passes through the cornea, it is refracted (about 75% of refraction occurs here) and passes through the aqueous humor (watery fluid between the cornea and lens), and the pupil. Tension of the suspensory ligaments is greatest between the relaxed ciliary muscles and the lens when viewing distant objects. This tension flattens the lens and little refraction occurs. Accommodation of the lens (minor refraction adjustments as the lens changes shape) occurs if objects are closer than 20 feet. As objects approach, the ciliary muscles contract and reduce the tension of the suspensory ligaments. This allows the lens to passively return to its somewhat round shape, which refracts light to a greater extent. At the same time, the pupillary constrictor muscles of the iris, under parasympathetic control, reduce the amount of light entering the posterior cavity. Light continues through the vitreous body (jelly-like substance), and strikes the retina.

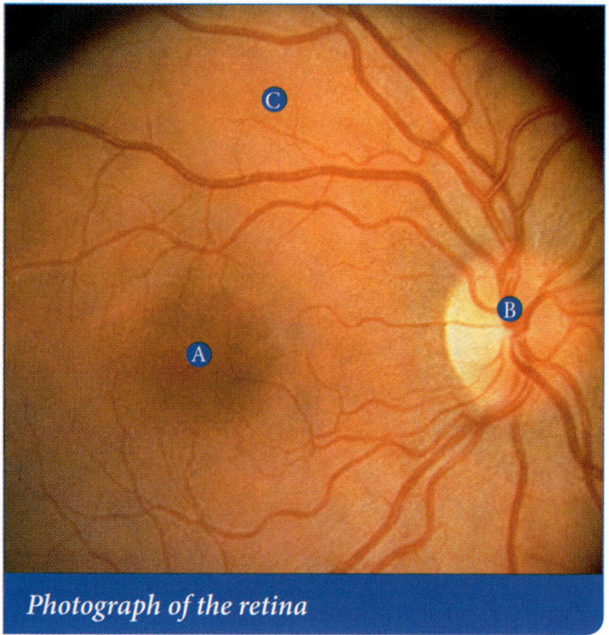

Photograph of the retina

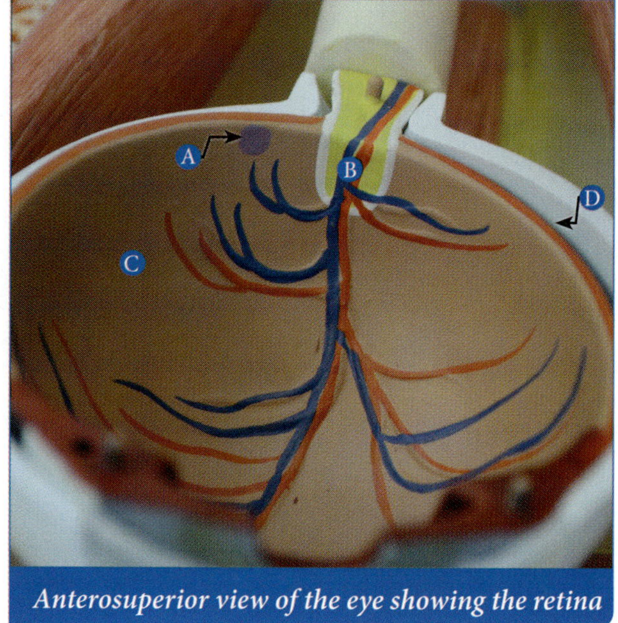

Anterosuperior view of the eye showing the retina

Ⓐ **Macula lutea** – Macula = spot; lutea = yellow. A small, yellow patch that an image is focused on when looking at a specific object. It contains a depression called the fovea centralis, where our highest visual acuity (resolution) occurs. It only contains cone cells, which perceive color, and is absent of rod cells (night vision). The perception of a faint star in peripheral vision, and the consequent disappearance when looking directly at it, is a vivid example of rod and cone cell distribution throughout the retina.

Ⓑ **Optic disk** – Optic nerve exits and blood vessels enter and exit

Ⓒ **Retina** – Light sensitive, transparent membrane

Ⓓ **Choroid** – Pigmented, highly vascular layer. Red-eye in a photograph results from a camera's flash reflecting off of blood vessels in this layer.

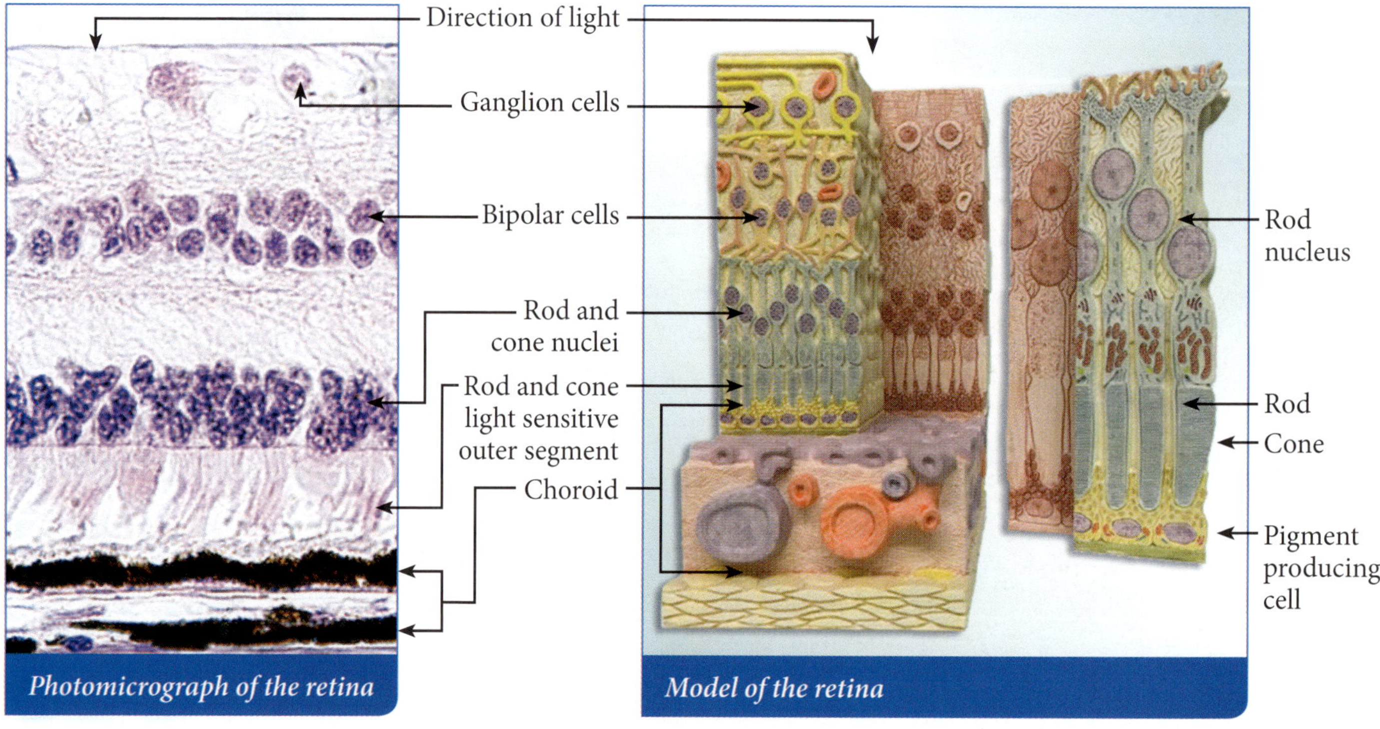

Photomicrograph of the retina

Model of the retina

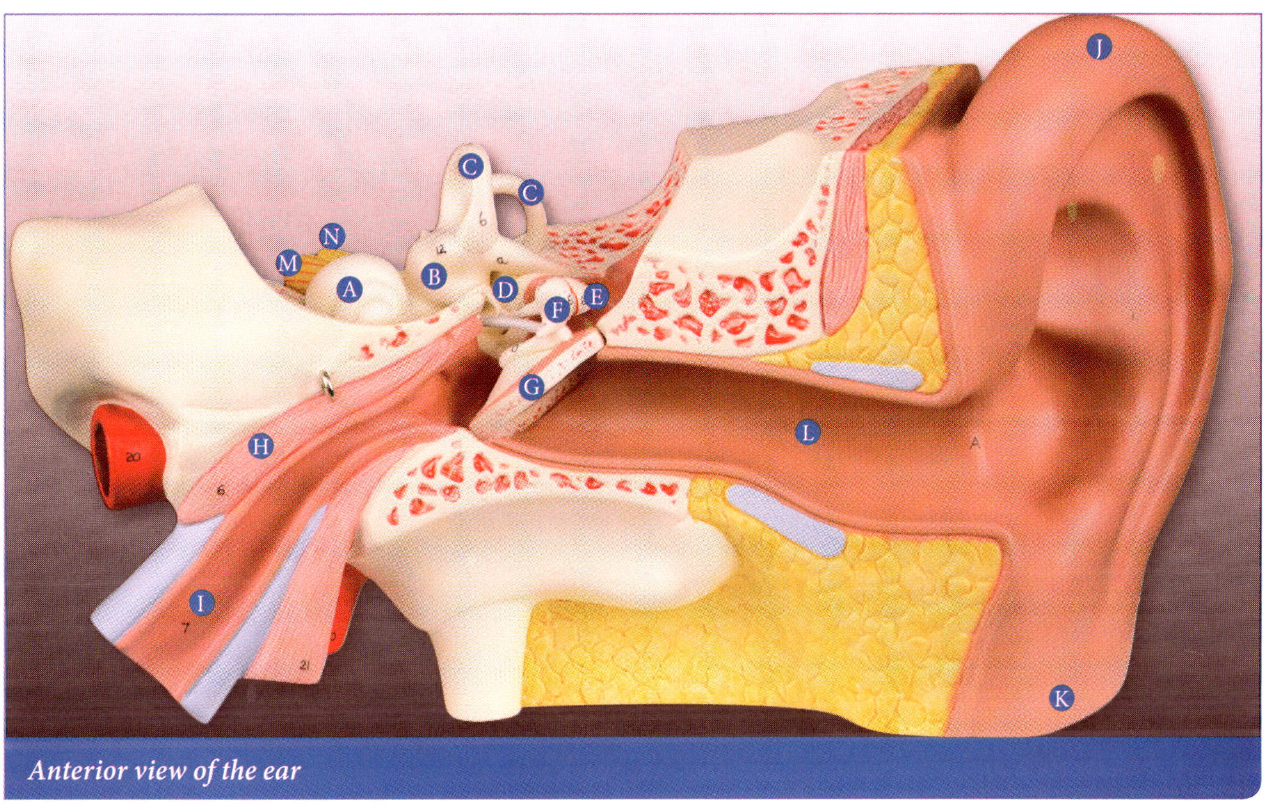

Anterior view of the ear

- Ⓐ Cochlea
- Ⓑ Vestibule
- Ⓒ Semicircular ducts
- Ⓓ Stapes
- Ⓔ Incus
- Ⓕ Malleus
- Ⓖ Tympanic membrane
- Ⓗ Tensor tympani muscle
- Ⓘ Auditory tube
- Ⓙ Helix of the Auricle
- Ⓚ Lobule of the Auricle
- Ⓛ Auditory canal
- Ⓜ Cochlear nerve
- Ⓝ Vestibular nerve

Equilibrium is maintained by the vestibule (contains saccule and utricle) and semicircular ducts. They send signals to the brain (via the vestibular nerve) in regard to body orientation, coordination, balance, linear acceleration, and rotation. Each is fluid filled and contains hair cells that bend in response to fluid movement.

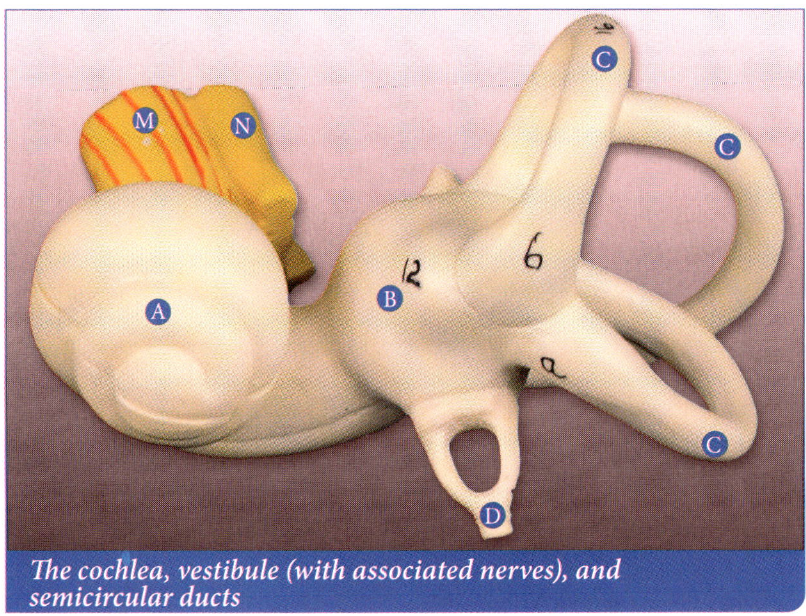

The cochlea, vestibule (with associated nerves), and semicircular ducts

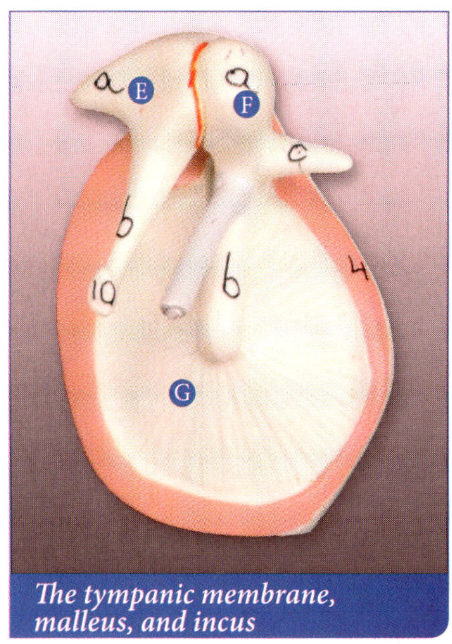

The tympanic membrane, malleus, and incus

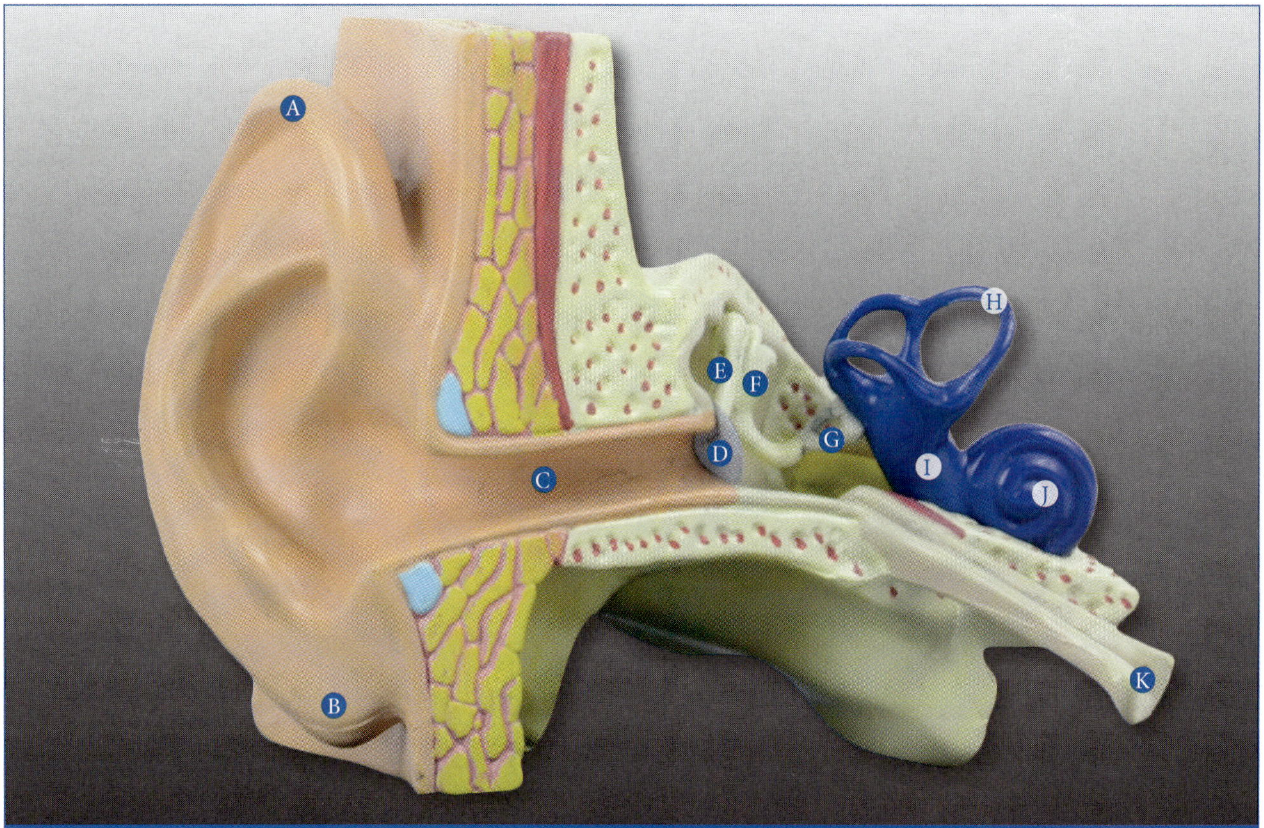

Anterior view of the ear (this model is actual size)

- **A** Helix of Auricle
- **B** Lobule of Auricle
- **C** Auditory canal
- **D** Tympanic membrane
- **E** Malleus
- **F** Incus
- **G** Stapes
- **H** Semicircular ducts
- **I** Vestibule
- **J** Cochlea
- **K** Auditory tube

Pressure waves traveling in air or water are funneled into the auditory canal with the aid of the auricle. Sound waves strike the tympanic membrane, causing it to vibrate at speeds of 20–20,000 Hz or cycles/second. The vibration is transferred to the cochlea's oval window via the auditory ossicles.

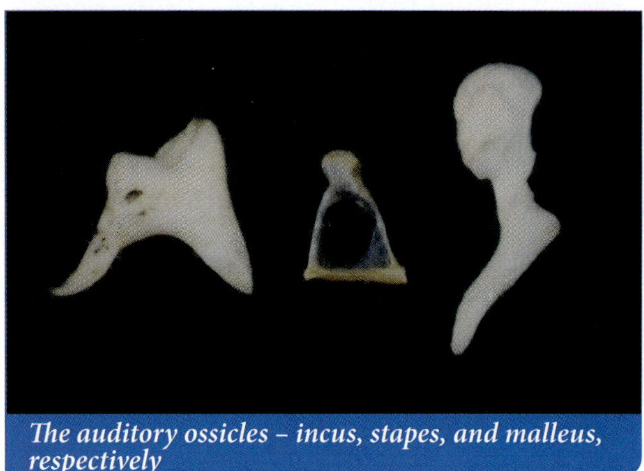

The auditory ossicles – incus, stapes, and malleus, respectively

Relative size of the auditory ossicles, the body's smallest bones

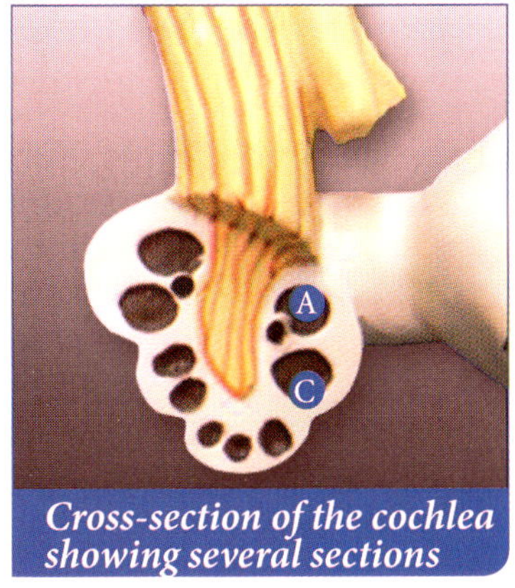

Cross-section of the cochlea showing several sections

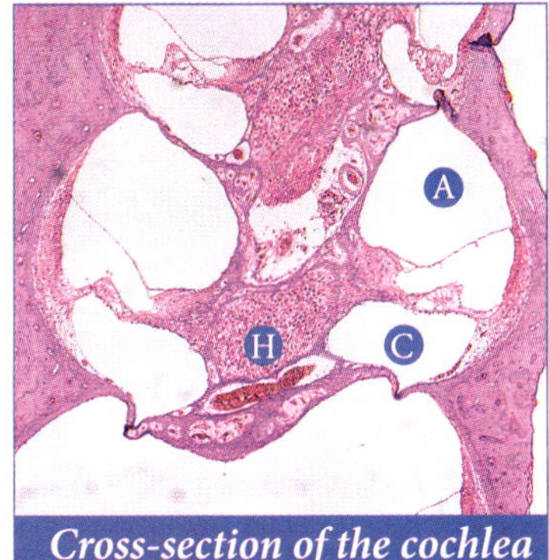

Cross-section of the cochlea showing several sections

The vibrating oval window sends pressure waves through perilymph fluid of the scala vestibule, apex of cochlea, and scala tympani, resulting in the bulging round window.

The above photograph and photomicrograph of the cochlea show several turns of the spiral, fluid filled tube. This snail-like structure is responsible for our perception of sound. On the right are enlarged views through one section of the cochlea.

Ⓐ Scala vestibuli

Ⓑ Cochlear duct

Ⓒ Scala tympani

Ⓓ Tectorial membrane

Ⓔ Hair cells

Ⓕ Basilar membrane

Ⓖ Vestibular membrane

Ⓗ Spiral ganglion

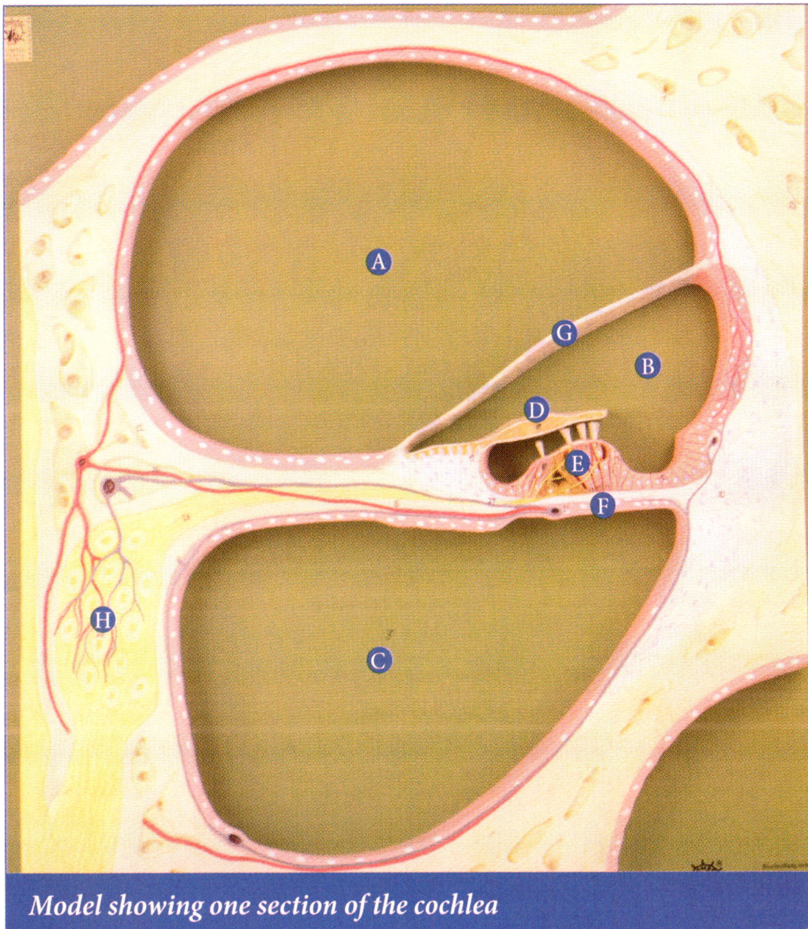

Model showing one section of the cochlea

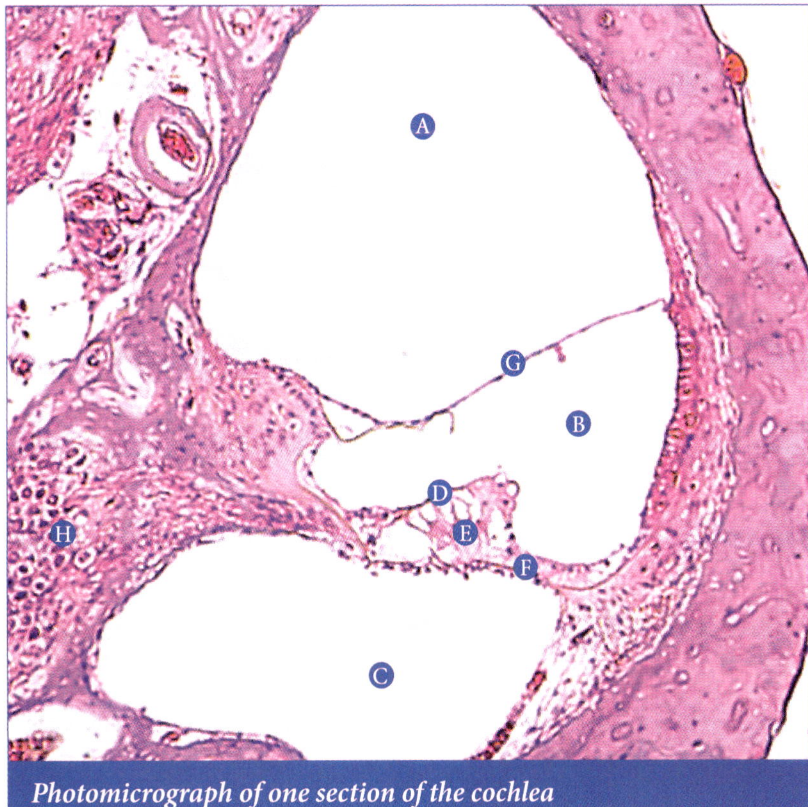

Photomicrograph of one section of the cochlea

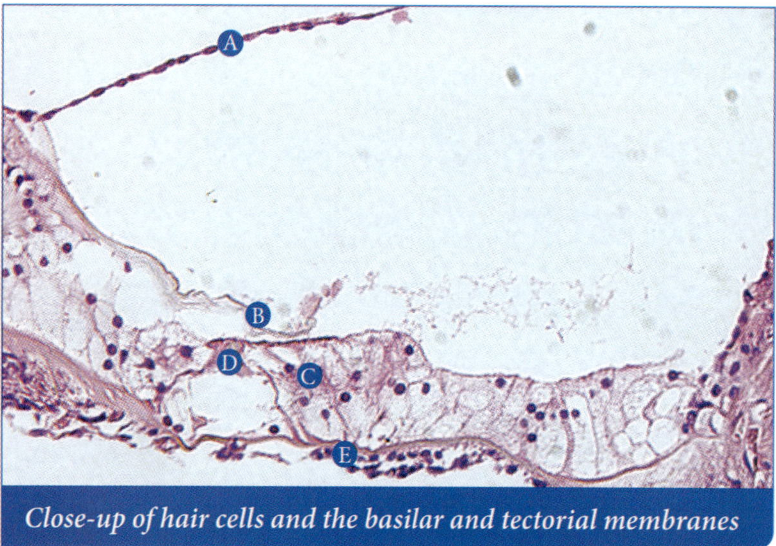

Close-up of hair cells and the basilar and tectorial membranes

Ⓐ   Vestibular membrane

Ⓑ   Tectorial membrane

Ⓒ   Outer hair cells

Ⓓ   Inner hair cells

Ⓔ   Basilar membrane

As pressure waves are generated through the scala tympani, the basilar membrane is pushed up, causing the inner hair cells to make contact with the tectorial membrane. Each stereocilia that is bent upon contact with the tectorial membrane will open potassium (K^+) gates, causing the cell to depolarize and release a neurotransmitter on a sensory dendrite. This will send an action potential along cochlear neurons of the cochlear nerve to the brain, where sound is perceived.

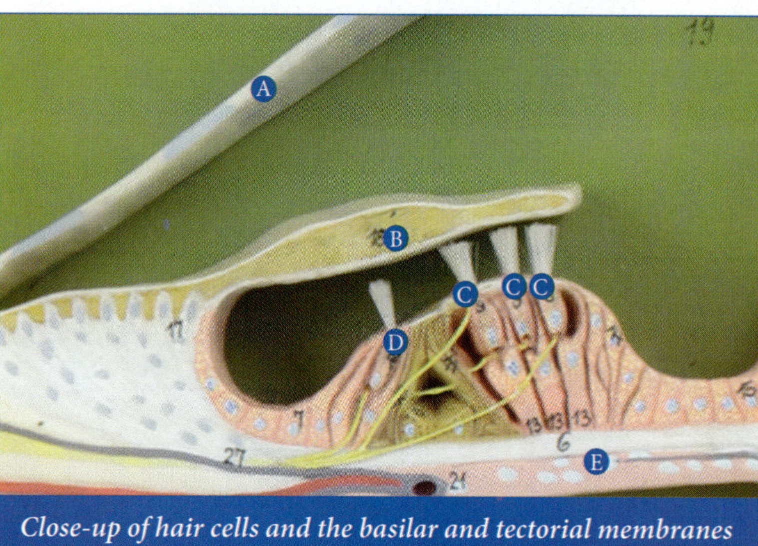

Close-up of hair cells and the basilar and tectorial membranes

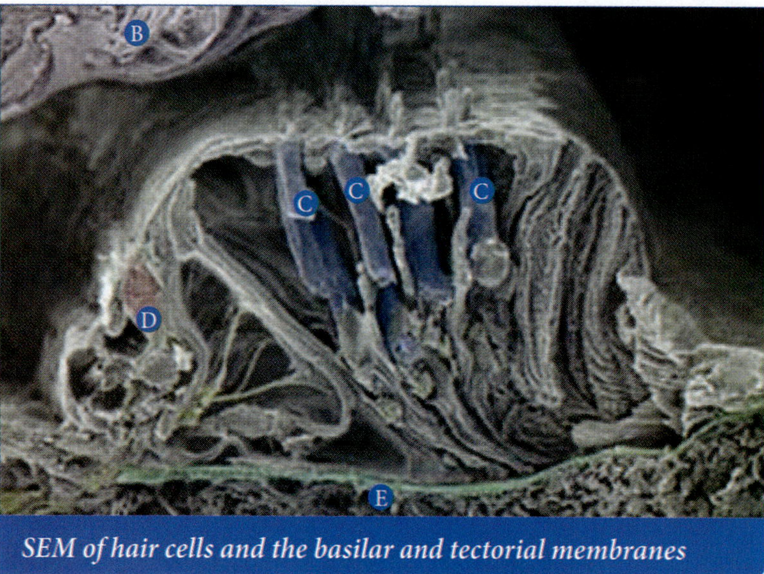

SEM of hair cells and the basilar and tectorial membranes

The above electronmicrograph of the cochlea shows three v-shaped rows of outer hair cells and one linear row of inner hair cells. Each cochlea has around 20,000 outer hair cells with 100 stereocilia on each cell. The outer hair cells are not involved in the perception of sound, but help keep the cochlea tuned. The stereocilia are embedded in the tectorial membrane and are innervated by sensory and motor neurons. This allows communication to the brain and control of the length of the stereocilia. Without outer hair cells, our hearing would be out of pitch. Hearing is possible due to the bending of the inner hair cells' stereocilia. Each cochlea has about 3,500 inner hair cells with about 50 stereocilia on each cell. There is only sensory innervation to the inner hair cells, not motor.

Perception

3 CHAPTER

Touch, Smell, Hearing, Dynamic Equilibrium, and Vision

The saying, "beauty is in the eye of the beholder," illustrates that perception is everything. Perception (from the Latin *percipio*) is defined as the process of attaining awareness or understanding of the environment by organizing and interpreting sensory information. Signals sent to the brain from specialized receptors help us interpret ourselves and the world around us. Each person has unique collections of input, association processes, and interpretation that result in individual personality and perspective. Many people may see the same event or read the same passage and have dramatically different viewpoints and opinions. Each of us has a unique outlook and interpretation of ourselves and the world.

Visual illusions demonstrate that the old adage, "seeing is believing," can be misleading. What we may believe to see and the events remembered may not be accurate. Most of us value eyewitness testimony in a court of law, when in fact, it has been proven to be highly fallible and unreliable. In one study, the percentage of correct identification ranged from 34–48% and the percentage of false identification 34–38%. In another study, the majority of eyewitnesses had wrongfully accused suspects when actually, DNA evidence proved their innocence. Some theme park rides use visual, auditory, motion, and tactile stimuli to make one believe he or she is flying through a rainforest or space, when in reality, he or she is confined in a dark building. Hopefully by understanding unique qualities and inaccuracies of human perception, we will continue to develop tolerance and acceptance.

Touch, smell, hearing & equilibrium

- Two-point discrimination test
- Olfactory discrimination
- The Weber test
- The Rinne tests
- The Barany test

Vision tests

- Near point determination
  - **Accommodation** of the lens
  - **Convergence** of the eye
- Color blindness (using Ishihara's plates)
- Snellen test
- Astigmatism test
- Measurement of visual field (peripheral vision)
- Determination of the blind spot
- Peripheral drift illusion
- Afterimage

One of the oldest tactile illusions is the Aristotle illusion. It is easy to perform. Cross your fingers, then touch a small spherical object such as a dried pea (or your nose), and it feels like you are touching two peas. This also works if you touch your nose.

Your fingertips are among the most sensitive parts of your body and this makes them surprisingly easy to fool. Take an ordinary comb and pencil and lay your index finger along the top of the comb, then run the pencil back and forth along the side of the teeth. Even though the teeth are moving from side to side in a wave-like motion, your finger will feel as if a raised dot is travelling up and down the comb.

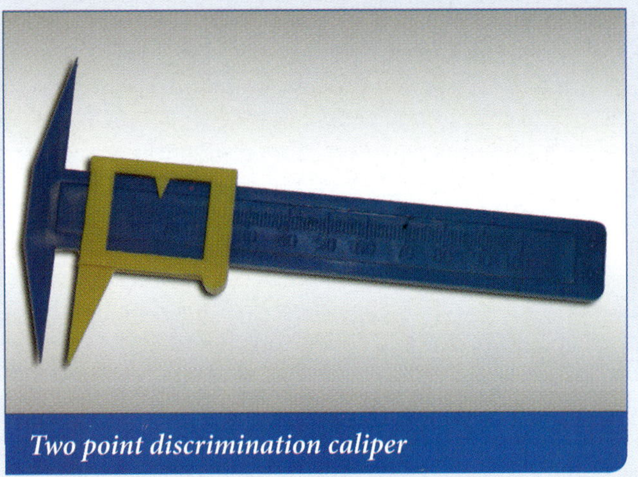

Two point discrimination caliper

The ability to distinguish two separate, simultaneous points of contact is known as two point discrimination. Two point discrimination is directly related to touch receptor density, which has regional variation. With your lab partner's eyes closed, use the caliper to determine the minimum distance they can perceive two separate points. It is important to touch each caliper point to the skin at the exact same time. It's best to begin with the caliper spread to a distance where the subject can easily determine two points of contact, then reduce until only one point can be felt. Follow the procedures and record your results on the Class Activity for this chapter.

Olfaction (smell) requires odor molecules to bind with one of 400 different types of receptors located on bipolar neurons' (olfactory cells) cilia. We can detect some odorants in quantities as low as a few parts per trillion and can detect up to a trillion odors. Adaptation (decreasing sensitivity) occurs quickly among olfactory cells and the CNS. About 50% loss of sensitivity occurs in about 2 seconds.

OLFACTORY DISCRIMINATION

Sniff the contents of each of the ten vials. After sampling each vial, fill in Columns 1 to 3 on the Class Activity for this chapter. If an odor is detected in the vial, place a check in Column 1. If the odor is recognized as familiar or as one that has been smelled in the past, place a check in Column 2. In Column 3, attempt to name each odor. If you are unable to identify the odor, leave this column blank.

After all ten vials have been sampled, the identification task is repeated for each vial. This time, however, use the odor selection list provided by your instructor to assist in identifying the smell of each odor. Enter your odor choices in Column 4.

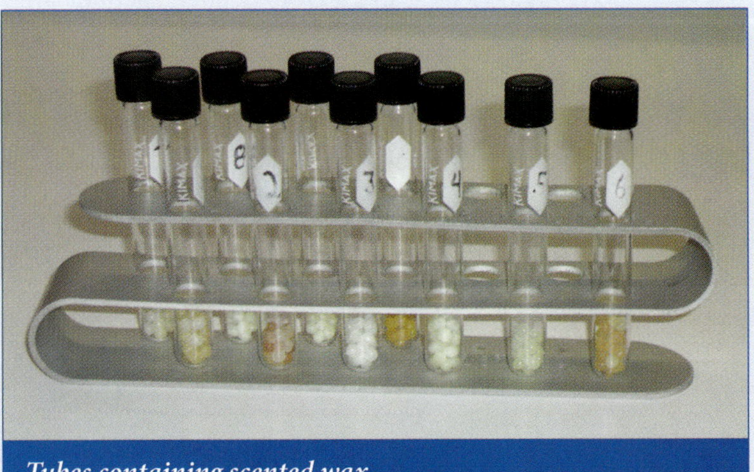

Tubes containing scented wax

TESTS FOR CONDUCTION DEAFNESS

Conduction deafness is restricted movement caused by damage to the tympanic membrane or ossicles. Deafness can also be caused by otitis externa and/or otitis media (infection of the outer and inner ear, respectively) or as simple as excess cerumen (ear wax).

The **Weber test** detects differences in conduction between each ear. If both ears are normal or both ears have the same amount of conduction loss, this test is inconclusive. Your lab partner should strike the tuning fork Ⓐ with the rubber mallet Ⓑ and place the base of the tuning fork Ⓒ on the middle of the forehead Ⓓ. If the sound is louder in one ear compared to the other, this is an indication of conduction deafness in the ear that is perceived to be louder. Record results on the Class Activity for this chapter.

The **Rinne test** is complementary to the Weber test. This is done by striking the tuning fork and placing the base on the mastoid process Ⓔ until the sound disappears. Then place the tines of the tuning fork just outside the ear canal. If the sound cannot be heard, then conduction deafness is possible. Record results on the Class Activity for this chapter.

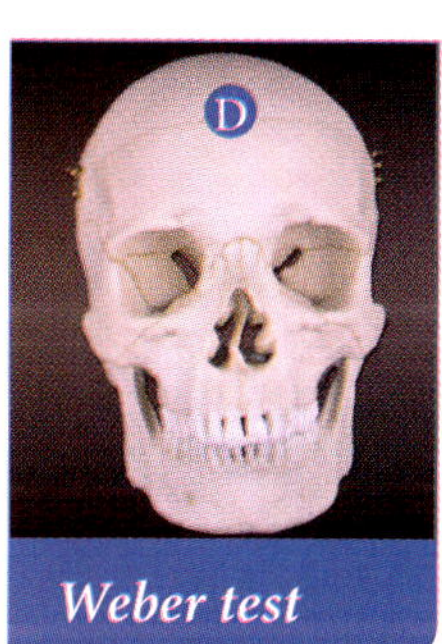

Weber test

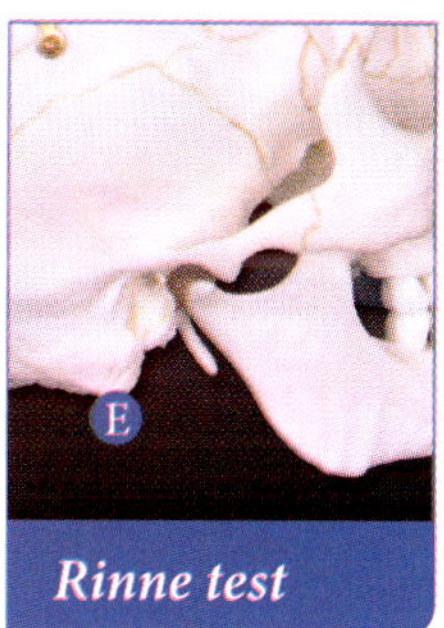

Rinne test

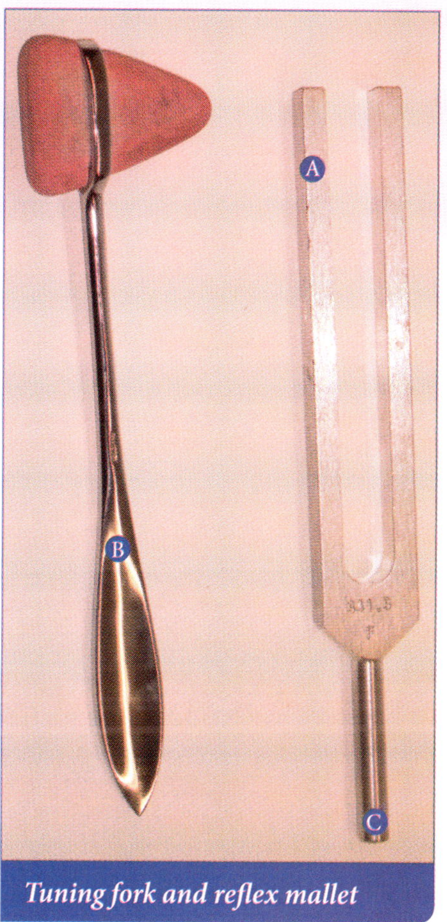

Tuning fork and reflex mallet

Dynamic equilibrium – movements in the fluid within the semicircular canals, utricle, saccule, and vestibule.

The **Barany test** examines an individual's responses to rotation and illustrates dynamic equilibrium. When a person, with eyes closed, is submitted to angular rotation, he or she will accurately signal the direction of rotation when movement begins. However, after a period of rotation at constant velocity, rotation is reported to have stopped. During the period of acceleration with eyes open at the beginning of the rotation, the person will also experience a nystagmus (side to side movement of the eyes) in the direction of rotation. When the sensation of rotation fades at constant velocity, the nystagmus also disappears.

If the chair is abruptly stopped at this point, the person will have a sensation of rotation in the direction opposite to that which was previously experienced. In addition, there will be a post rotatory nystagmus, also in the direction opposite to the previous rotation. Although rotation has ceased, endolymph fluid is still moving in the semicircular canals, and the central nervous system cannot correctly interpret the information. An example of fluid movement after rotation ceases is performed by spinning an egg, bringing it to an abrupt stop with one finger, quickly releasing the finger, and watching the egg begin to spin on its own.

In an area free of obstruction, one student sits on a lab chair (the subject) and others are responsible for rotating and observing the student in response to rotation. With the subject's eyes closed, rotate the chair slowly. Can the subject accurately detect the direction of rotation? Keep slowly rotating for ten seconds. Can the subject tell which way they are spinning? Have the subject open their eyes, stop rotation, and observe any eye movements. Next, rotate the subject slowly with eyes open. Do their eyes move back and forth (nystagmus)? In which direction? Record results on the Class Activity for this chapter.

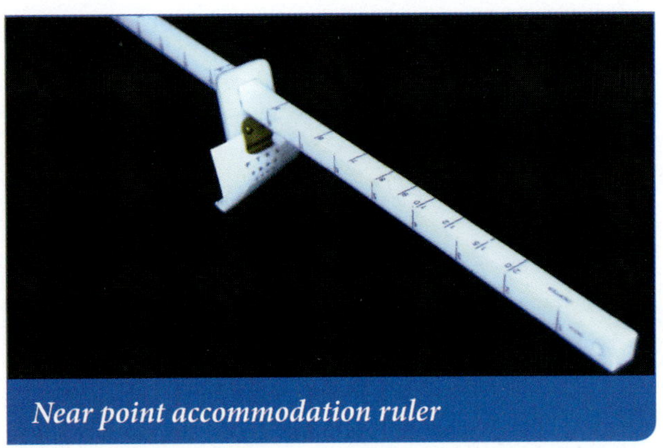

Near point accommodation ruler

Accommodation of the lens occurs as we focus on objects 20 feet away and closer. The closer the object is, the more round the lens becomes, due to contraction of ciliary muscles and subsequent reduction in tension of suspensory ligaments.

Convergence of the eyes occurs as an object approaches. During convergence, the visual axis of each eye is trained on the object, causing "cross eyes" when one looks at the object inches away. See pages 2-4 and 2-5 for models associated with accomodation and convergence.

To determine near point vision, hold the near point accommodation ruler to your forehead. Cover one eye and look at the print on the sliding attachment. Focus on the letters and slide the attachment until the letters become blurry. Repeat using the other eye and record results on the Class Activity for this chapter in centimeters.

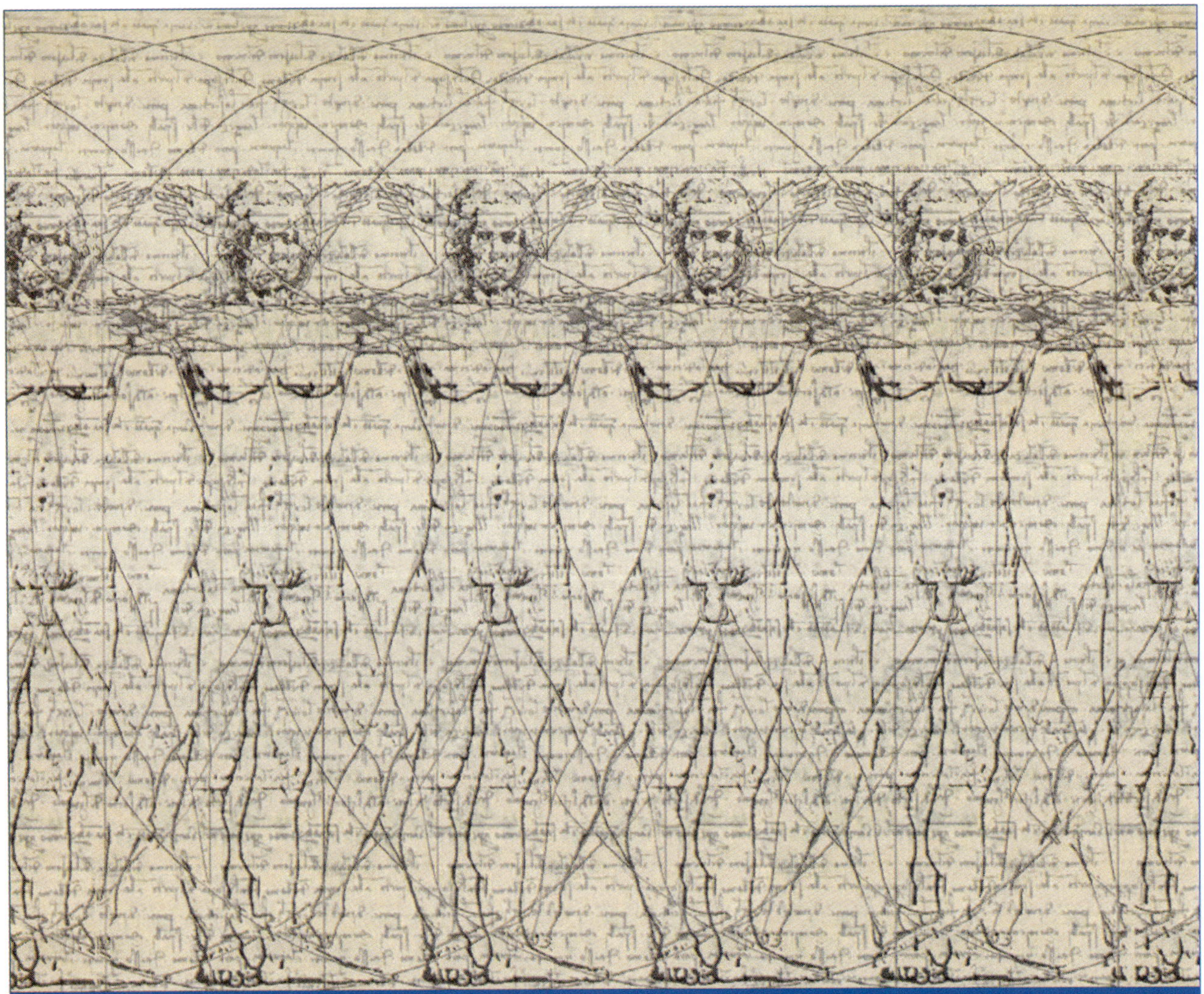

A Stereogram of Vitruvian Man by Leonardo da Vinci

Convergence (medial rotation of eyes) allows each eye to align to the center of focus (visual axis) on an object that is inches away. If both eyes do not converge, but maintain the visual axis of looking at a distant object, the above stereogram can produce an optical illusion. By maintaining the visual axis of looking at a distant object, each eye focuses on a separate image, the brain combines them, and a 3-D image appears.

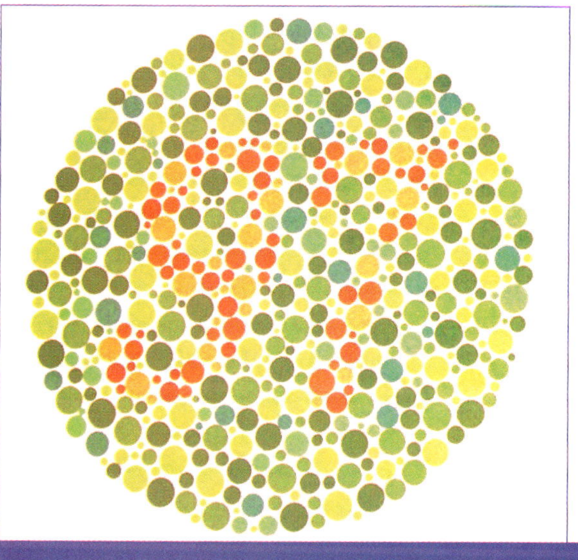

Ishihara plate for red/green color blindness

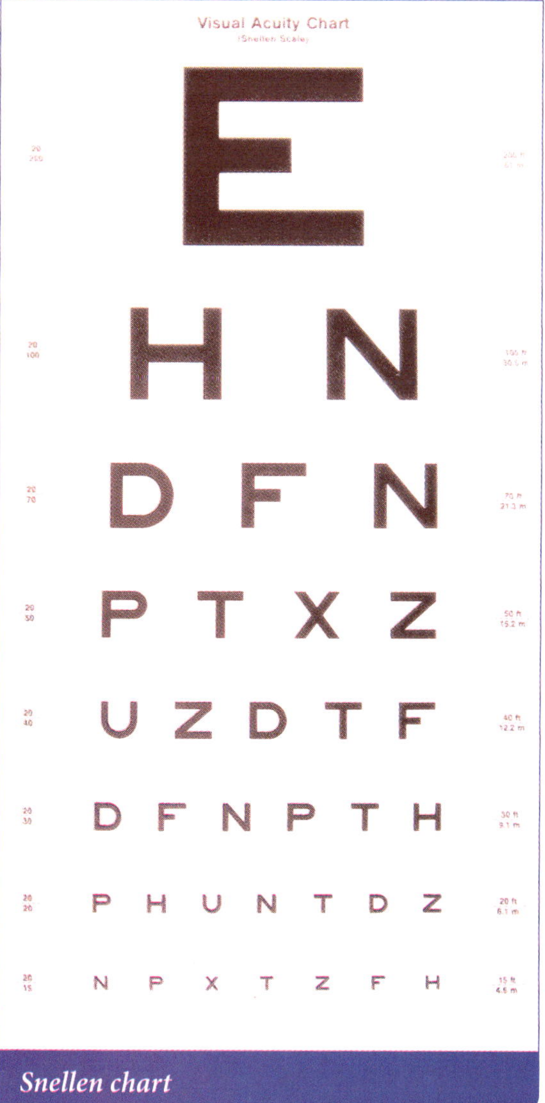

Snellen chart

Look at the above Ishihara plates. People with normal color vision see the numbers 97 and 16. Red/green color blind people do not see any numbers. Many genes involved in forming color receptors (cones) are on the X chromosome, of which males have only one copy (XY), compared with XX chromosomes of females, with two chances of receiving functional genes. About 7% of the male population and only 0.5% of females are color blind.

The Snellen chart measures visual acuity (sharpness and clarity). At 20 feet, a person with perfect vision, 20/20, can read the second to the bottom line.

Astigmatisms arise from asymmetries of the cornea. Malformation of the cornea can be detected when darker lines along a particular axis in the chart below are perceived.

Astigmatism chart

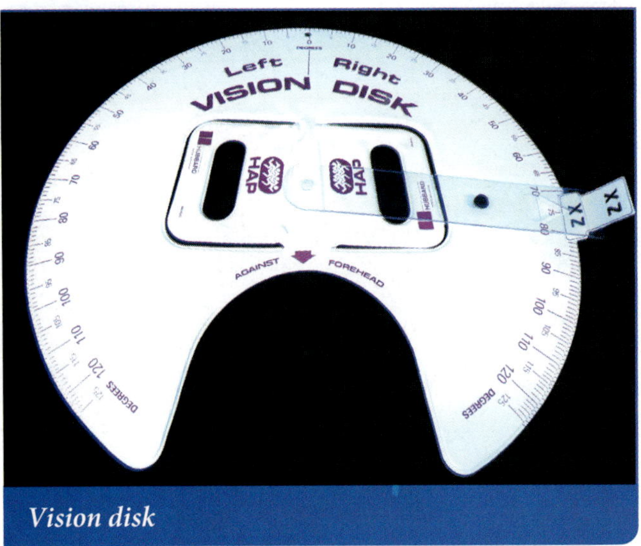

"Rotating wheels" demonstrate the peripheral drift illusion, copyright A. Kitaoka 2003 (September 2, 2003).

Vision disk

Alternating bands of the above image overlap in peripheral vision of both eyes to create this unusual illusion.

The peripheral drift illusion allows observers to see circles appear to rotate clockwise and counterclockwise, even though we know they are not moving.

The vision disk measures the visual field of each eye. Hold the disk to your forehead, and with only one eye open, focus on the triangular plastic straight ahead. Your lab partner will move the sliding ruler towards the center of vision until the object appears in your peripheral vision. Record the angle of first perception for each eye on the Class Activity for this chapter.

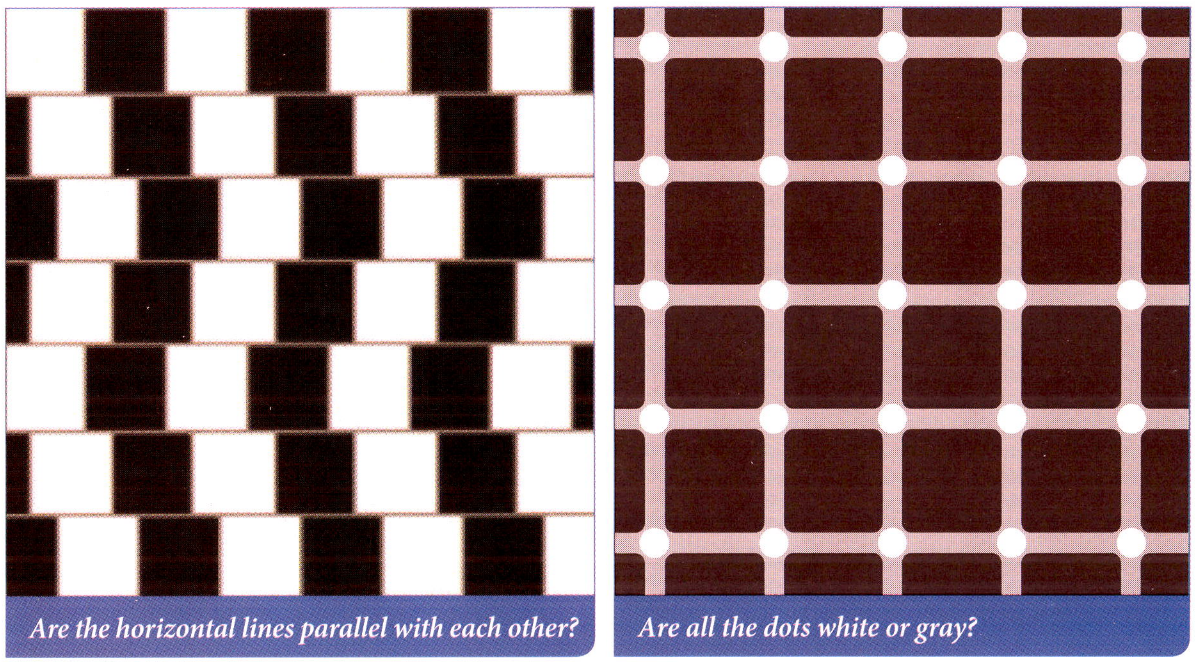

Are the horizontal lines parallel with each other?

Are all the dots white or gray?

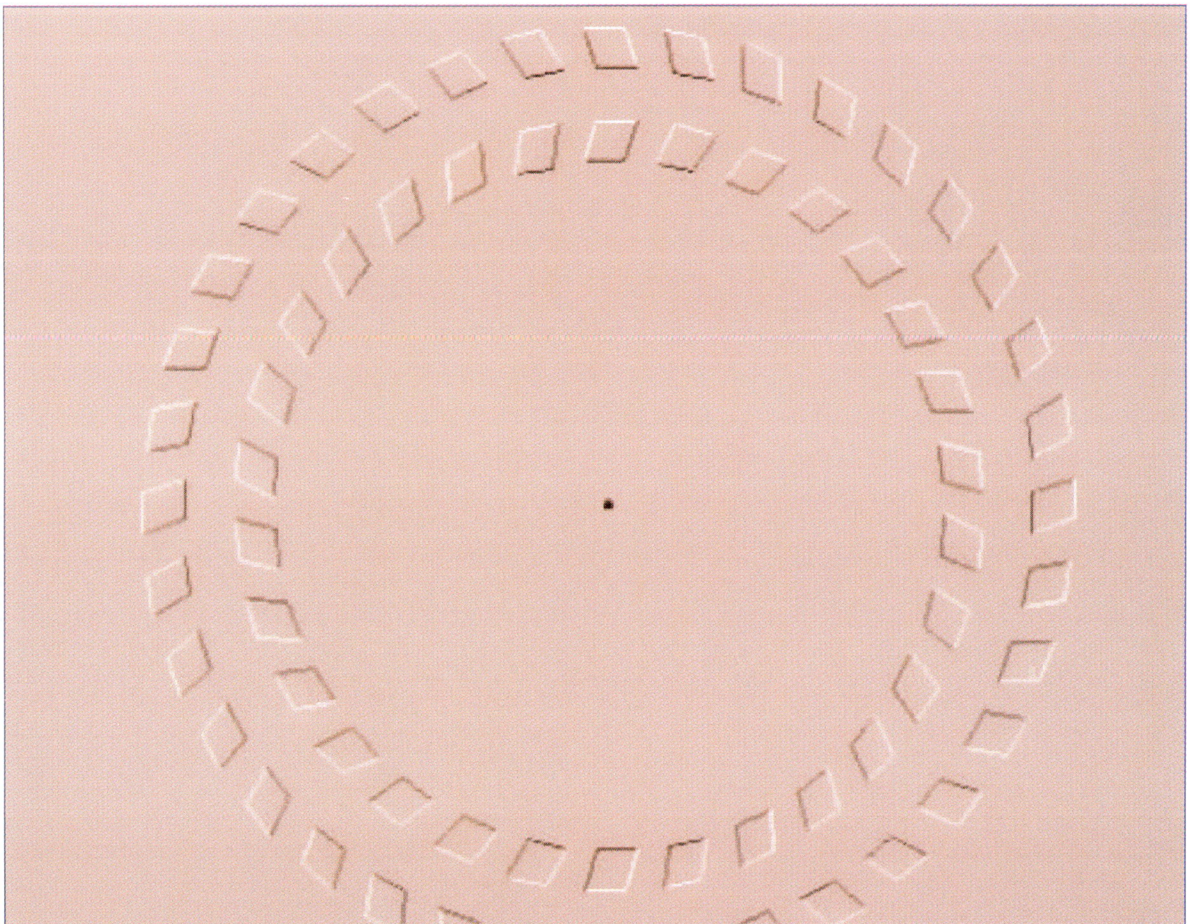

Stare at the dot in the center and move your head closer to, then further from, the page.

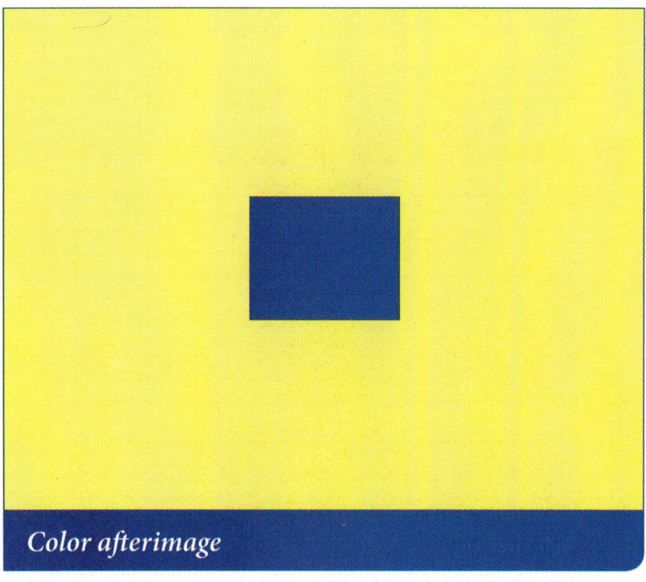

Color afterimage

Stare at the blue square for 30 seconds. Then look to the blank area to the right. Blink your eyes periodically to help see the afterimage.

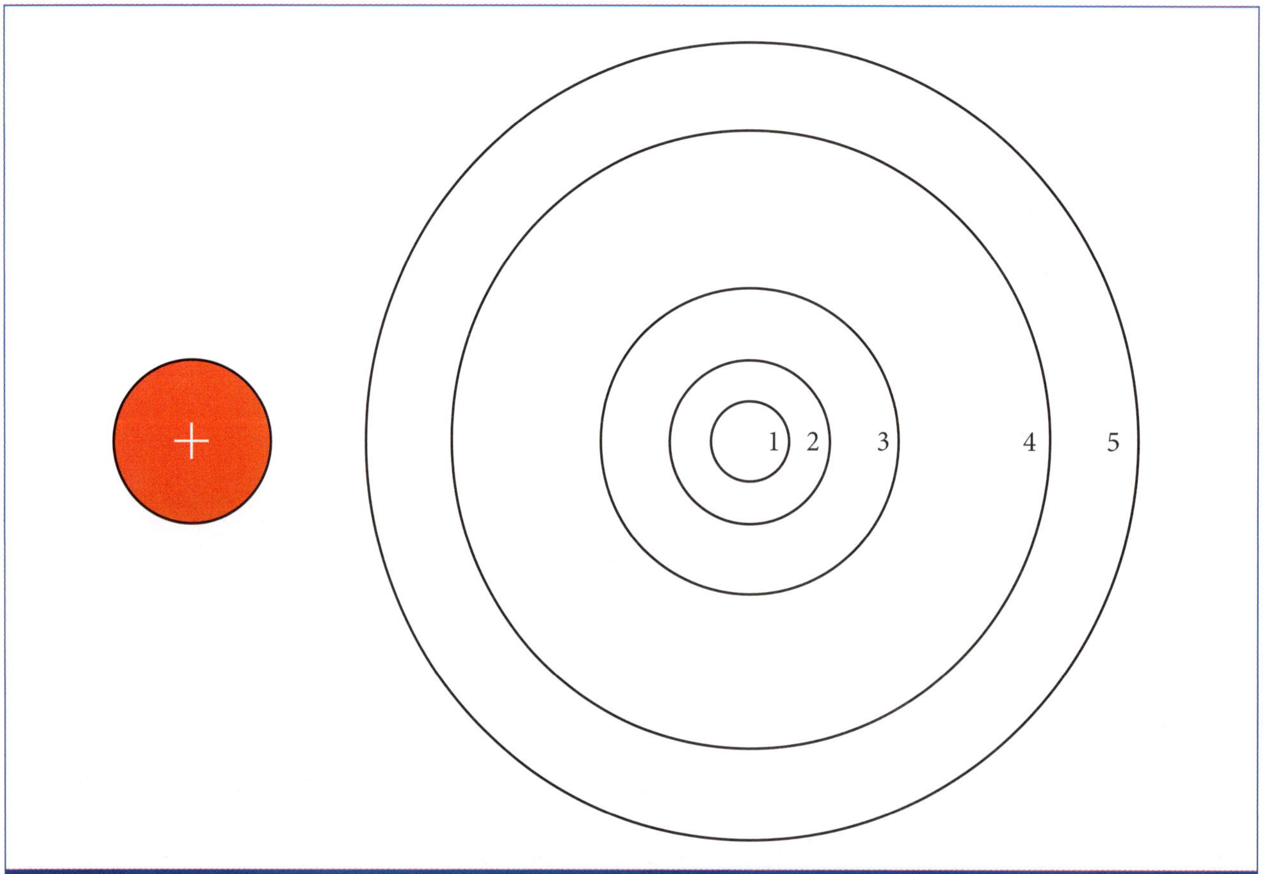

Afterimage relative size

Stare at the red circle for 30 seconds. Now, look at the numbered circles at the same distance. Move closer until the afterimage only fills the 1 circle. Move back until 3, 4, and 5 are filled. Blink your eyes periodically to help see the afterimage. Record your distance away from the page for circles 1, 3, 4, and 5.

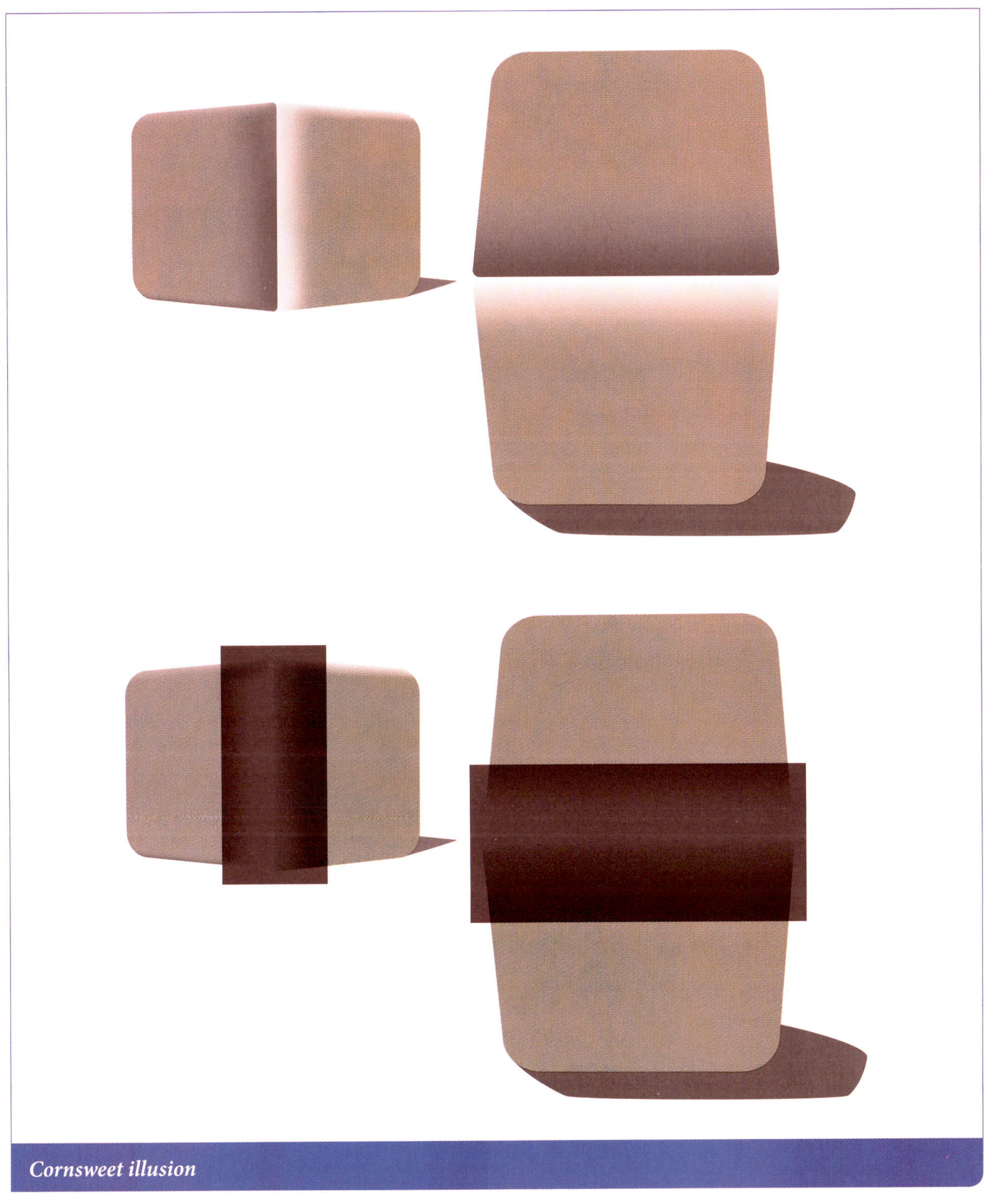

Cornsweet illusion

This is an example of the corn sweet illusion. On the top your eye is tricked into thinking that the centers of the grays have different shades. When you remove the gradients in the center of the two shapes, as seen on the bottom, you can see that the grays are actually the same shade.

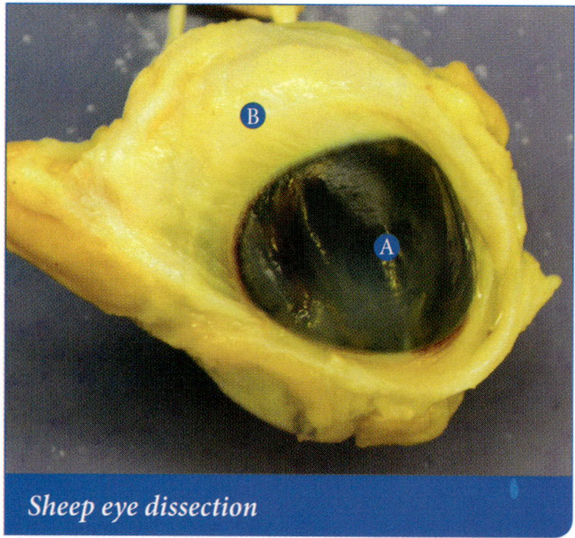

Sheep eye dissection

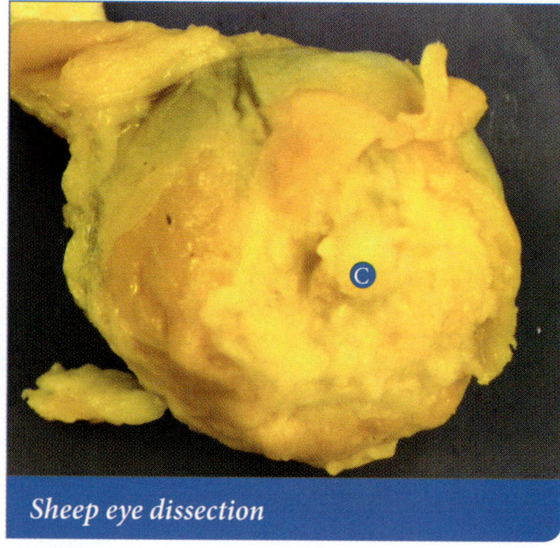

Sheep eye dissection

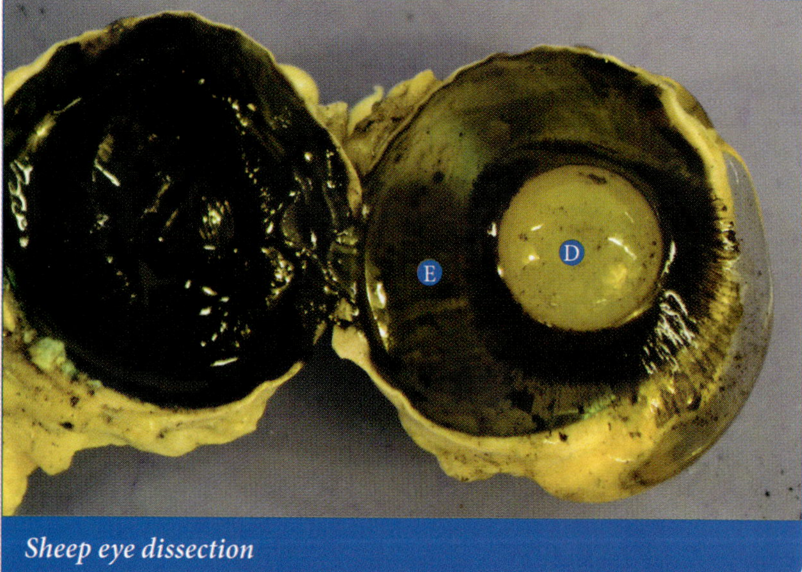

Sheep eye dissection

- **A**   Cornea
- **B**   Sclera
- **C**   Optic nerve
- **D**   Lens
- **E**   Vitreous body
- **F**   Retina
- **G**   Tapetum lucidum – not found in human eyes. Nocturnal animals possess this shiny layer to improve night vision.

Using scissors, carefully cut the eye in half along the coronal plane of section.

Sheep eye dissection

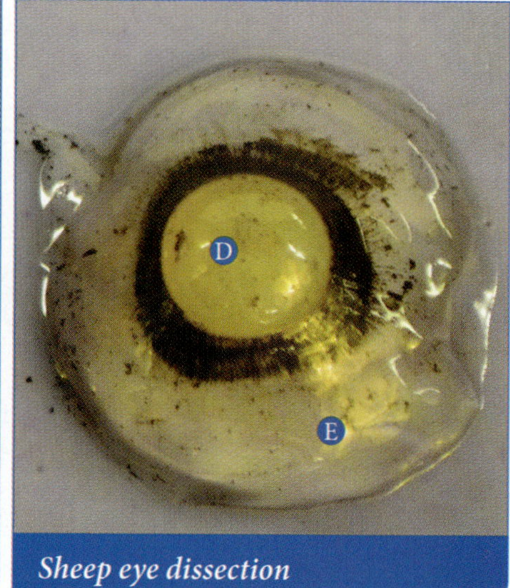

Sheep eye dissection

4 Blood

Plasma and the Formed Elements

Blood is a powerful symbol in human culture. It can be used to define relationships (blood relative) and can evoke emotional responses (fainting or nausea). We only have about five liters of blood. A little bit of blood looks like a lot! Blood carries food and oxygen to our cells and takes wastes to the kidneys. Its pH, like most body fluid, is 7.35–7.45. By volume, just over half of blood is plasma, the watery portion with dissolved substances. The remainder is blood cells, mostly erythrocytes, or red blood cells (RBCs). There are about five million RBCs/µl and about 50 µl in one drop, so there are about 250 million RBCs in one drop of blood! Erythrocytes are responsible for delivering oxygen to tissues and converting carbon dioxide into bicarbonate ions and protons (a reversible reaction) in the blood stream. Leukocytes (white blood cells) can emigrate, or leave the bloodstream, to an infected or injured area, and are responsible for our immunity.

Plasma

- **Water** – Of the 55% plasma in whole blood, most of it is water.

- **Plasma proteins** – Important in maintaining blood osmolarity (colloid osmotic pressure). About 7% of plasma consists of protein.

  - *albumin* (60%) – Transports many types of molecules and buffers blood pH.

  - *globulins* (35%) – Clotting enzymes, transport and immunity.

  - *fibrinogen* (5%) – Involved in coagulation (forming a clot).

  Thrombin (a protease enzyme) cleaves fibrinogen into shorter strands of insoluble fibrin. Fibrin strands are linked together by Factor XIII into polymer strands that form web-like netting to trap platelets and cells.

- **Nutrients** – Glucose, amino acids, fatty acids, lipoproteins, and vitamins.

- **Electrolytes** – Sodium, chloride, calcium, bicarbonate, potassium.

- **Wastes** – Urea, uric acid, creatinine, creatine, and bilirubin.

- **Gases** – Carbon dioxide, oxygen, and nitrogen.

- **Hormones** – Endocrine hormones are transported in blood.

Formed elements

- **Erythrocytes**, or red blood cells (RBCs) are by far the most abundant blood cell. RBCs make up about 45% of the volume of whole blood and are responsible for transporting oxygen, and converting carbon dioxide to bicarbonate and protons and visa versa. Hemoglobin is the oxygen carrying protein that is responsible for the red appearance of blood.

- **Leukocytes**

  - **Neutrophils** – 60–70% of all leukocytes

  - **Eosinophils** – 2–4% of all leukocytes

  - **Basophils** – 0.5–1%

  - **Lymphocytes**

  - **Monocytes**

- **Platelets** – Involved in forming a blood clot. They are fragments of membrane and cytoplasm of a megakaryocyte.

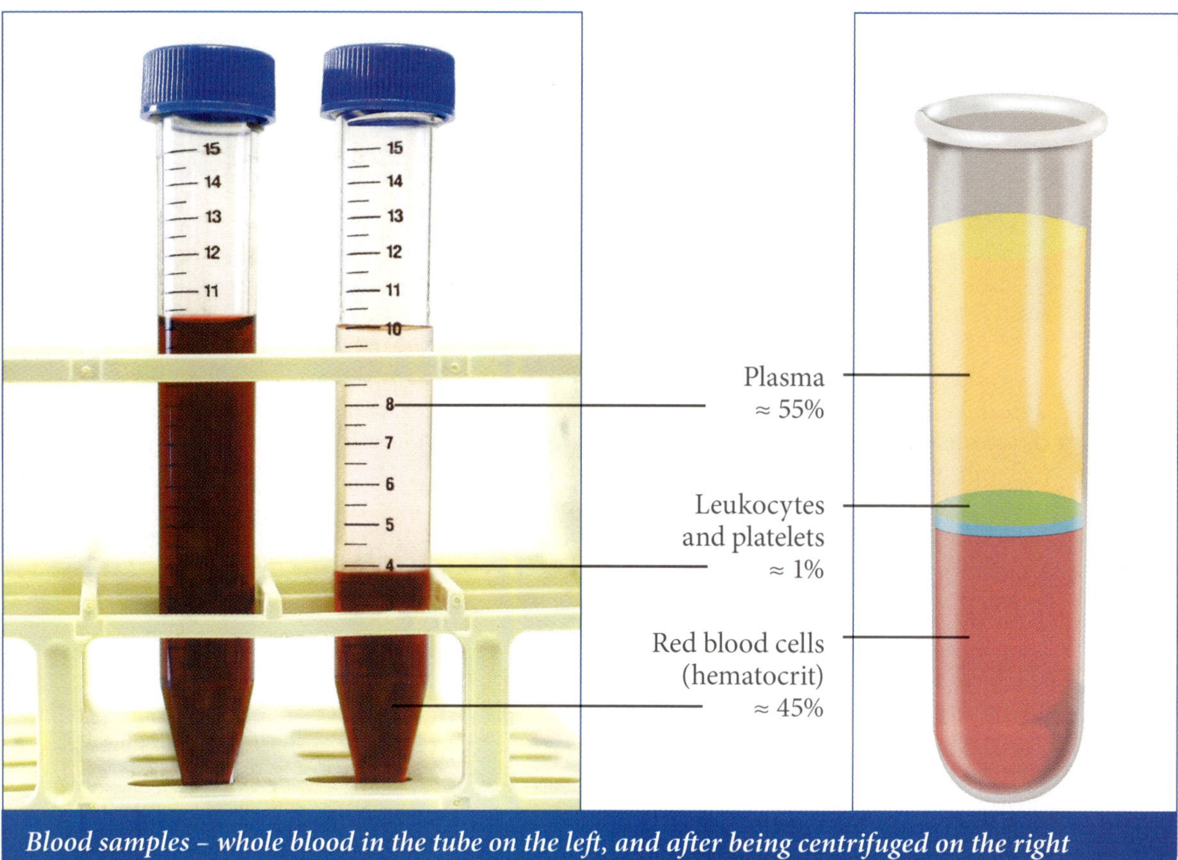

Blood samples – whole blood in the tube on the left, and after being centrifuged on the right

Whole blood is made of two primary components: plasma and cells. Each contributes to roughly half the volume of whole blood. When blood is put into a test tube containing heparin (an anticoagulant that interrupts fibrin formation) and spun in a centrifuge, cells are packed into the bottom half and plasma floats on top.

Plasma – A little over half the volume of whole blood, 55%, is plasma. This can vary over the course of a day depending on hydration levels. After drinking 32 ounces of water, the plasma percentage will increase (this results in a decrease of the percentage of red blood cells of whole blood, but not the number of red blood cells) and osmolarity of whole blood will drop slightly. If the body is deprived of water for a day, the percentage of plasma will decrease and osmolarity of whole blood will increase.

92%   Water – Of the 55% plasma in whole blood, most of it is water.

6%   Plasma proteins – Important in maintaining blood osmolarity (colloid osmotic pressure)
     albumin (60%) – Transports many types of molecules and buffers blood pH
     globulins (35%) – Three types based on size. Clotting enzymes, transport, and immunity
     fibrinogen (5%) – Involved in coagulation (forming a clot)

2%   Nutrients – Glucose, amino acids, fatty acids, lipoproteins, and vitamins

Electrolytes – Sodium, chloride, calcium, bicarbonate, potassium

Wastes – Urea, uric acid, creatinine, creatine, and bilirubin

Gases – Carbon dioxide, oxygen, and nitrogen

Hormones – All endocrine hormones travel in the plasma

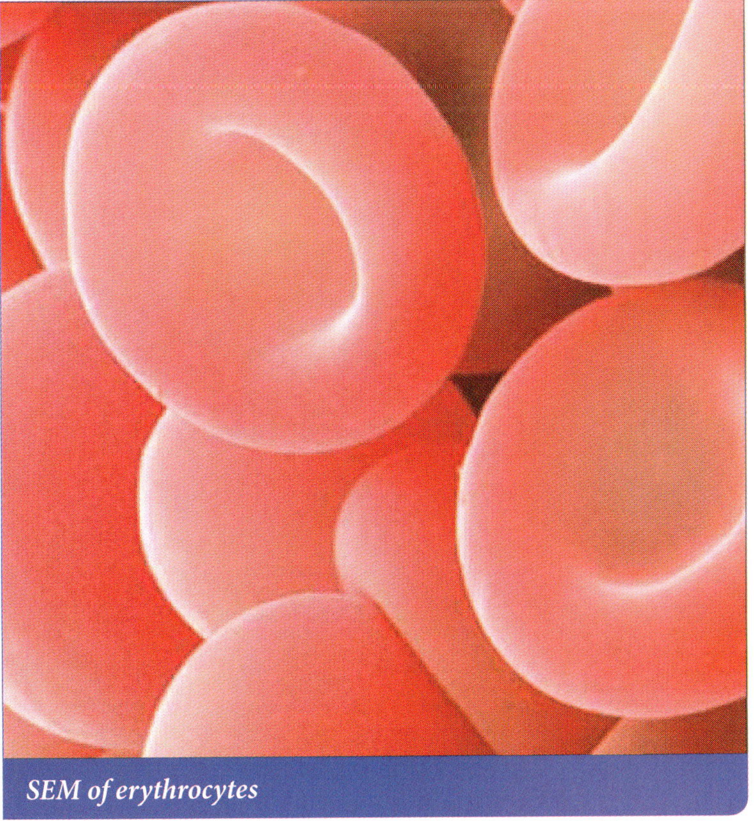

SEM of blood

Erythrocytes, or red blood cells (RBCs), are by far the most abundant blood cell. They have a diameter of 7 μm and are described as a biconcave disk, a shape that increases surface area and flexibility. RBCs make up about 45% of the volume of whole blood and are responsible for transporting oxygen and converting carbon dioxide to bicarbonate and protons. Every second of your life, you produce about 2.3 million RBCs, or 200 billion a day! RBCs circulate about three months, die, and are broken down in the spleen. Hemoglobin is the oxygen carrying protein that is responsible for the red appearance of blood. Hemoglobin molecules make up about one third of the cytoplasm (250 million molecules) of a red blood cell. This quaternary structured protein can carry four molecules of oxygen. 4 O_2 molecules/hemoglobin molecule $\times$ 250,000,000 hemoglobin molecules/RBC = 1 RBC. One RBC can transport one billion oxygen molecules!

SEM of erythrocytes

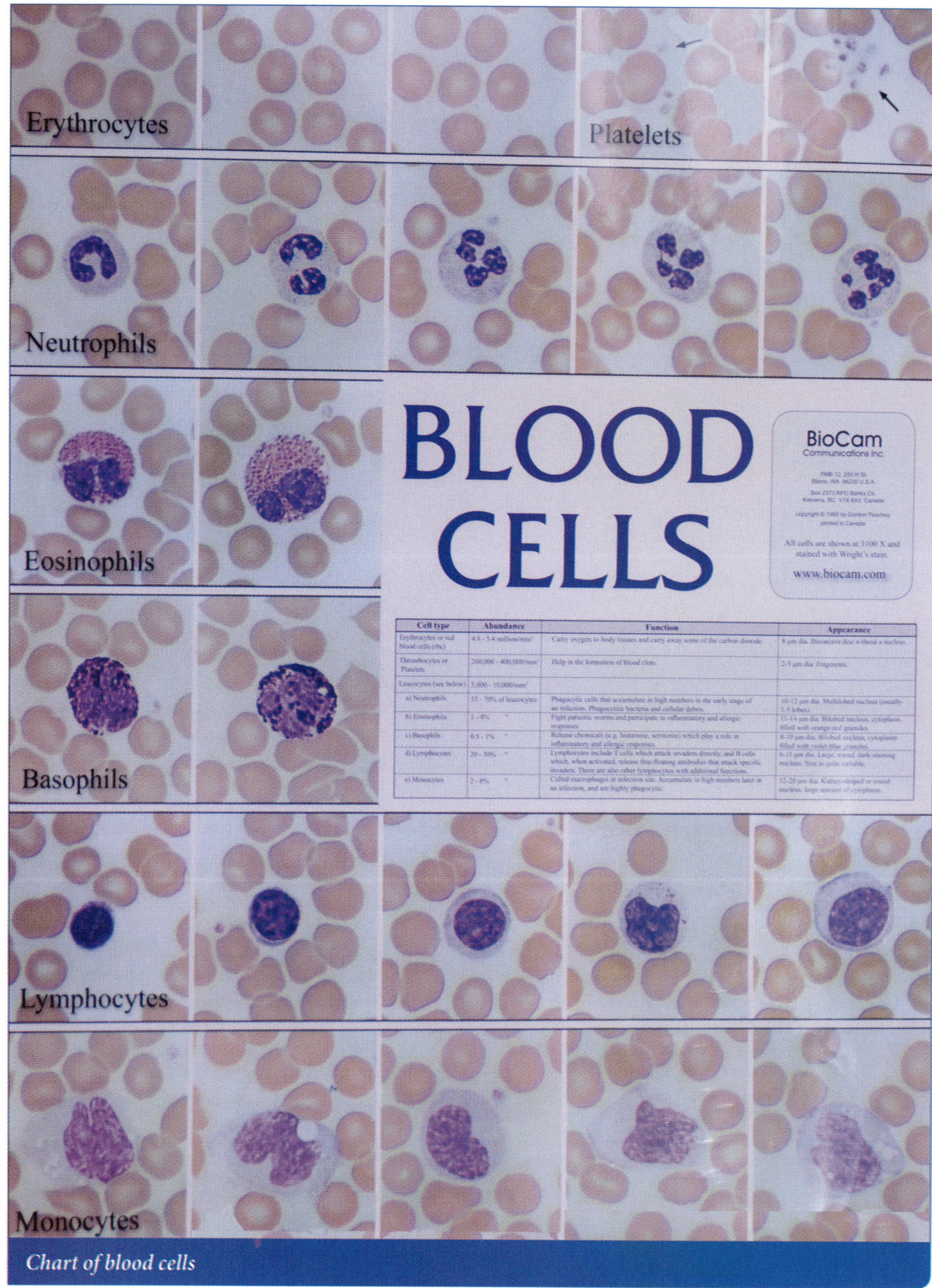

Chart of blood cells

Granular leukocytes end with the suffix, "phil," meaning attraction (in this case, specific cytoplasmic granules attract the stain used in preparing the blood slide).

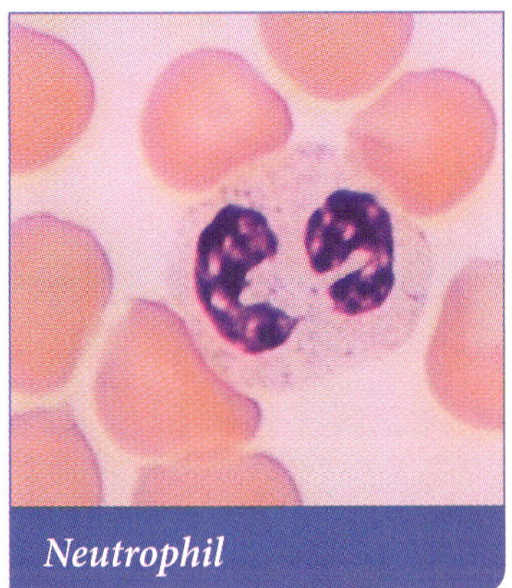

Neutrophil

60–70% of all leukocytes

Bacteria killers. Perform phagocytosis of bacteria and releases hydrogen peroxide, hypochlorite, and superoxide into tissue, killing many bacteria and themselves.

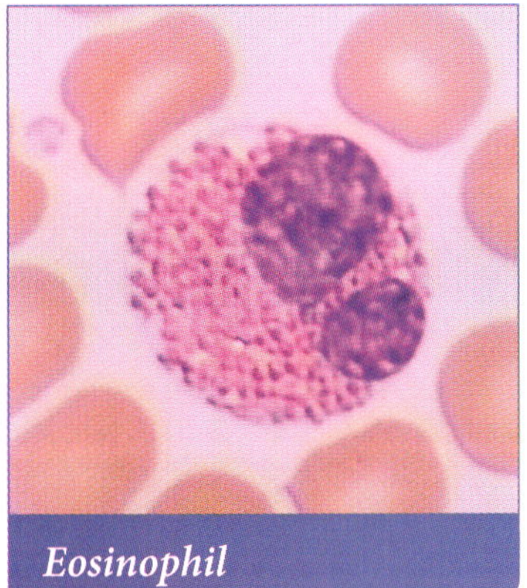

Eosinophil

2–4% of all leukocytes

Produce hydrogen peroxide and superoxide, phagocytize antibody tagged antigen, promote basophil action, and release enzymes that break down histamine.

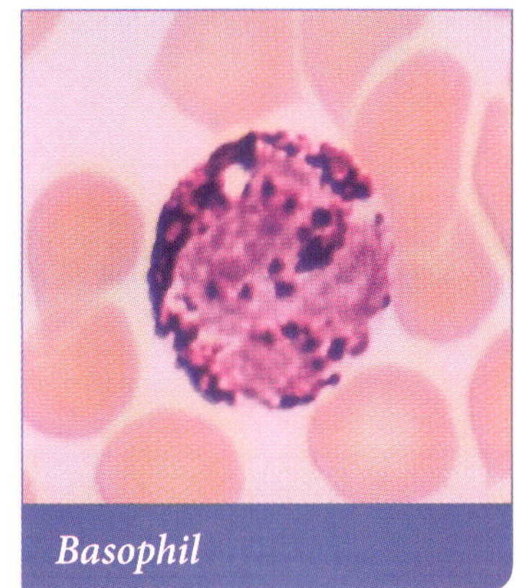

Basophil

0.4–1% of all leukocytes

Secrete histamine (vasodilator) and heparin (anticoagulant) to increase and maintain blood flow, and leukotrienes that attract and promote neutrophils and eosinophils.

Agranular leukocytes end with the suffix, "cyte."

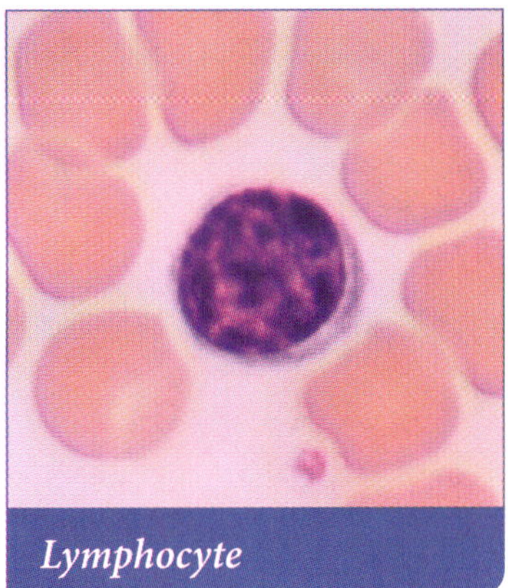

Lymphocyte

25–30% of all leukocytes

Have a variety of types and functions. T cells and B cells are responsible for specific immunity (antibody production in response to non-self antigens). NK (natural killer) cells hunt and kill bacteria, cells with viruses, and cancer cells.

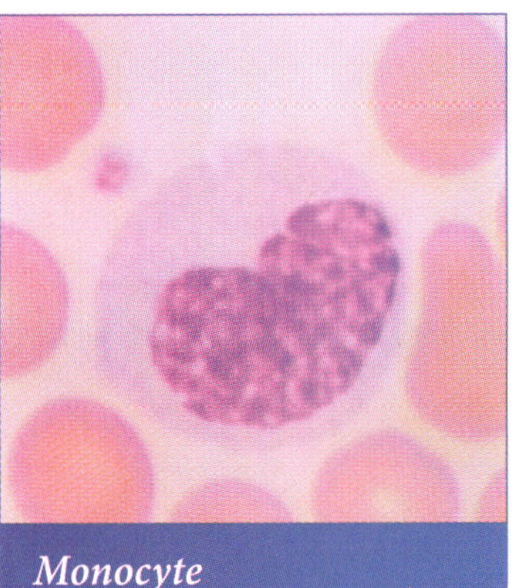

Monocyte

3–8% of all leukocytes

Leave the circulatory system and become wandering or fixed macrophages. Macrophages are phagocytosis machines that reside in connective tissue throughout the body but are no longer classified as leukocytes.

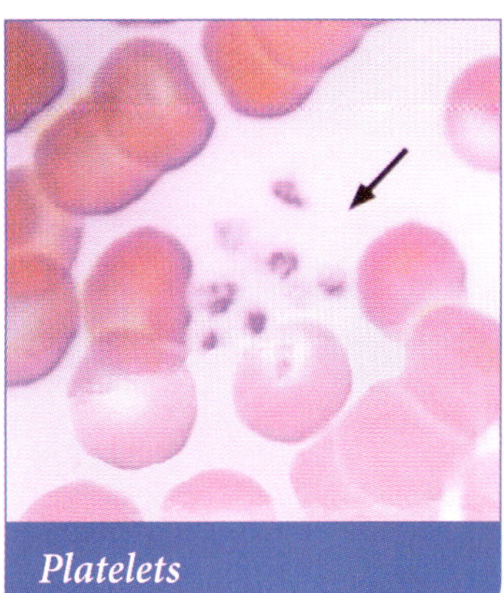

Platelets

Platelets are abundant (250,000,000/µl) but very small (2–3 µm in diameter).

Arise from a huge cell called a megakaryocyte that lives in red bone marrow. Projections extend into sinusoid areas and break off. These fragments of membrane and cytoplasm are platelets.

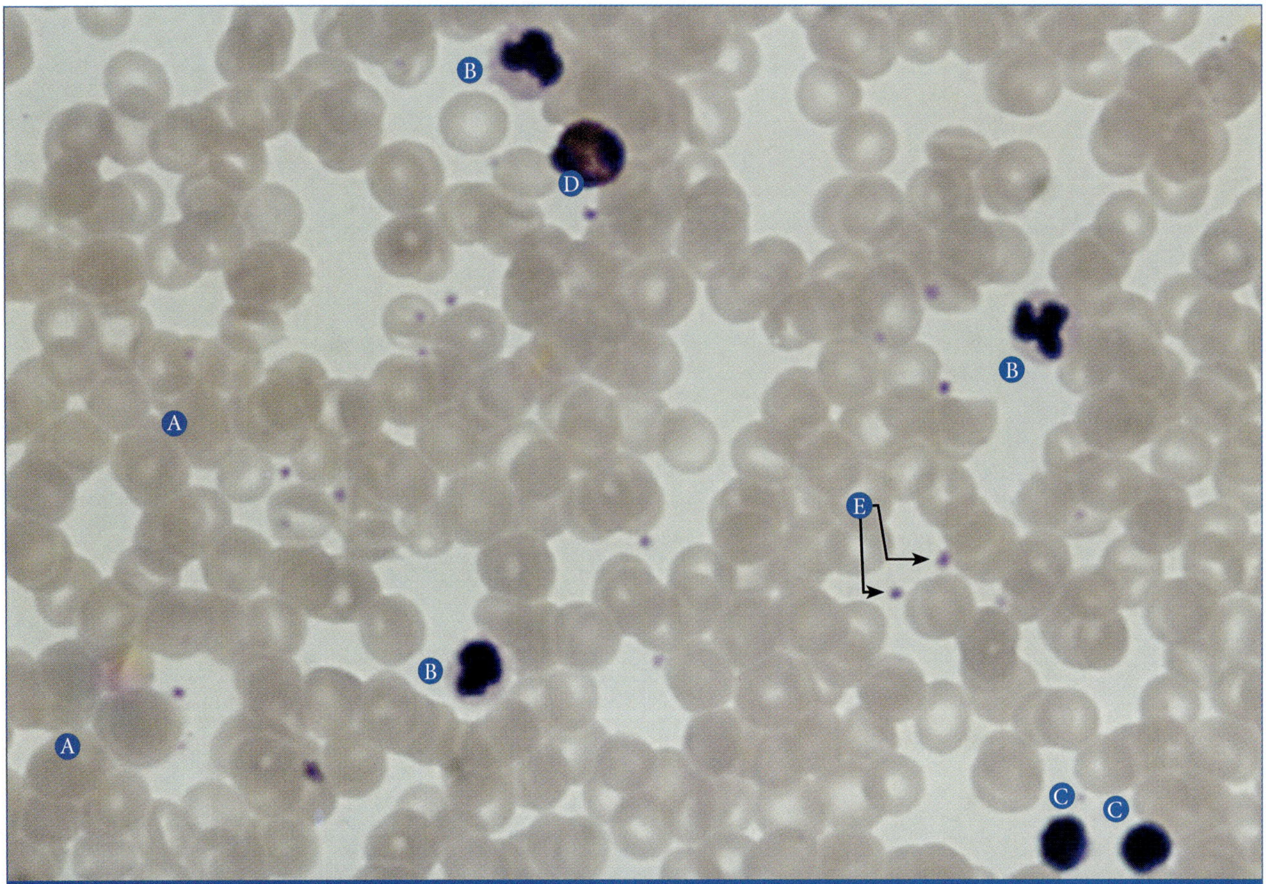

Photomicrograph of blood smear

Ⓐ   Erythrocytes (majority of blood cells—99%)

Ⓑ   Neutrophils

Ⓒ   Lymphocytes

Ⓓ   Eosinophil

Ⓔ   Platelets

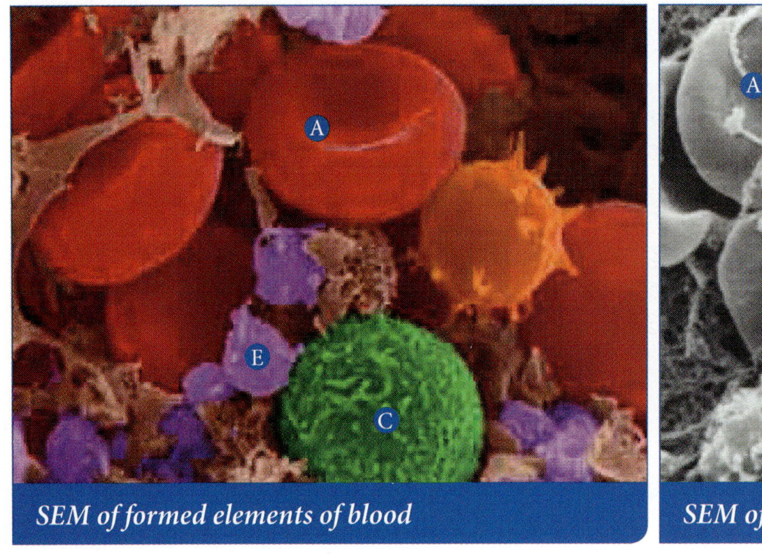

SEM of formed elements of blood

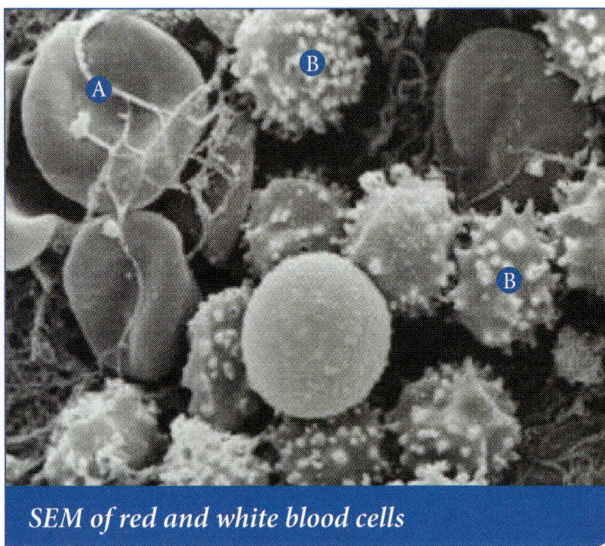

SEM of red and white blood cells

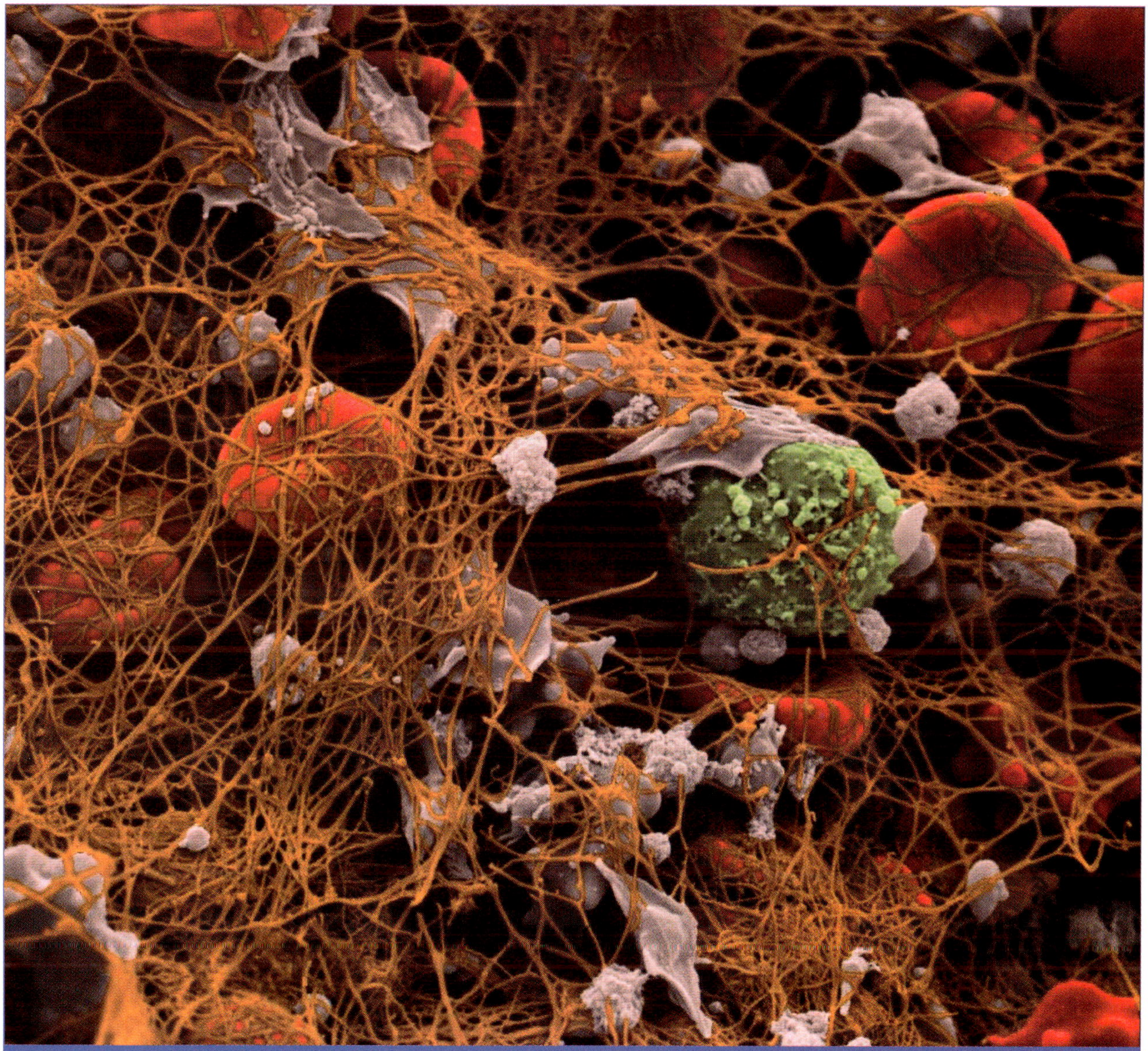

SEM of clotting blood

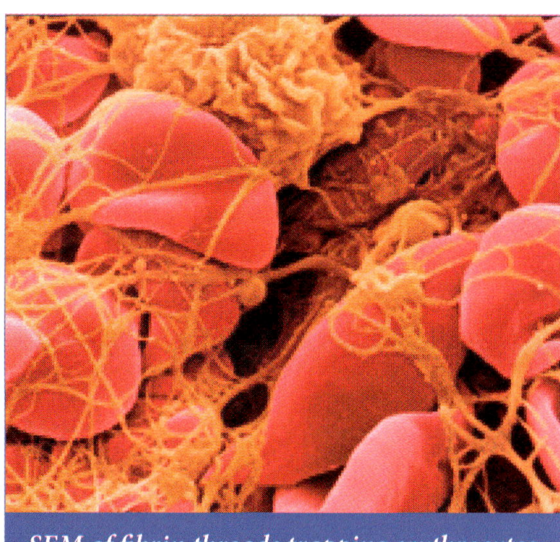

SEM of fibrin threads trapping erythrocytes

Shown here are two colorized SEMs of the last stage of hemostasis, coagulation (blood clotting). Red blood cells are shown in red, the leukocyte is green, platelets are gray, and the web-like yellow strands are fibrin, the protein that traps platelets and cells, forming a clot.

Both extrinsic and intrinsic mechanisms of coagulation lead to the activation of Factor X, which when combined with activated Factor V, calcium, and PF_3 (platelet factor 3), becomes prothrombinase, an enzyme. This enzyme converts prothrombin into thrombin. Thrombin (a protease enzyme) cleaves fibrinogen into shorter strands of insoluble fibrin. Fibrin strands are linked together by Factor XIII into polymer strands that link with each other to form web-like netting to trap platelets and cells. All of the above molecules are found in the plasma at all times.

The Heart

Cardiac Structures and Autorythmic Electrical Activity

The heart is a muscular pump that begins beating during the fourth week of embryonic development and never stops until death. The heart receives blood during the relaxation phase (diastole) from the lungs via the pulmonary veins, and from the body via the vena cava into the left and right atrium, respectively. During systole (contraction) equal amounts of blood are ejected from each ventricle. The right ventricle delivers blood at a relatively low blood pressure (40 mmHg) to the lungs and the pulmonary circuit, and the left ventricle supplies the systemic circuit, or the rest of the body, at higher pressure (120 mmHg). A normal blood pressure reading of 120/80 mmHg is achieved by left ventricular systole (120 mmHg) and diastole (80 mmHg).

Chambers and valves of the heart

- Right atrium
- Right auricle
- Tricuspid (right AV) valve
- Chordae tendineae
- Papillary muscle
- Right ventricle
- Pulmonary semilunar valve
- Left atrium
- Left auricle
- Bicuspid (mitral or left AV) valve
- Left ventricle
- Aortic semilunar valve

Blood vessels

- Superior vena cava
- Inferior vena cava
- Pulmonary trunk
- Pulmonary artery
- Pulmonary veins
- Aorta
- Aortic arch
- Brachiocephalic trunk
- Right common carotid artery
- Right subclavian artery
- Left common carotid artery
- Left subclavian artery

Blood vessels of the myocardium

- Anterior interventricular artery
- Great cardiac vein
- Right coronary artery
- Circumflex artery
- Coronary sinus
- Posterior interventricular artery
- Anterior interventricular vein

Electrocardiogram (ECG)

- P-wave
- QRS complex
- T-wave
- Atrial depolarizarion
- Atrial repolarizarion
- Ventricular depolarizarion
- Ventricular repolarizarion

Cardiac muscle

- Cardiocytes
- Intercalated disks
- Mitochondria
- Striations
- Actin and myosin filaments

Ⓐ Intercalated disks

Ⓑ Nuclei

Ⓒ Mitochondria

Ⓓ Actin and myosin filaments

The photomicrograph to the right shows many cardiocytes and intercalated disks. Notice the light and dark bands (striations) along the cardiocytes.

Photomicrograph of cardiac muscle fibers 400×

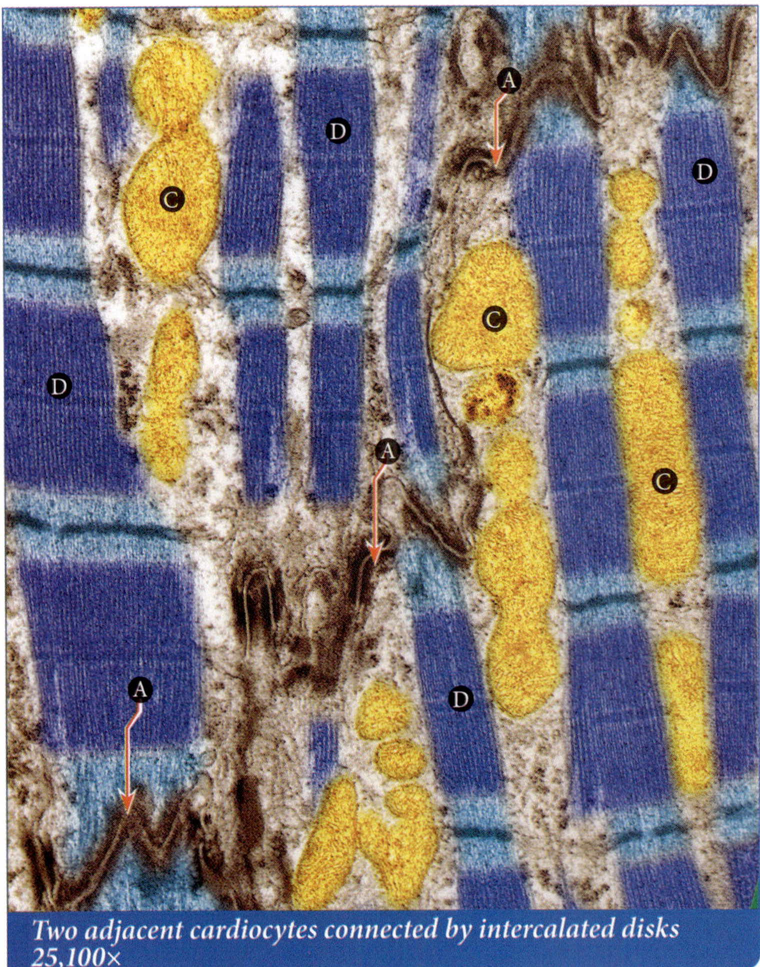

Two adjacent cardiocytes connected by intercalated disks 25,100×

The heart is primarily made of cardiac muscle (myocardium). A cardiocyte (cardiac muscle cell) is striated (alternating bands of actin and myosin), involuntary, and contains one central nucleus. Large mitochondria contribute approximately 25% of a cardiocyte's volume.

Intercalated disks form between adjacent cardiac muscle cells and contain many gap junctions. Gap junctions are openings where ions can travel from one cell to another, allowing depolarization and contraction of cardiac tissue virtually simultaneously.

The TEM to the left is an extreme close-up of two adjacent cardiocytes. Gap junctions connect the two cells with tube-like protein complexes that allow ion flow between the cells. These can be seen as intercalated disks (arrows). Myosin filaments are colored dark blue, actin filaments are light blue, mitochondria are yellow, and the intercalated disk (arrows) is a dark, wavy line.

- Ⓐ Superior vena cava
- Ⓑ Right auricle
- Ⓒ Pulmonary trunk
- Ⓓ Pulmonary artery
- Ⓔ Pulmonary veins
- Ⓕ Left auricle
- Ⓖ Aorta
- Ⓗ Aortic arch
- Ⓘ Brachiocephalic trunk
- Ⓙ Right common carotid artery
- Ⓚ Right subclavian artery
- Ⓛ Left common carotid artery
- Ⓜ Left subclavian artery
- Ⓝ Anterior interventricular artery
- Ⓞ Great cardiac vein
- Ⓟ Right coronary artery

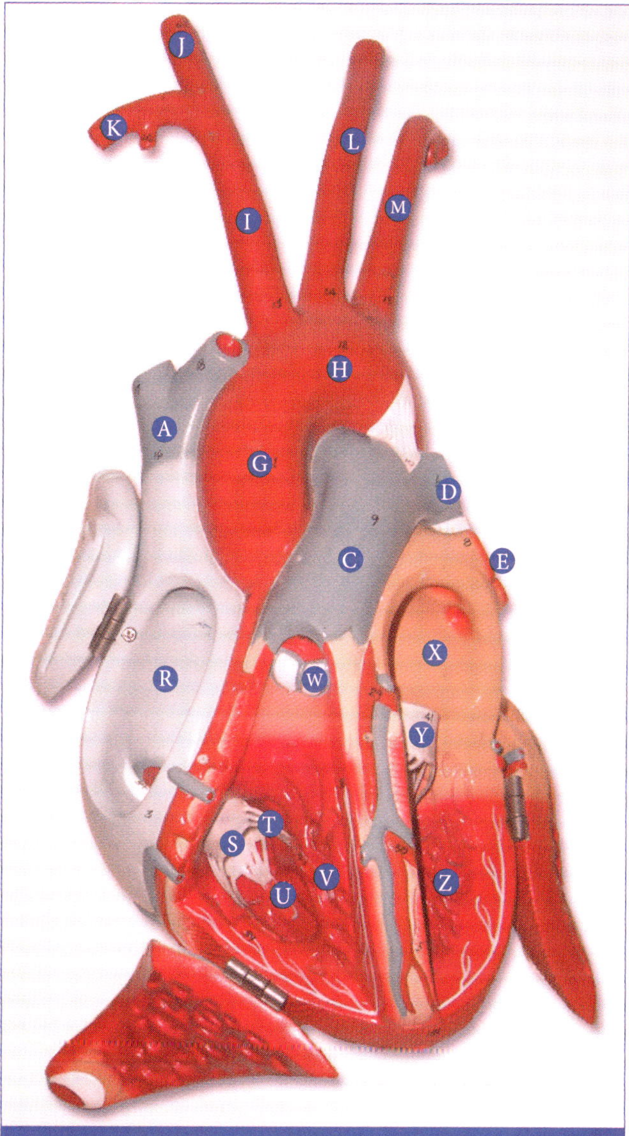

Anterior view of heart model with flaps opened

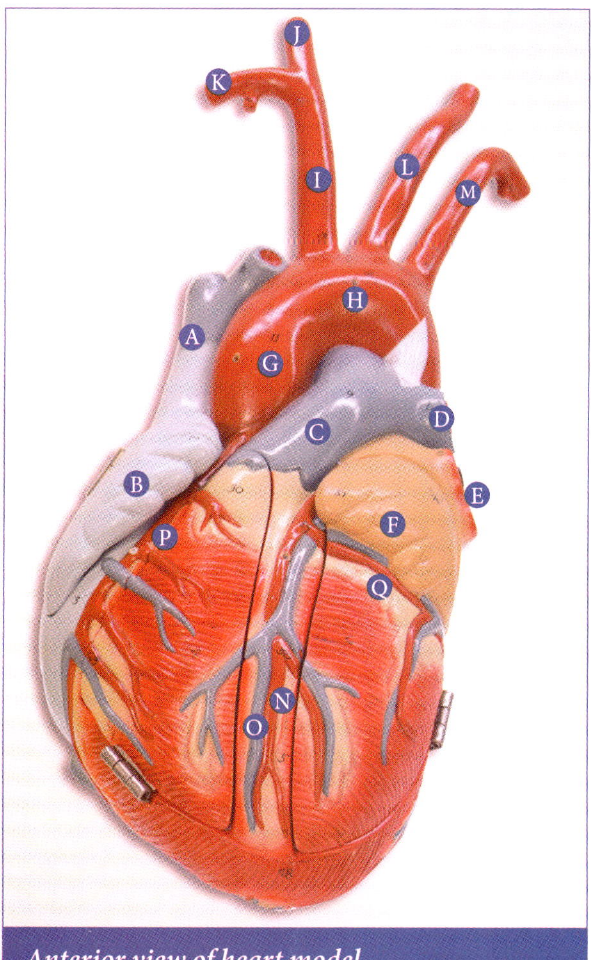

Anterior view of heart model

- Ⓠ Circumflex artery
- Ⓡ Right atrium
- Ⓢ Tricuspid (right AV) valve
- Ⓣ Chordae tendineae
- Ⓤ Papillary muscle
- Ⓥ Right ventricle
- Ⓦ Pulmonary semilunar valve
- Ⓧ Left atrium
- Ⓨ Bicuspid (mitral or left AV) valve
- Ⓩ Left ventricle

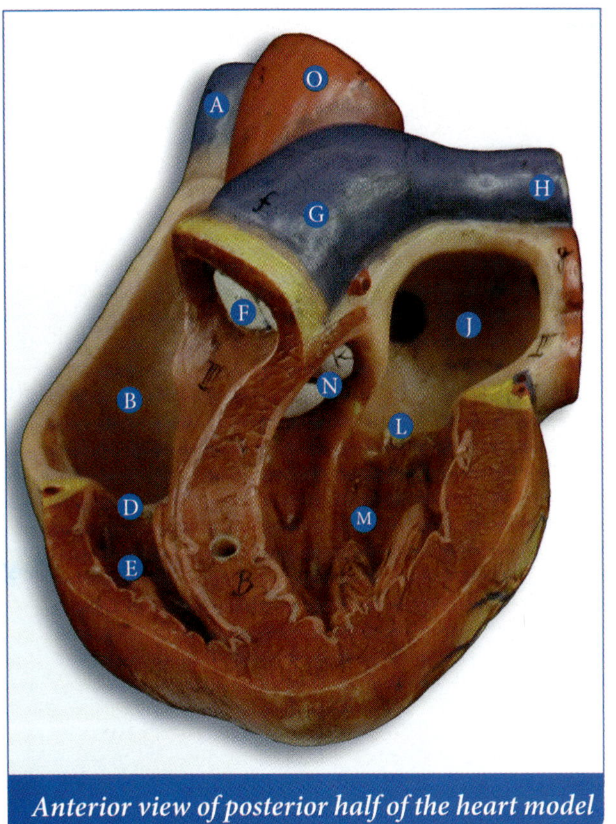

Anterior view of posterior half of the heart model

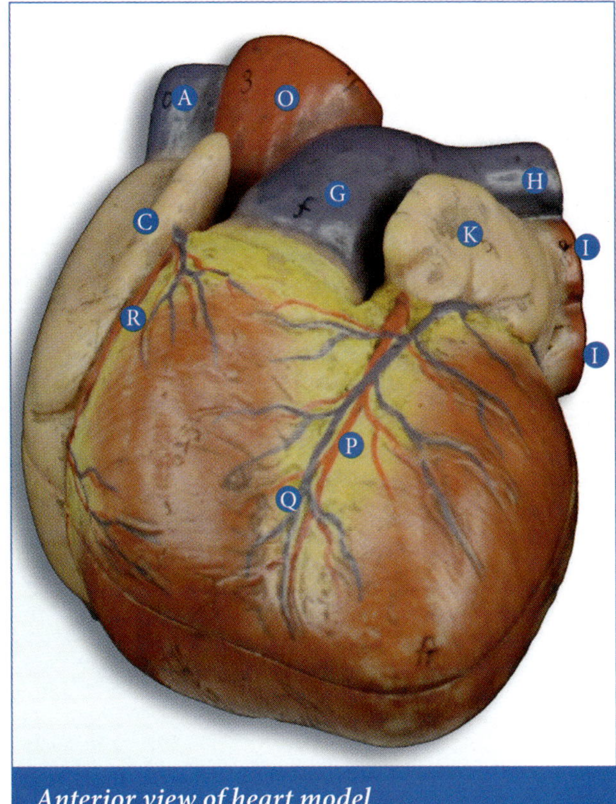

Anterior view of heart model

Posterior view of heart model

A Superior vena cava

B Right atrium

C Right auricle

D Tricuspid (right AV) valve

E Right ventricle

F Pulmonary semilunar valve

G Pulmonary trunk

H Pulmonary artery

I Pulmonary veins

J Left atrium

K Left auricle

L Bicuspid (mitral or left AV) valve

M Left ventricle

N Aortic semilunar valve

O Aorta

P Anterior interventricular artery

Q Great cardiac vein

R Right coronary artery

S Coronary sinus

A. Superior vena cava
B. Inferior vena cava
C. Right auricle
D. Pulmonary trunk
E. Pulmonary artery
F. Pulmonary veins
G. Left auricle
H. Aorta
I. Brachiocephalic trunk
J. Left common carotid artery
K. Left subclavian artery
L. Descending aorta

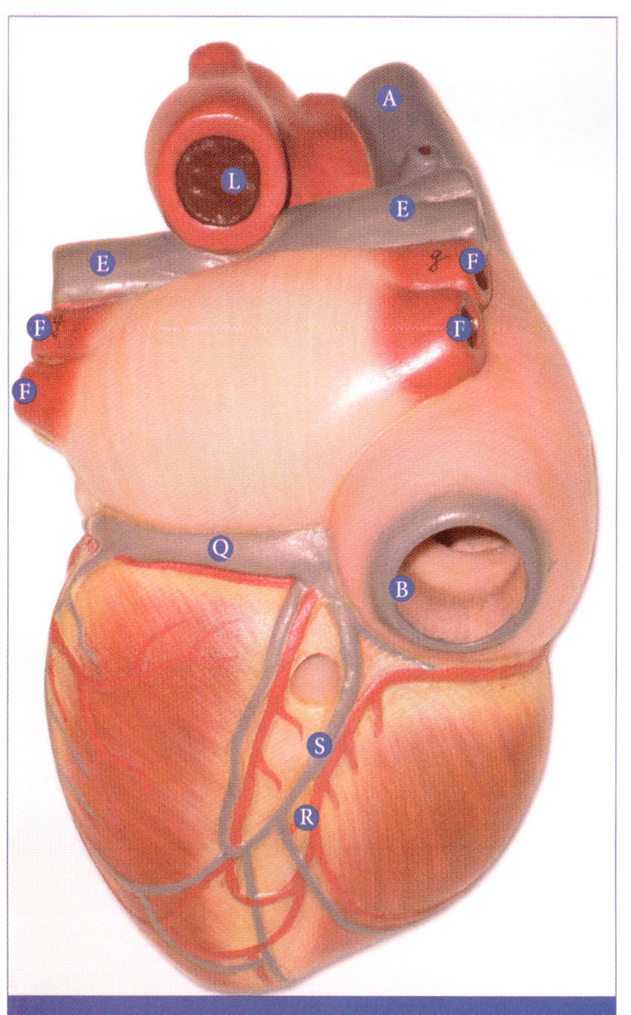

Anterior view of heart model

Posterior view of heart model

M. Anterior interventricular artery
N. Great cardiac vein
O. Right coronary artery
P. Circumflex artery
Q. Coronary sinus
R. Posterior interventricular artery
S. Posterior interventricular vein

A. Superior vena cava
B. Inferior vena cava
C. Right atrium
D. Right auricle
E. Right ventricle
F. Pulmonary trunk
G. Pulmonary artery
H. Pulmonary veins
I. Left atrium
J. Left auricle
K. Left ventricle
L. Aorta
M. Descending aorta
N. Anterior interventricular a.
O. Great cardiac vein
P. Right coronary artery
Q. Circumflex artery
R. Esophagus
S. Trachea

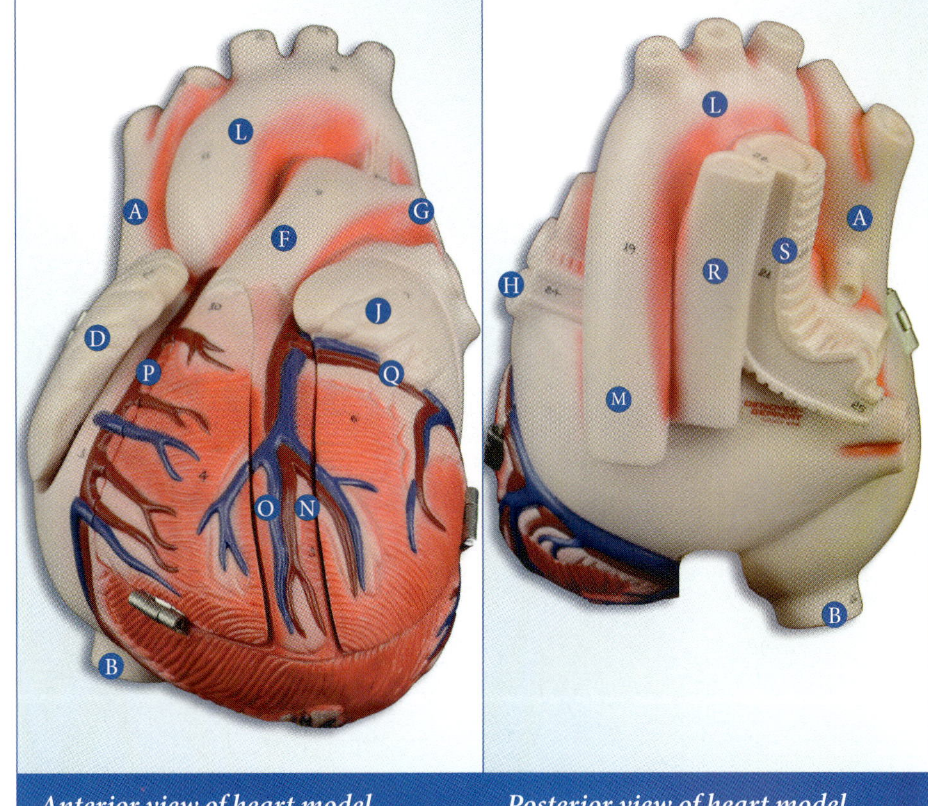

Anterior view of heart model

Posterior view of heart model

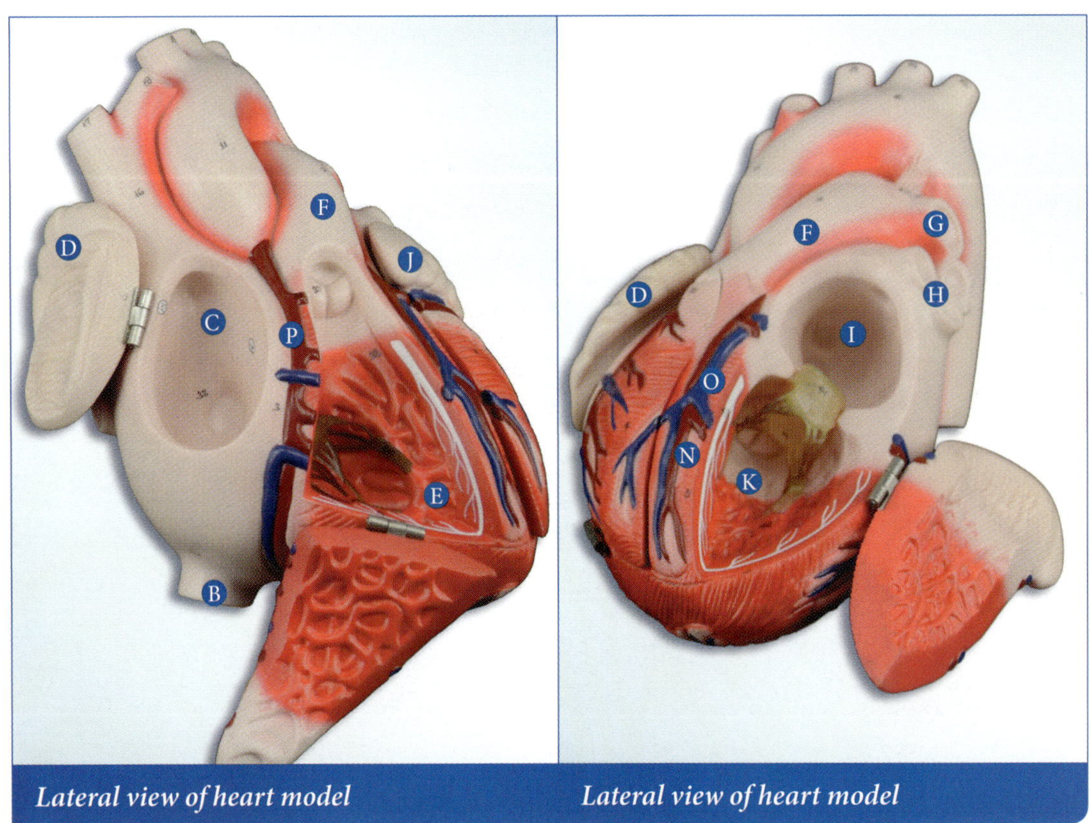

Lateral view of heart model

Lateral view of heart model

Ⓐ Superior vena cava

Ⓑ Inferior vena cava

Ⓒ Right auricle

Ⓓ Pulmonary trunk

Ⓔ Pulmonary artery

Ⓕ Pulmonary veins

Ⓖ Left auricle

Ⓗ Aorta

Ⓘ Anterior interventricular artery

Ⓙ Great cardiac vein

Ⓚ Right coronary artery

Ⓛ Circumflex artery

Ⓜ Coronary sinus

Ⓝ Posterior interventricular artery

Ⓞ Posterior interventricular vein

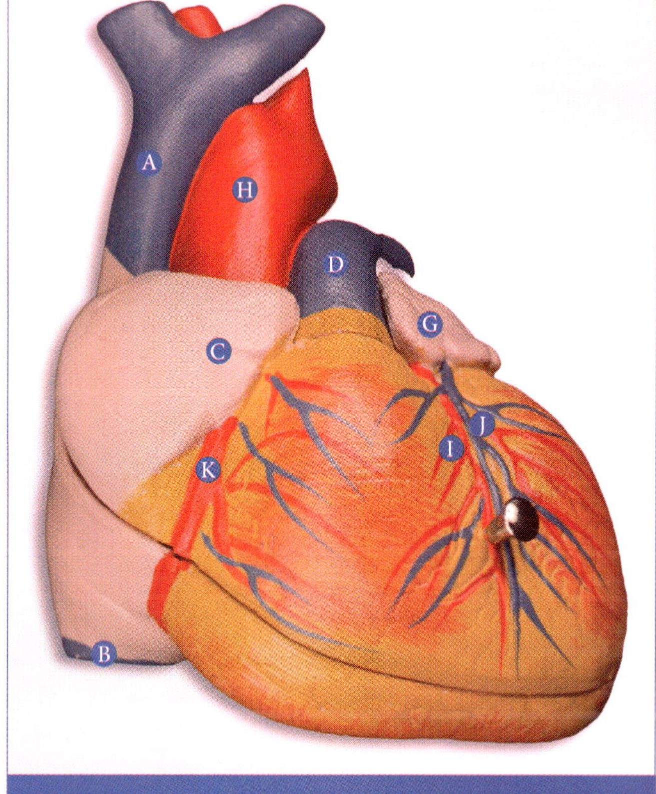

Anterolateral view of heart model

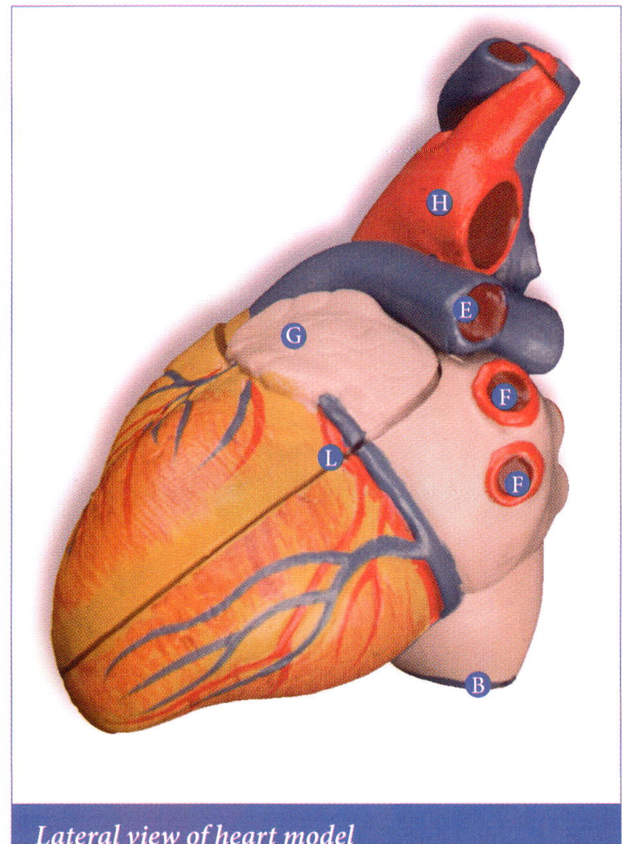

Lateral view of heart model

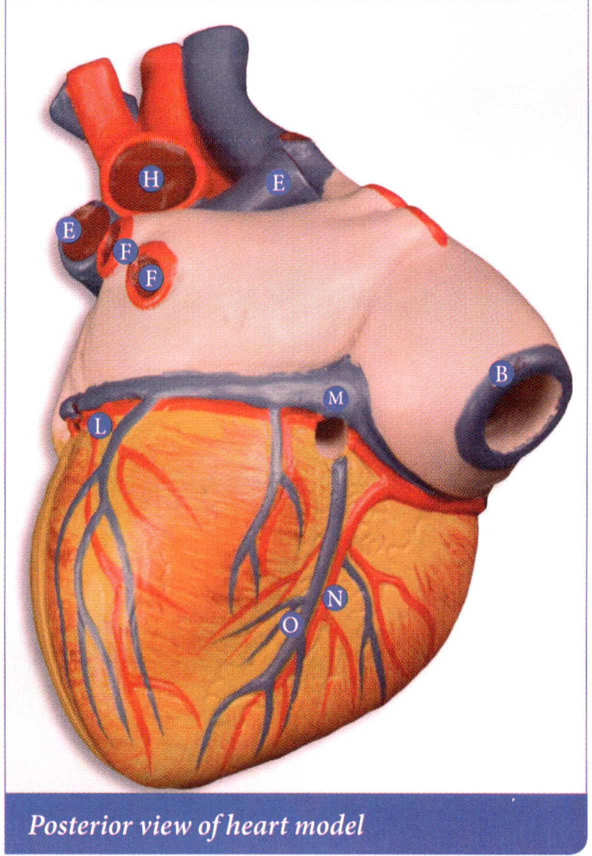

Posterior view of heart model

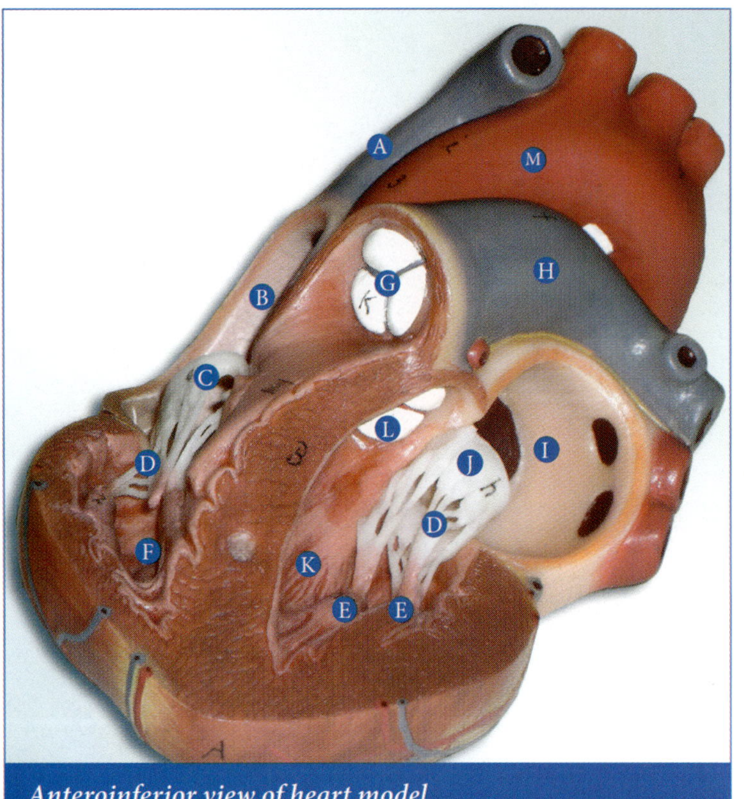

Anteroinferior view of heart model

As the heart enters diastole, blood begins to fill the chambers passively (elastic recoil and pressure differential). During atrial systole, additional blood is forced into each ventricle. As the ventricles contract (ventricular systole), pressure builds and exceeds the pressure in the atria, causing the bicuspid and tricuspid to slam shut. This creates turbulent blood flow and vibrations of the heart wall, which can be heard as the first heart sound, Lubb. Equal amounts of blood are ejected from each ventricle. Ventricular ejection continues until the pressure in the aorta is greater than the pressure in the left ventricle, forcing the aortic semilunar valve to close, creating turbulent flow and vibrations that are heard as the second heart sound, Dupp. The pulmonary semilunar valve also closes at this time but because of relatively low blood pressure in the pulmonary circuit, little vibration and low sound volume is produced.

A   Superior vena cava

B   Right atrium

C   Tricuspid (right AV) valve

D   Chordae tendineae

E   Papillary muscle

F   Right ventricle

G   Pulmonary semilunar valve

H   Pulmonary trunk

I   Left atrium

J   Bicuspid (mitral or left AV) valve

K   Left ventricle

L   Aortic semilunar valve

M   Aorta

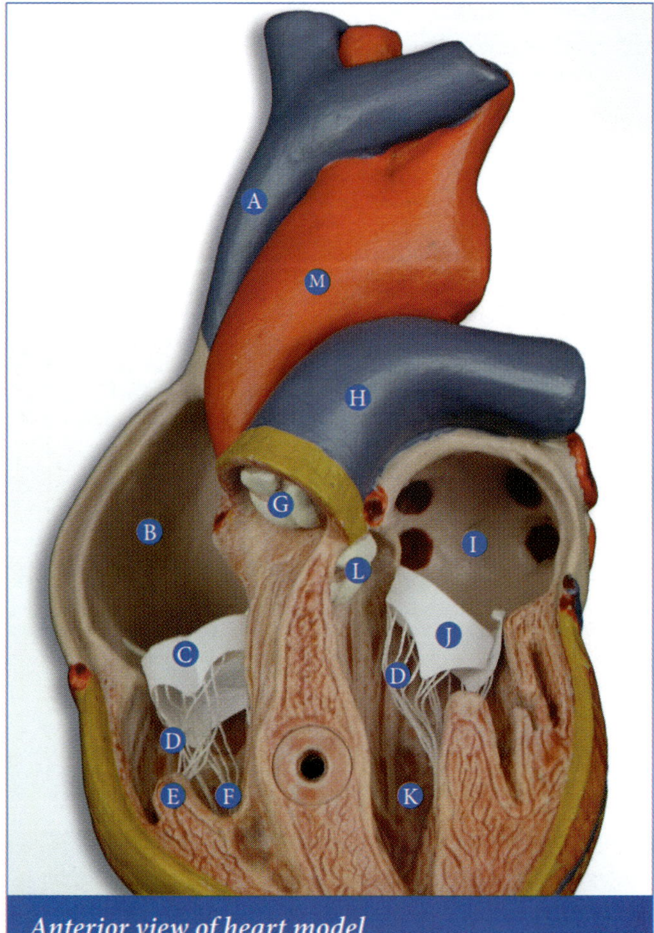

Anterior view of heart model

Heart sound 1, known as Lubb, is created by vibrations caused by the closing of the bicuspid **F** (mitral or left AV) valve and to a lesser extent, the tricuspid **A** (right AV) valve.

Heart sound 2, known as Dupp, is created by vibrations caused by the closing of the aortic semi-lunar valve **H** and to a lesser extent, the pulmonary semilunar valve **D**.

The left side of the heart is where most of the sound of a heartbeat is generated due to relatively high blood pressure, about 120/80 mmHg, compared to the right side, about 40 mmHg.

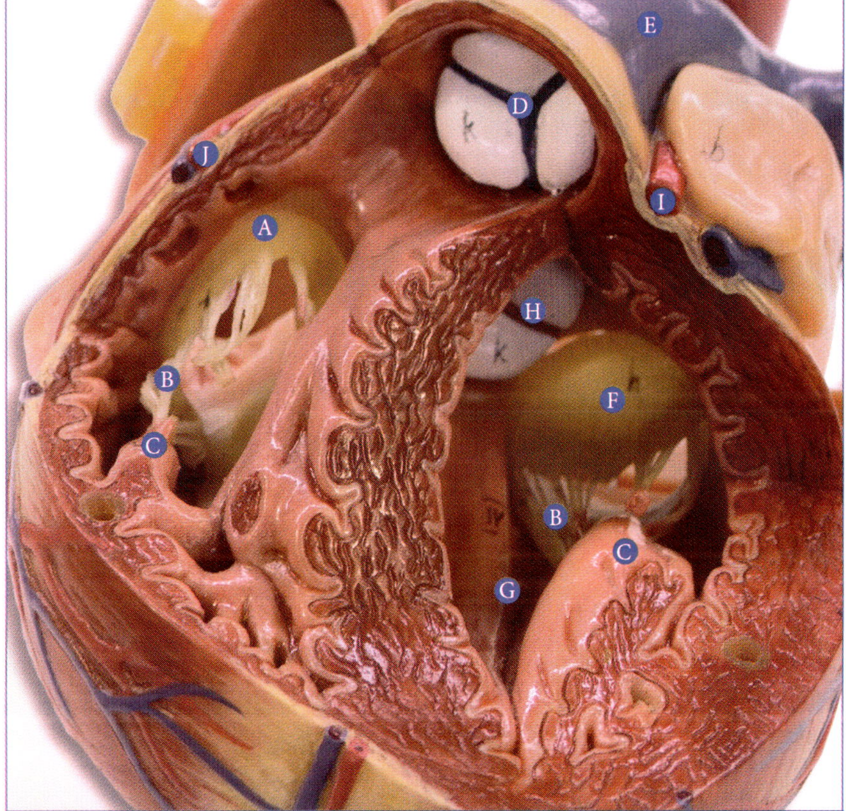

Anteroinferior view of the heart showing all four valves

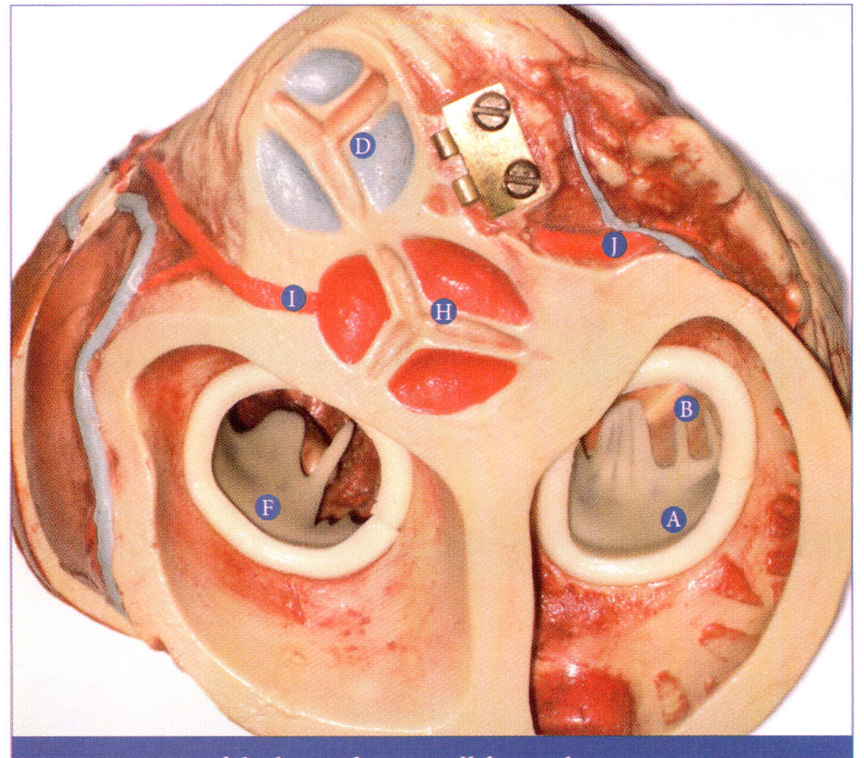

Superior view of the heart showing all four valves

A Tricuspid (right AV) valve

B Chordae tendineae

C Papillary muscle

D Pulmonary semilunar valve

E Pulmonary trunk

F Bicuspid (mitral or left AV) valve

G Left ventricle

H Aortic semilunar valve

I Left Coronary artery

J Right Coronary artery

Ⓐ Right atrium

Ⓑ Tricuspid (right AV) valve

Ⓒ Chordae tendineae

Ⓓ Papillary muscle

Ⓔ Right ventricle

Ⓕ Pulmonary trunk

Ⓖ Pulmonary arteries

Ⓗ Pulmonary veins

Ⓘ Left atrium

Ⓙ Bicuspid (mitral or left AV) valve

Ⓚ Left ventricle

Ⓛ Aorta

Ⓜ Right coronary artery

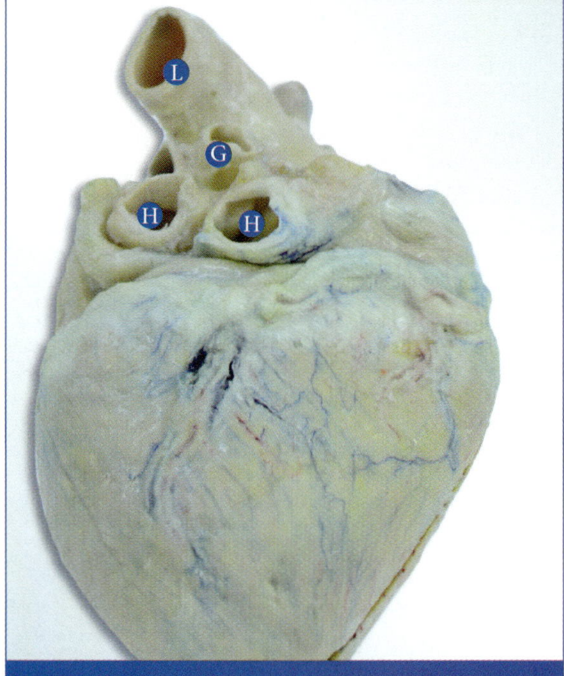

Posterior view of a plastinated sheep's heart

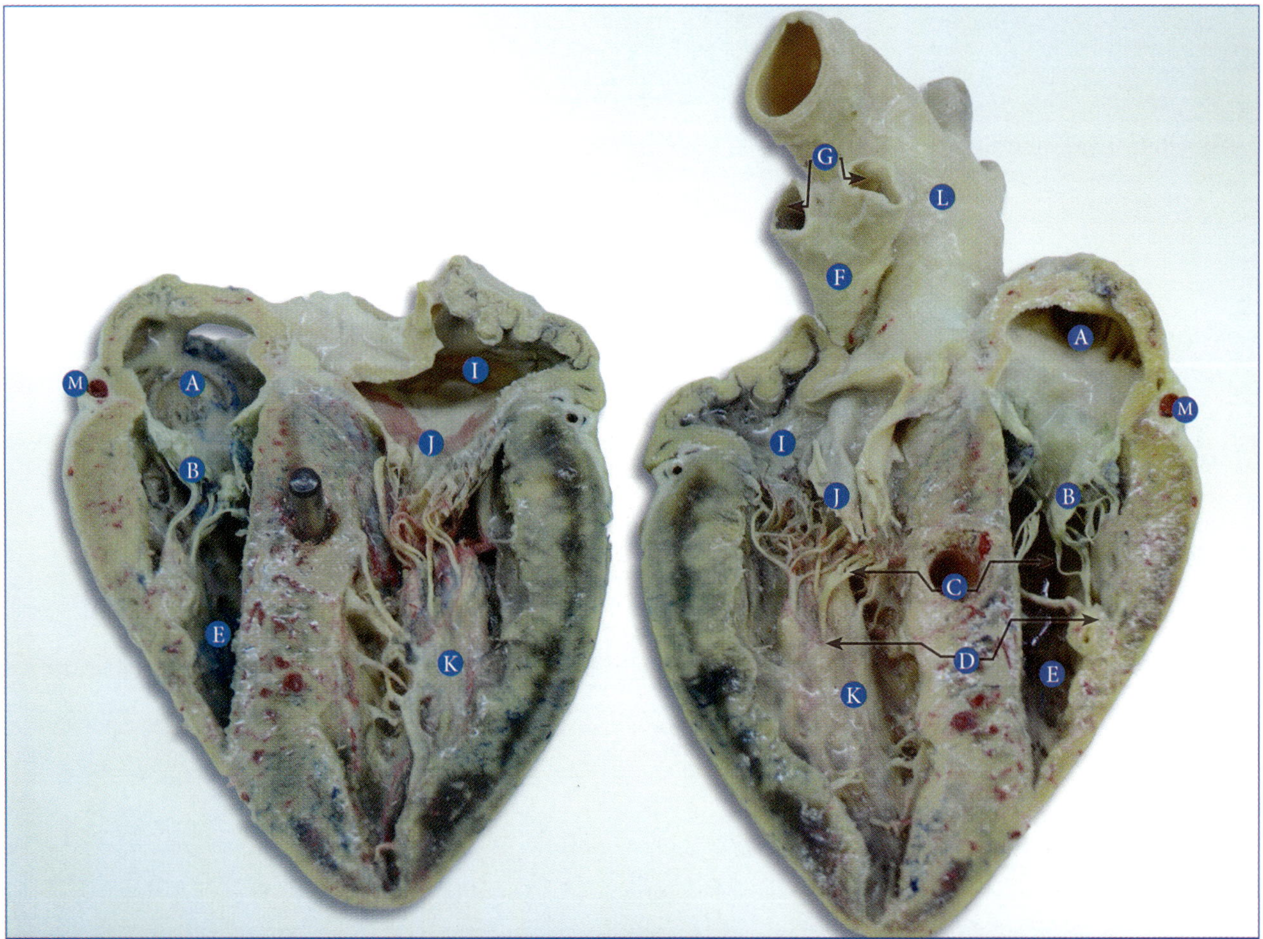

Coronal section of a plastinated sheep's heart

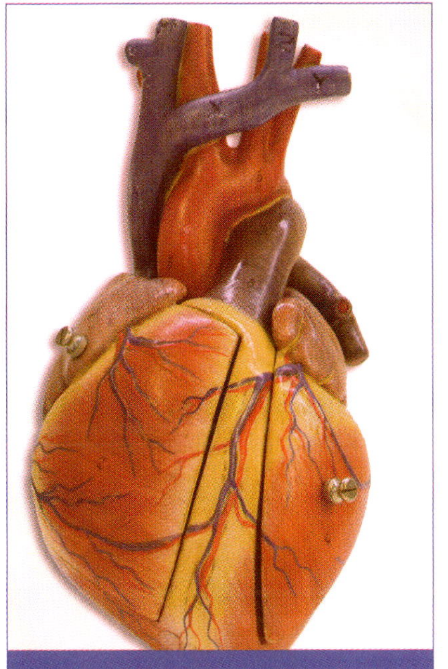

Anterior view of the fetal heart

The fetal heart has two unique structures that would be fatal in the adult heart: the foramen ovale and the ductus arteriosus. These structures allow most blood to bypass the lungs, since the lungs are fluid filled (amniotic fluid) and do not provide gas exchange. After parturition (child birth) and the first breath is taken, a drop in pressure occurs in the right atrium. A momentary flow of blood from the left atrium to the right atrium causes two flaps to shut and seal the foramen ovale, separating the two atria and eliminating the right to left shunt (bypass). This same pressure change causes the ductus arteriosus to collapse, allowing blood flow to the lungs via the pulmonary arteries.

- **A** Foramen ovale
- **B** Ductus arteriosus
- **C** Aorta
- **D** Pulmonary trunk
- **E** Pulmonary arteries
- **F** Pulmonary veins
- **G** Left atrium
- **H** Bicuspid (mitral or left AV) valve
- **I** Left ventricle

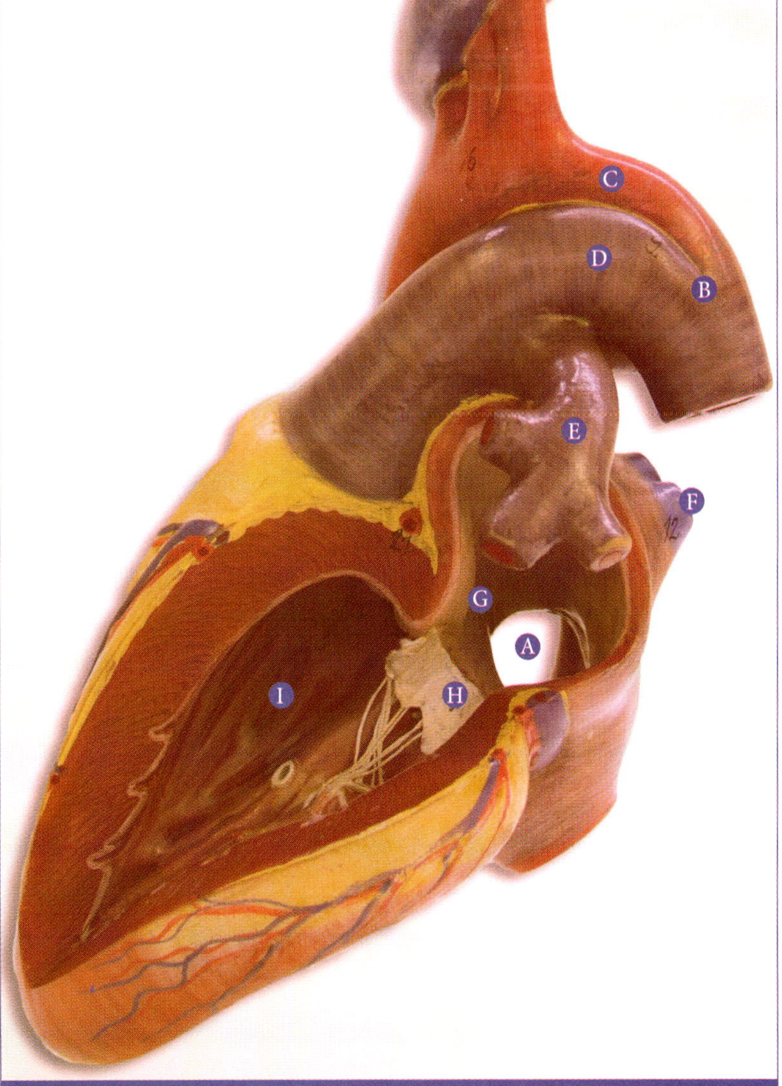

Lateral view of the fetal heart showing the foramen ovale and ductus arteriosus

The heart has an intrinsic electrical conduction system that causes a rhythmic contraction of the myocardium.

It begins when the sinoatrial (SA) node spontaneously depolarizes from an influx of sodium ions, bringing resting membrane potential to threshold. When the SA node (the pacemaker) fires, the atrial cardiocytes depolarize (transmitted through gap junctions of intercalated disks) and contract almost instantly. The depolarization wave spreads across the atria and to the atrioventricular (AV) node, causing it to depolarize and send a signal down the bundle branches which give rise to Purkinje fibers and cause the ventricular myocardium to contract. The ventricles contract about 200 milliseconds (2/10 of a second) after the SA node fires, allowing the ventricles to become further filled with blood from atrial contraction.

An electrocardiogram (ECG or EKG, the German spelling, electrokardiogram) is a recording made from the electrical activity of the heart. Electrodes are placed on the skin and connected to an electrocardiograph. As nodes fire and the myocardium depolarizes, slight changes in voltage can be detected throughout the body. The recording of a heartbeat produces unique waves that show atrial depolarization (P-wave), ventricular depolarization and atrial repolarization (QRS complex), and ventricular repolarization (T-wave).

A typical heartbeat consists of the following: The SA node fires, generating the P-wave. Milliseconds after the P-wave, both atria contract and add more blood into each ventricle. About 200 milliseconds after the SA node fires, the AV node depolarizes and delivers the signal to ventricular myocardium to depolarize, causing the QRS complex. Milliseconds after the QRS complex, each ventricle contracts and ejects blood from the heart, the right ventricle to the lungs and the left ventricle to the rest of the body. Shortly after ventricular systole, the T-wave appears as the ventricles repolarize.

The SA node intrinsically fires at a rate of about 100 action potentials, or beats per minute. The medulla oblongata contains nuclei of sympathetic and parasympathetic divisions of the autonomic nervous system that alter heart rate. At rest, this intrinsic rhythm is influenced by cranial nerve X, the vagus. Parasympathetic vagal tone reduces the resting heart rate to 60–80 beats per minute by releasing acetylcholine which bind to muscarinic receptors on the SA node, increasing time between each spontaneous depolarization. Sympathetic activity releases norepinephrine from sympathetic postganglionic neurons that increases heart rate and cardiac contractility. Concurrent inhibition of vagal tone is necessary. Release of catecholamines from the adrenal medulla further enhances cardiac output.

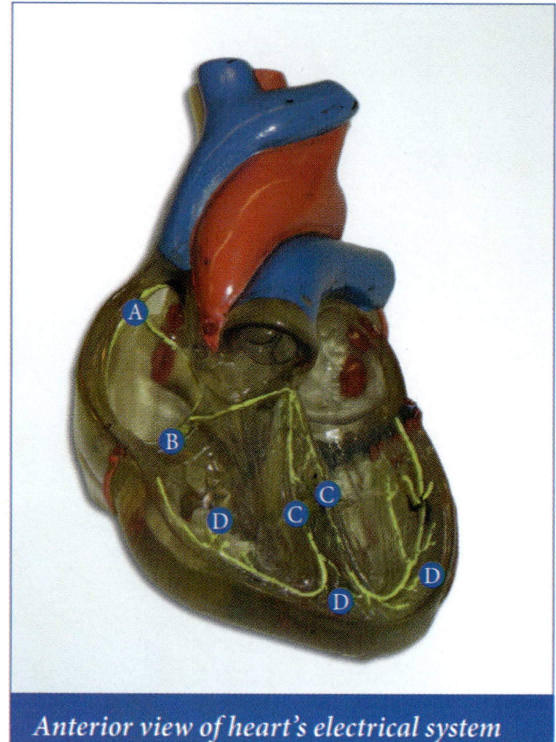

Anterior view of heart's electrical system

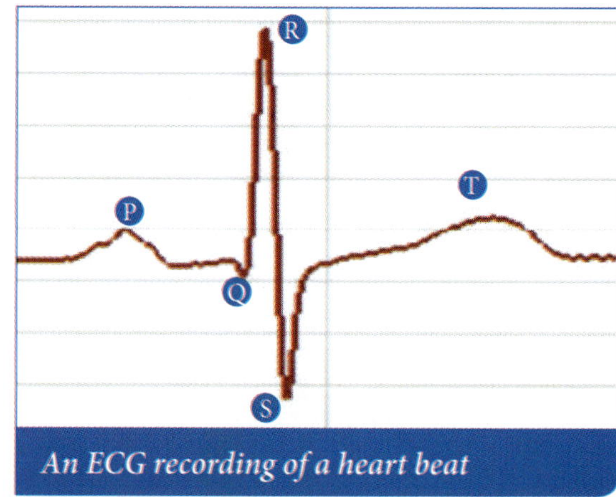

An ECG recording of a heart beat

- **A** Sinoatrial (SA) node
- **B** Atrioventricular (AV) node
- **C** Bundle branches
- **D** Purkinje fibers
- **P** P-wave
- **Q**
- **R** QRS complex
- **S**
- **T** T-wave

ECG AND HEART RATE USING IWORX PHYSIOLOGY DATA ACQUISITION

The subject will place one disposable adhesive electrode disk on each wrist and on the right leg. Snap the red lead wire onto the right wrist, the black lead wire on the left wrist, and the green lead wire to the right leg. The subject must remain as still as possible, as any movement will be detected by the ECG recording as noise.

Click "Record," located in the upper right of the screen, as the subject sits still. Let the recording run for 10–15 seconds. Click "Autoscale" on the ECG recording, then "Autoscale" on the heart rate recording. This should produce a recording that looks something like this:

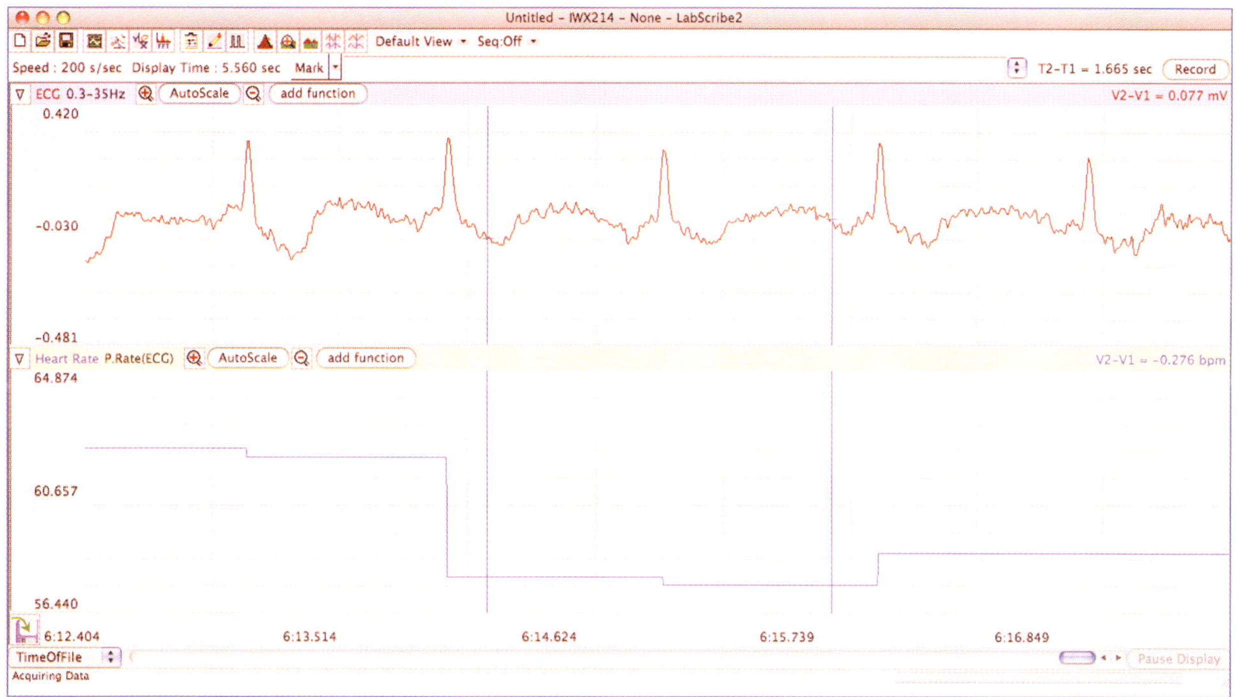

After a few minutes of base line recording, the subject should hold their breath for 15 seconds. Notice the changes on the ECG and Heart rate. Click "Stop." Use the Double Display Time icon to adjust the Display Time of the main window to show the six progressive heart beats on the main window. Click on the "Analysis" window icon in the toolbar from the LabScribe toolbar below.

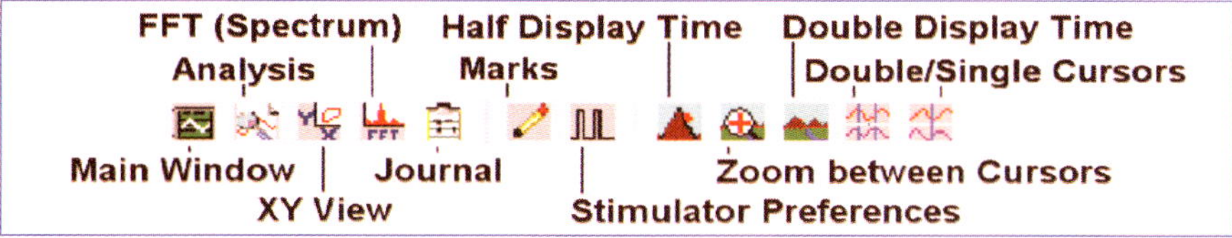

Look at the Function Table that is above the uppermost channel displayed in the Analysis window. The values for Abs. Area, V2-V1, and T2-T1 on each channel are seen in the table across the top margin of each channel.

Place one cursor on the first heart beat's R wave on the ECG channel. Place the other cursor on the sixth heart beat's R wave and record the values for T2-T1 on the ECG channel on the group's Data Sheet. Multiply time for six heart beats by ten. This will give you the heart rate in minutes.

T2-T1 ______________________

Heart Rate (beats / minute) ______________________________

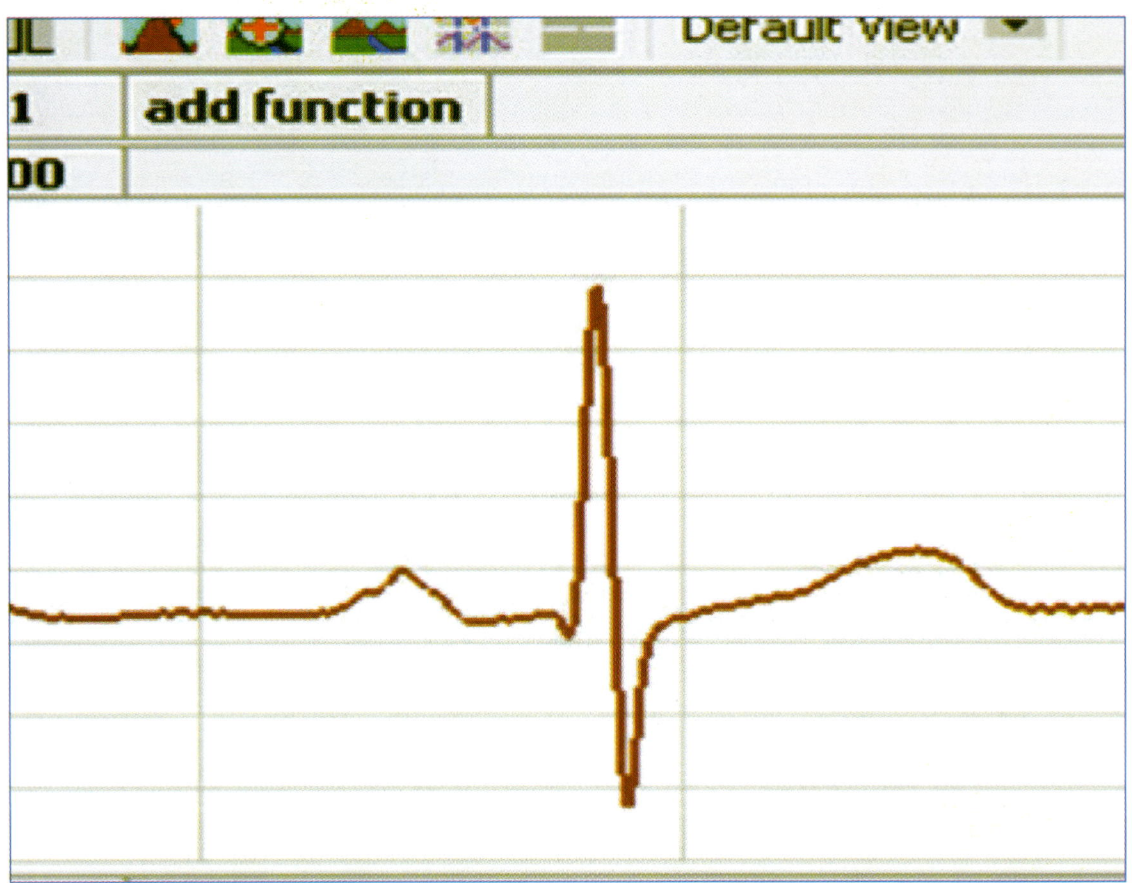

Label the above figure in regard to P, QRS, and T waves. Where does atrial and ventricular depolarization and repolarization occur?

Blood Vessels

Transport and Exchange

Blood vessels range in size and number. Arteries and veins transport blood away from and back to the heart, respectively. They only serve in the distribution of blood, not exchange. Resistance arteries control systemic blood pressure by varying their diameter by contraction and relaxation of smooth muscle in the tunica media. Capillaries are the vessels in which exchange of gases, nutrients and wastes occur. Of our five liters of blood, only about 5% is in our capillary beds at any given moment, yet all exchange is performed between blood and tissues in capillaries. We have an estimated 60,000 miles of blood vessels in our bodies.

Vessels and their composition

- Vein
- Medium size (distributing) artery
- Arteriole
- Capillary
- Tunica intima
  - Endothelium
  - Basement membrane
  - Internal elastic lamina
- Tunica media (smooth muscle)
- External elastic membrane
- Tunica externa
- Adipose tissue
- Valves – ensure one directional flow

Arteries of the head – a. = artery

- Common carotid a.
- Vertebral a.
- External carotid a.
- Internal carotid a.
- Maxillary a.
- Facial a.
- Superficial temporal a.
- Transverse facial a.
- Opthalmic a.
- Occipital a.
- Basilar a.
- Posterior cerebral a.
- Cerebellar a.
- Middle cerebral a.
- Posterior communicating a.
- Anterior cerebral a.
- Anterior communicating a.

Arteries of the hand

- Radial a.
- Ulnar a.
- Superficial palmar arch
- Common palmar digital a.
- Proper palmar digital a.
- Deep palmar arch
- Principal pollicis a.
- Palmar metacarpal a.

Conducting and distributing arteries

- Aorta
- Carotid a.
- Subclavian a.
- Axillary a.
- Brachial a.
- Radial a.
- Descending aorta
- Renal a.
- Common iliac a.
- External iliac a.
- Femoral a.
- Popliteal a.
- Posterior tibial a.
- Anterior tibial a.

Veins – v. – vein

- Superficial temporal v.
- Occipital v.
- Facial v.
- Internal jugular v.
- r. Brachiocephalic v.
- Subclavian v.
- Axillary v.
- Basilic v.
- Cephalic v.
- l. Brachiocephalic v.
- Renal v.
- Inferior vena cava
- Common iliac v.
- Internal iliac v.
- External iliac v.
- Great saphenous v.
- Femoral v.
- Small saphenous v.
- Dorsal venous arch
- Dorsal metacarpal v.
- Intercapitular v.

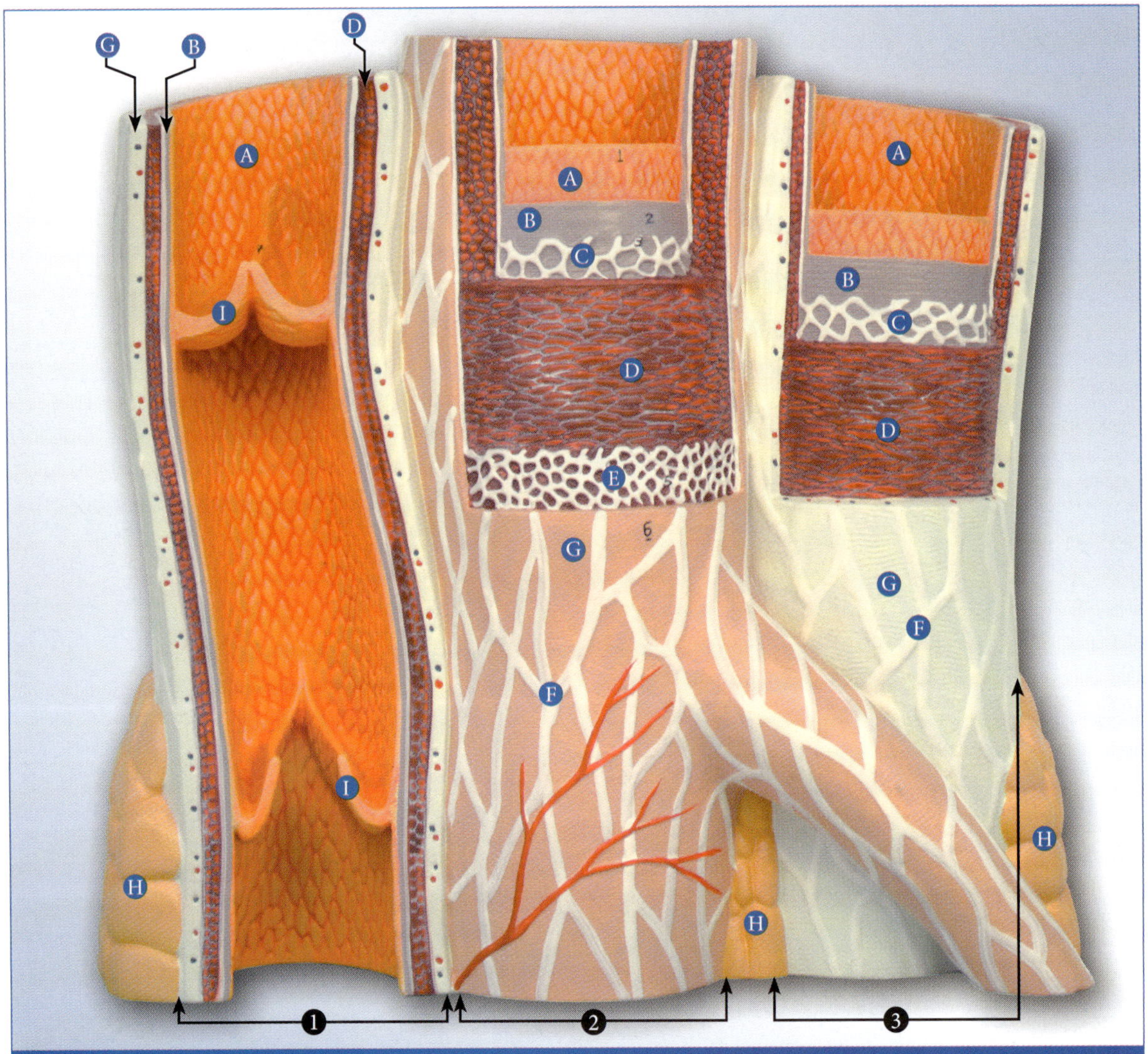

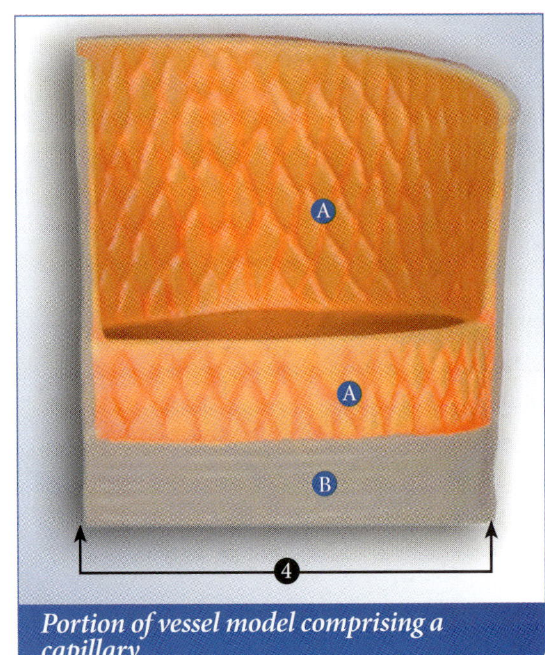

Model showing a vein, distributing artery, and an arteriole

❶ Vein

❷ Medium size (distributing) artery

❸ Arteriole

❹ Capillary

Ⓐ Endothelium

Ⓑ Basement membrane — Tunica intima

Ⓒ Internal elastic lamina

Ⓓ Tunica media (smooth muscle)

Ⓔ External elastic membrane

Ⓕ Collagen fibers

Ⓖ Tunica externa

Ⓗ Adipose tissue

Ⓘ Valves – ensure one directional flow towards heart

Portion of vessel model comprising a capillary

Arteries and veins transport blood to and from tissues, respectively. Arteries carry blood away from the heart. The largest is the aorta, which gives rise to other large, conducting arteries that feed regions of the body. Conducting arteries are compliant (able to stretch) and elastic (recoil back into shape). This allows blood pressure to remain relatively constant downstream. Medium sized or distributing arteries take blood to specific glands, organs, or regions. Small resistance arteries are too numerous to name and possess a relatively thick tunica media. This smooth muscle layer can cause changes in vessel diameter, which control overall systemic blood pressure. Change in the diameter of resistance arteries dramatically changes blood flow, which changes blood pressure. A resistance artery with a diameter of 3 mm allows 80 times the blood flow compared to an artery with a diameter of 1 mm. A small change in diameter causes a big change in flow and pressure. The smallest of the resistance arteries are thin walled and are called arterioles.

The venous system carries blood back to the heart after it returns from capillary beds. Post capillary vessels are venules that merge into medium sized veins, then large veins, and back into the heart as the largest veins, the inferior and superior vena cava. Because pressure is low in veins

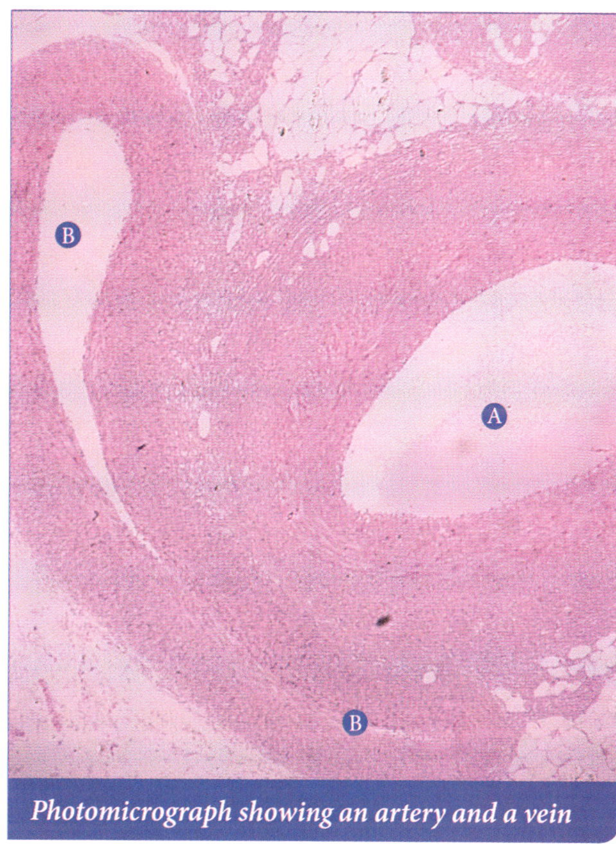

Photomicrograph showing an artery and a vein

(5–15 mmHg), one way valves help force blood back to the heart with help from muscle contraction (skeletal muscle pump).

SEM of an artery and a vein

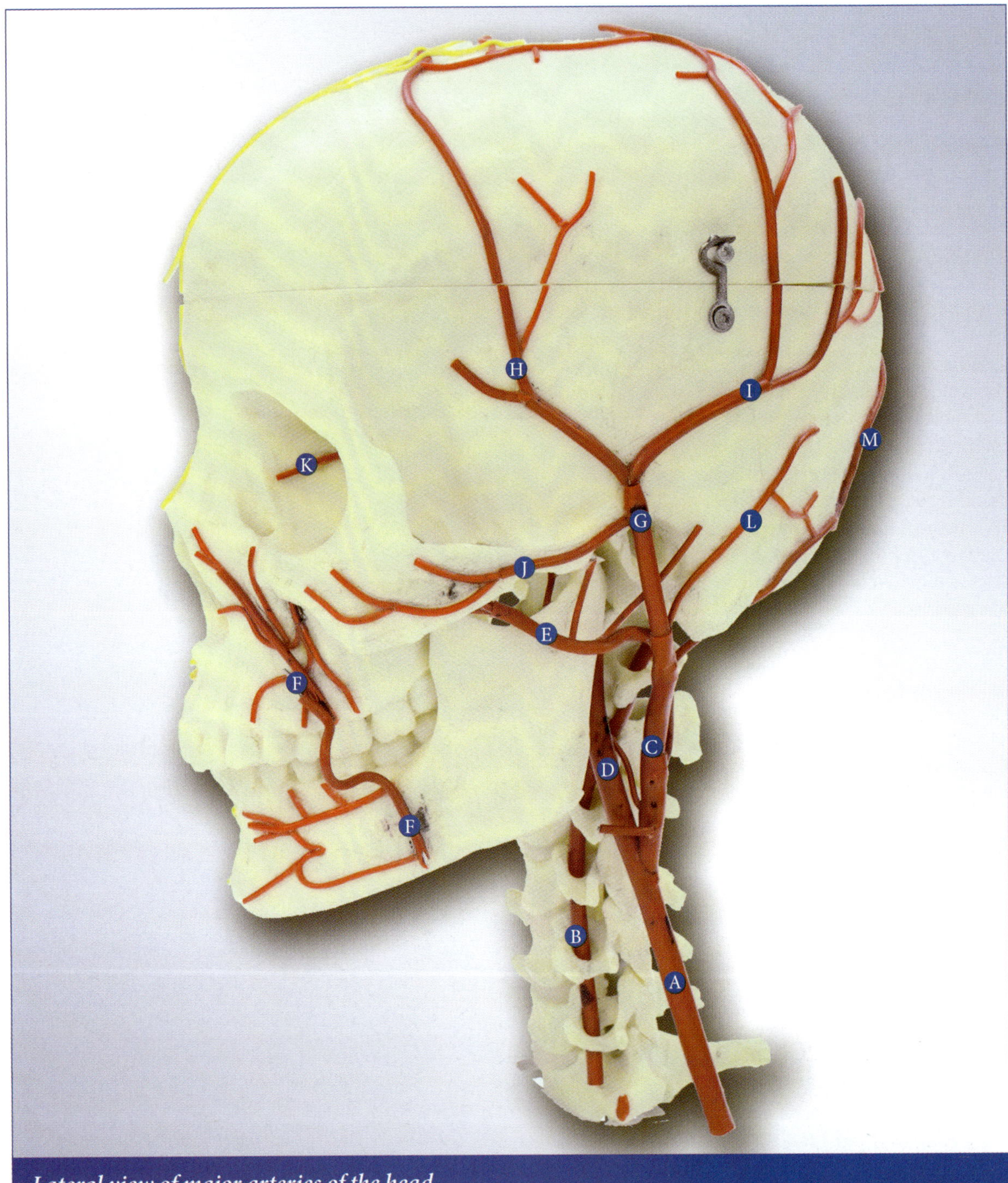

Lateral view of major arteries of the head

A	Common carotid a.	**F**	Facial a.	**J**	Transverse facial a.		
B	Vertebral a.	**G**	Superficial temporal a.	**K**	Opthalmic a.		
C	External carotid a.	**H**	Frontal branch of superficial temporal a.	**L**	Posterior auricular a.		
D	Internal carotid a.	**I**	Parietal branch of superficial temporal a.	**M**	Occipital a.		
E	Maxillary a.						

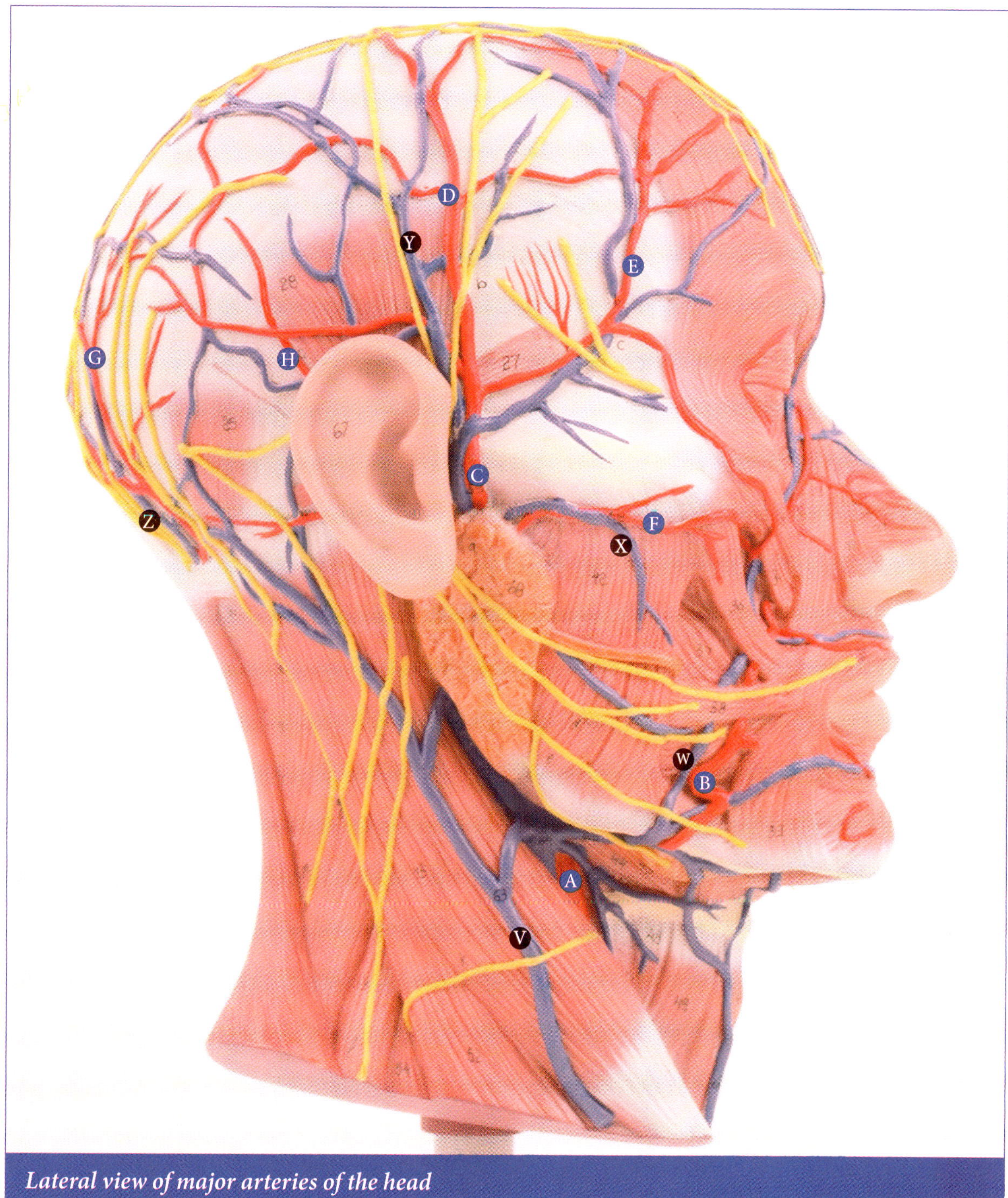

Lateral view of major arteries of the head

A Common carotid a.	**F** Transverse facial a.	**Y** Parietal branch of superficial temporal v.
B Facial a.	**G** Occipital a.	**Z** Occipital v.
C Superficial temporal a.	**H** Posterior auricular a.	
D Parietal branch of superficial temporal a.	**V** Jugular v.	
E Frontal branch of superficial temporal a.	**W** Facial v.	
	X Transverse facial v.	

The brain is supplied by four main arteries. Two carotid and two vertebral arteries ensure constant blood flow to both hemispheres of the cerebrum at all times. A temporary occlusion (blockage) of one of these arteries will not deprive the half of the brain it appears to supply. The cerebral arterial circle (circle of Willis) is fed by all four arteries and distributes blood to all regions of the brain.

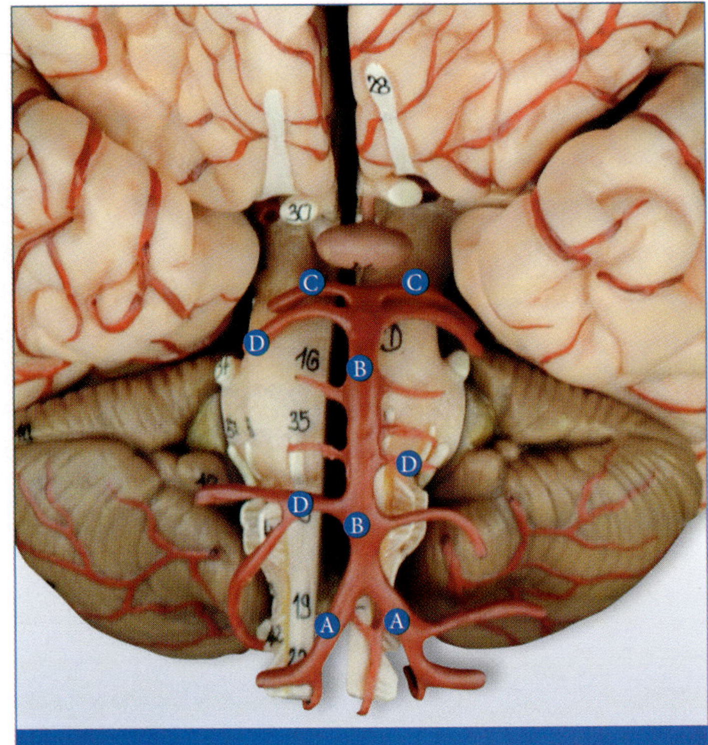

Inferior view of the arteries of the brain

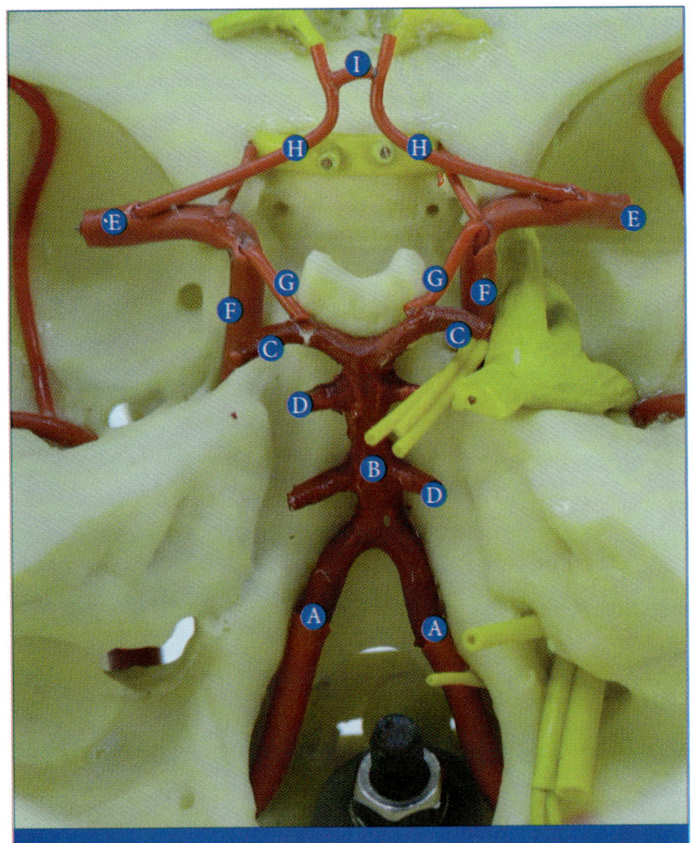

Superior view of cranial cavity with arteries of the brain

ARTERIES THAT SUPPLY THE BRAIN

- **A**   Vertebral a.
- **B**   Basilar a.
- **C**   Posterior cerebral a.
- **D**   Cerebellar a. (several are labeled)
- **E**   Middle cerebral a.
- **F**   Internal carotid a.
- **G**   Posterior communicating a. ← Cerebral arterial circle (circle of Willis)
- **H**   Anterior cerebral a.
- **I**   Anterior communicating a. ←

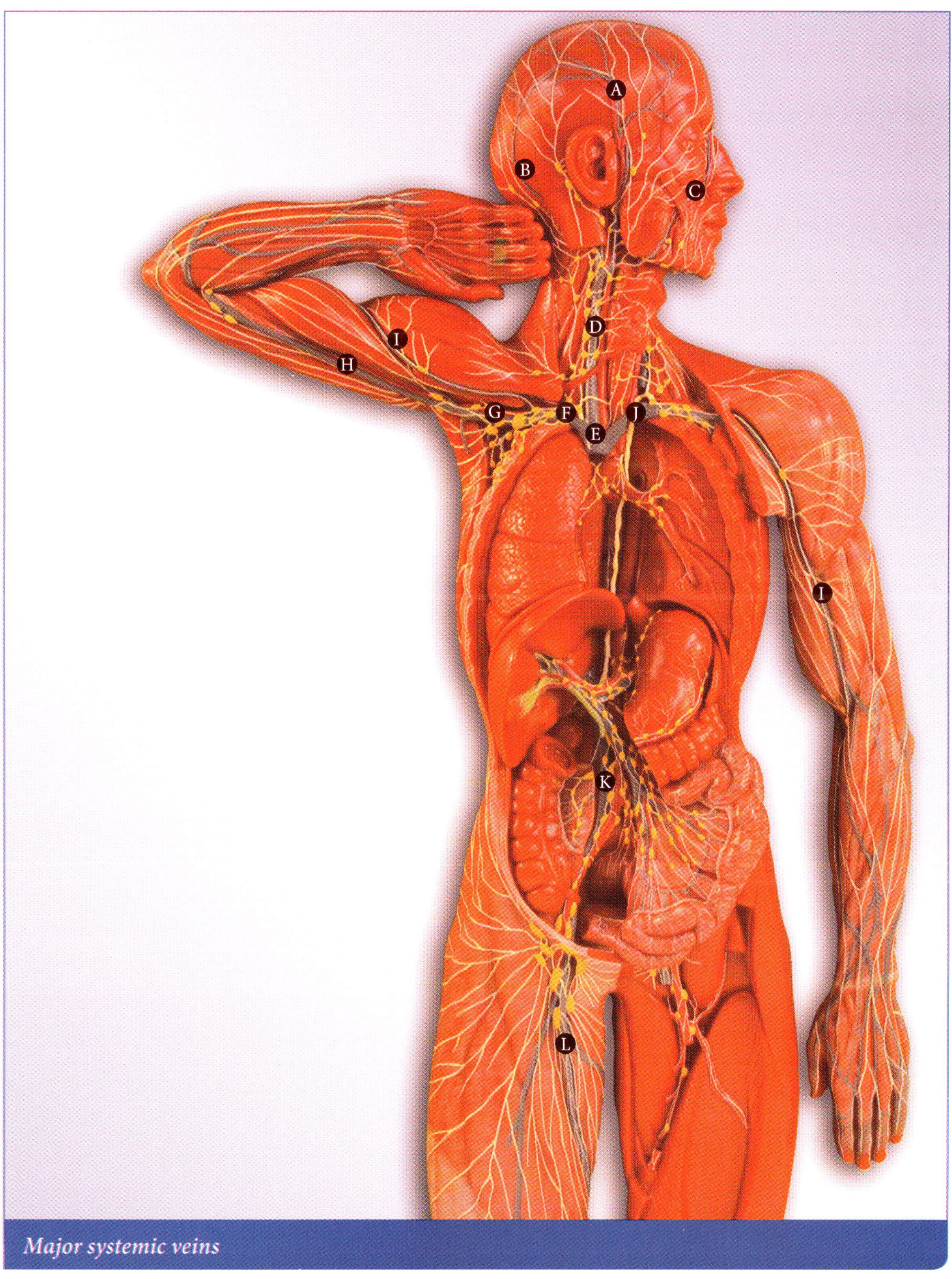

Major systemic veins

Ⓐ	Superficial temporal v.	Ⓔ	r. Brachiocephalic v.	Ⓘ	Cephalic v.
Ⓑ	Occipital v.	Ⓕ	Subclavian v.	Ⓙ	l. Brachiocephalic v.
Ⓒ	Facial v.	Ⓖ	Axillary v.	Ⓚ	Common iliac v.
Ⓓ	Internal jugular v.	Ⓗ	Basilic v.	Ⓛ	Great saphenous v.

VEINS

A r. Brachiocephalic v.

B Internal jugular v.

C Subclavian v.

D Axillary v.

E Basilic v.

F Cephalic v.

G Renal v.

H Inferior vena cava

I Internal iliac v.

J l. Brachiocephalic v.

ARTERIES

Q Aorta

R Carotid a.

S Subclavian a.

T Axillary a.

U Brachial a.

V Radial a.

W Descending aorta

X Renal a.

Y Common iliac a.

Z External iliac a.

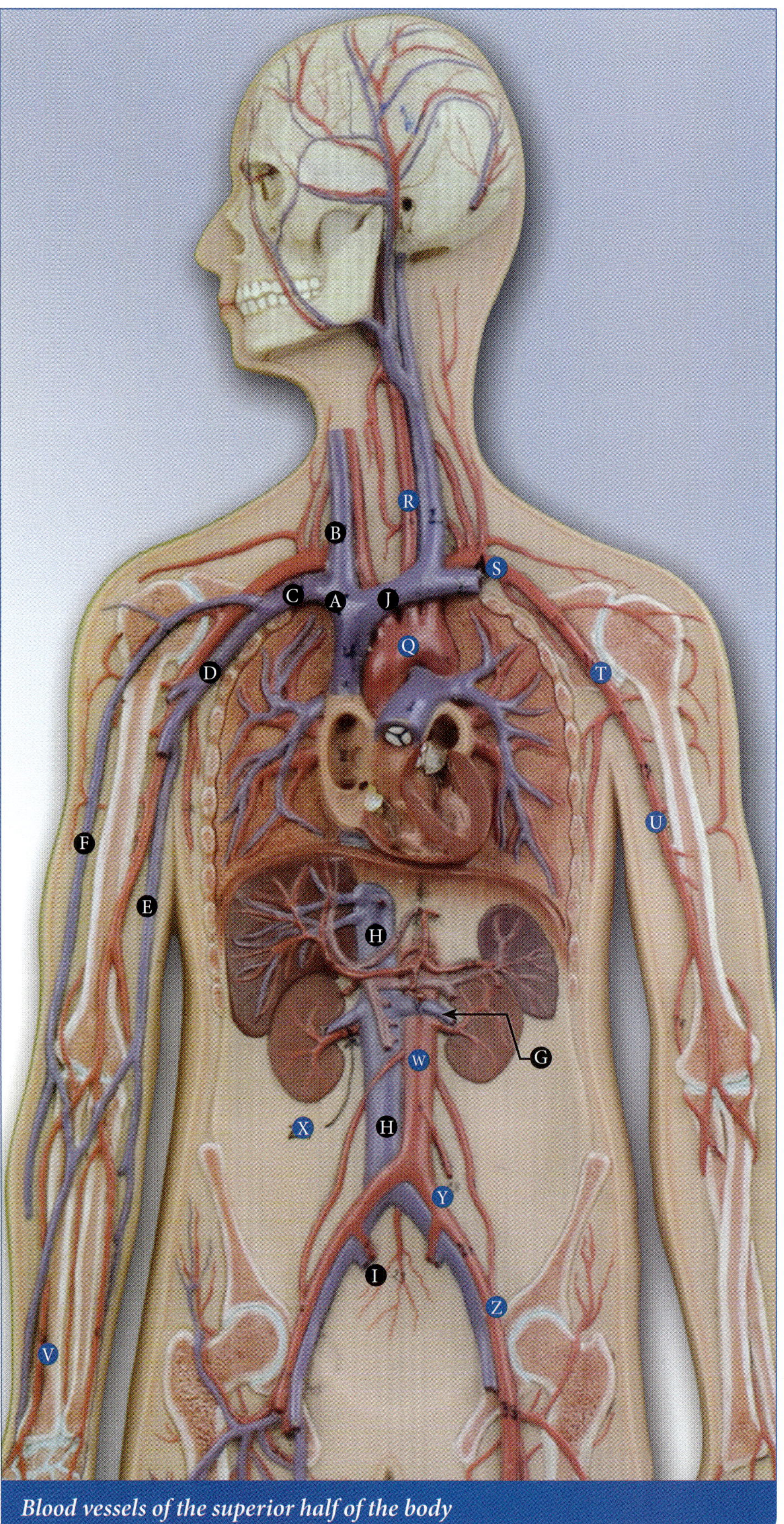

Blood vessels of the superior half of the body

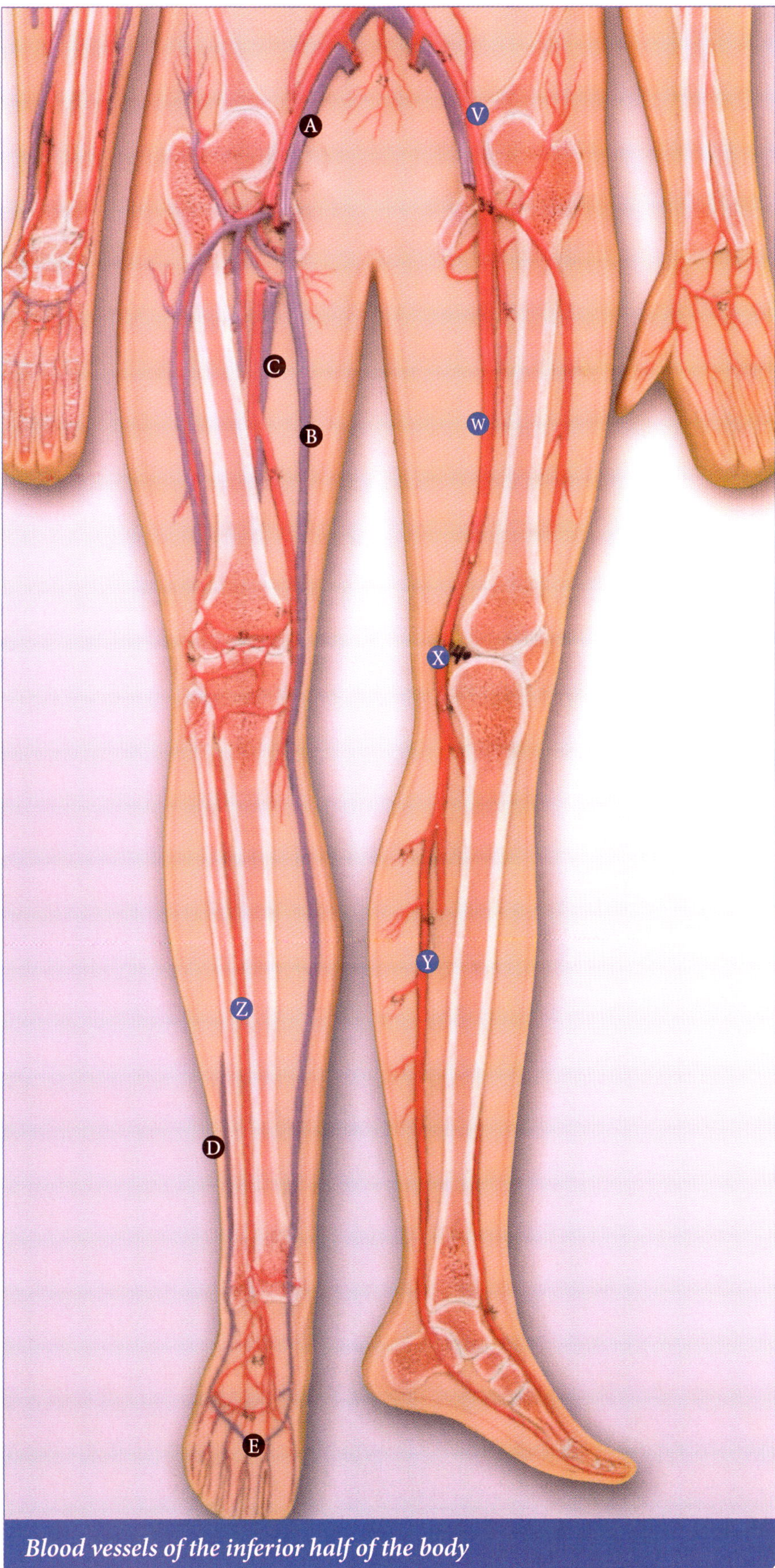

Blood vessels of the inferior half of the body

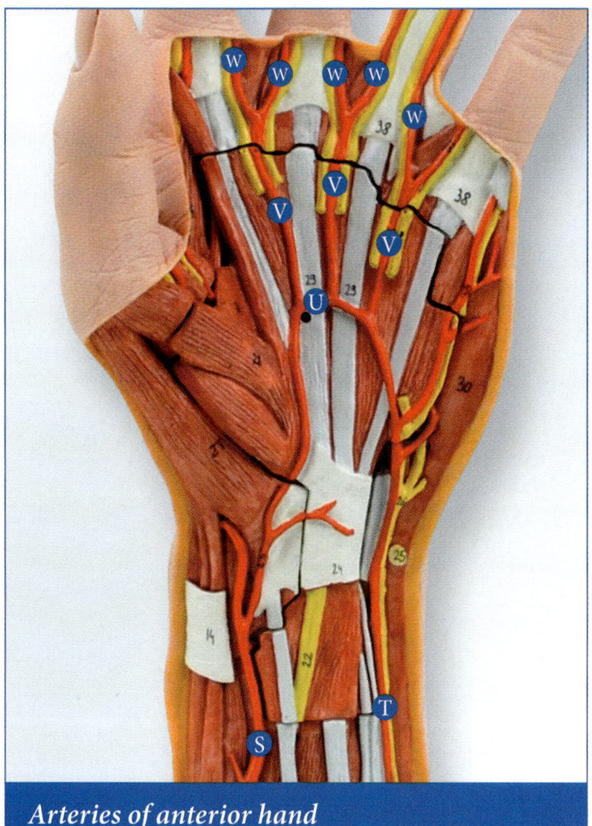

Arteries of anterior hand

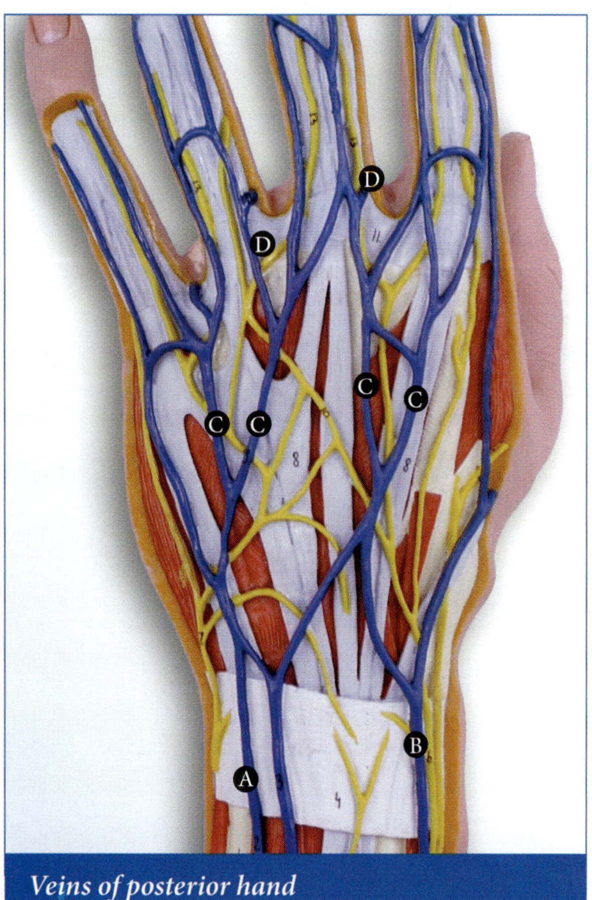

Veins of posterior hand

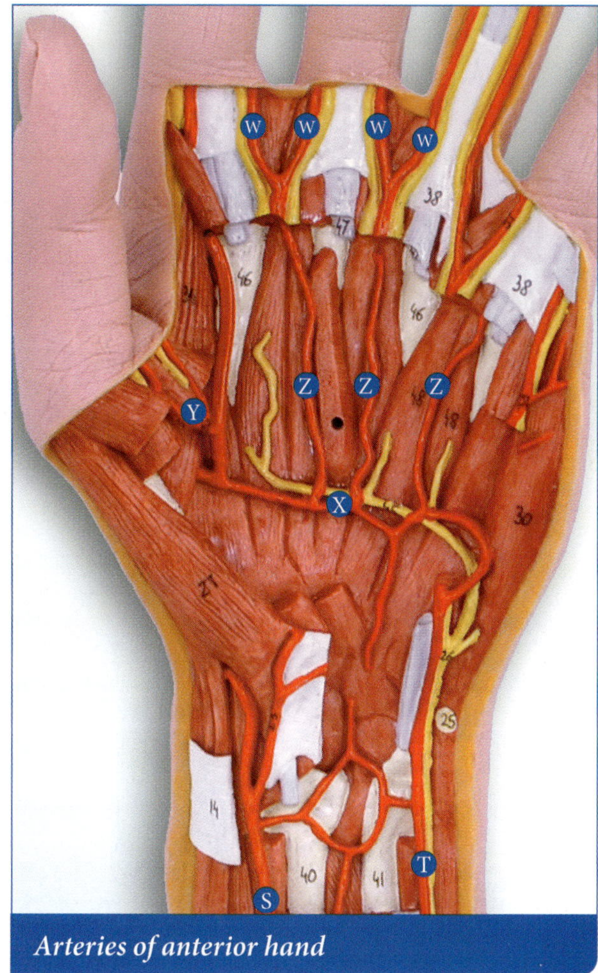

Arteries of anterior hand

VEINS

A Basilic v.

B Cephalic v.

C Dorsal metacarpal v.

D Intercapitular v.

ARTERIES

S Radial a.

T Ulnar a.

U Superficial palmar arch

V Common palmar digital a.

W Proper palmar digital a.

X Deep palmar arch

Y Principal pollicis a.

Z Palmar metacarpal a.

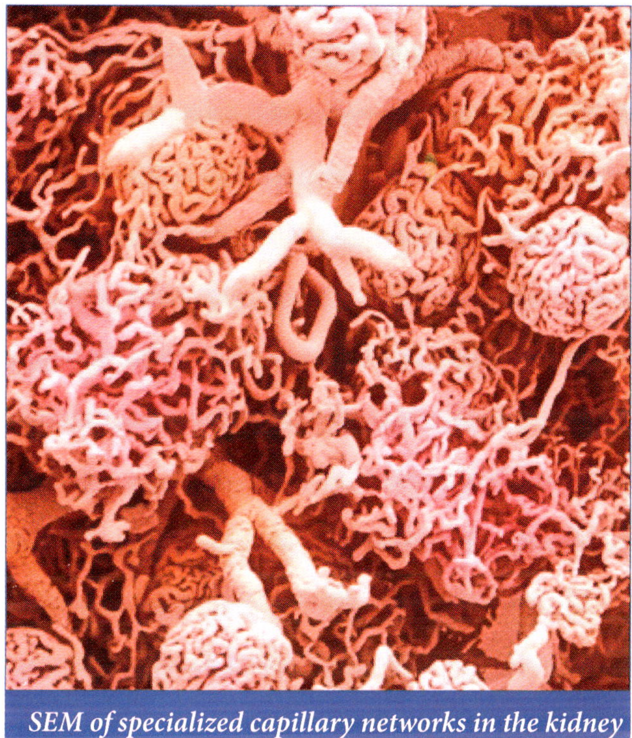

SEM of specialized capillary networks in the kidney

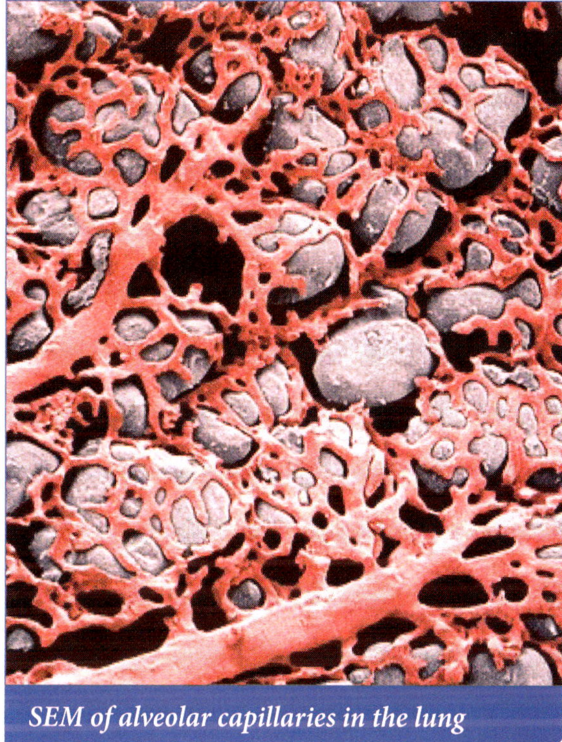

SEM of alveolar capillaries in the lung

The above left SEM of the kidney shows specialized fenestrated capillary beds (round ball-like structures) that function only in filtration. Reabsorption does not take place in this capillary bed.

The above right SEM shows the alveolar capillary network of the lung. Here absorption is the major flow of fluid, with hardly any filtration. CO_2 escapes the blood stream into alveolar air, and oxygen diffuses into the blood from the alveoli via the alveolar capillary network.

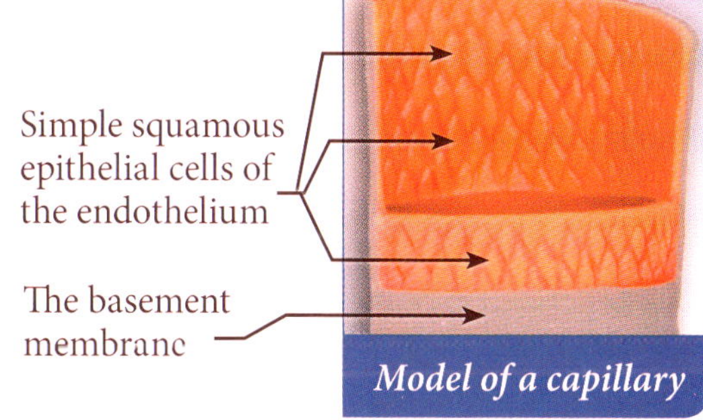

Model of a capillary

Capillaries are the vessels that provide filtration and reabsorption of blood plasma (along with dissolved nutrients and wastes) and diffusion of the blood gases, carbon dioxide and oxygen. Capillaries have very thin walls (only simple squamous epithelial cells and basement membrane) that are about 0.3 μm thick. This allows rapid exchange between the circulatory system and tissues. There are about a billion capillaries with a total surface area of a football field in the human body. Virtually all nutrient delivery and waste removal occurs through capillary walls. Gas exchange is dependent upon the thin capillary wall and large surface area of numerous capillaries. Capillaries can be continuous (tightly bound cells with very narrow clefts), fenestrated (small pores), or sinusoid (blood filled spaces).

Notice the narrow diameter of the capillary in the SEM to the right. Red blood cells (erythrocytes) must line up, single file, to pass through. This slows blood flow and allows diffusion and fluid exchange to take place.

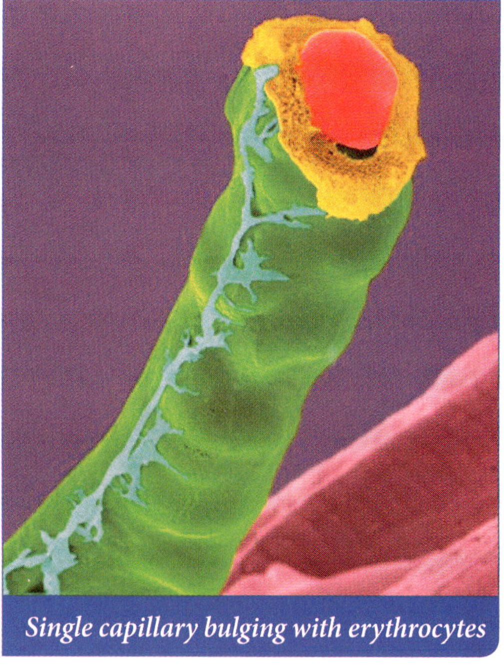

Single capillary bulging with erythrocytes

Lymphatic

Fluid Recovery, Lipid Absorption, and the Immune System

The lymphatic system is involved in fluid recovery, lipid absorption, and immunity. Fluid recovery is important in maintaining proper water balance in tissues. If flow of interstitial fluid through lymphatic vessels or nodes is interupted, edema, a swelling of tissues caused by excess fluid, occurs. Lipids are transported into the central lacteal (lymphatic capillary) of a villus of the small intestine and not directly into blood capillaries. Immunity involves leukocytes and organs to fight off invaders. Non-specific resistance involves physical barriers such as skin and a variety of leukocytes and macrophages that produce antimicrobial agents and phagocytize invaders without prior exposure to a pathogen. Specific immunity relies on the abilty of lymphocytes to produce trillions of unique antigen receptors. These receptors bind to specific regions (epitope) of invading pathogens or toxins. In response to a second exposure of a particular pathogen, memory B cells differentiate into plasma cells and produce antibodies (binding portion of antigen receptor) which bind to, and neutralize the invader.

Components of the lymphatic system

- Tissue (interstitial) fluid
- Lymphatic capillary
- Lymphatic fluid, or lymph
- **Lacteal** (lymphatic capillary with large openings involved in lipid absorption)
- **Lymph nodes** – Filter lymphatic fluid recovered from tissue and house B and T lymphocytes, plasma cells, macrophages and reticular cells.
- **Red bone marrow** – Bones of the rib cage and pelvis contain hemopoietic marrow that produce each of the formed elements of blood, one of which, leukocytes, are the cell type involved in resistance to pathogens.
- **Palatine tonsils** – Located in the lateral region of the oropharynx, help reduce airborne and ingested pathogens.
- **The spleen** – The largest of the lymphatic organs, contains macrophages that engulf invaders and dying erythrocytes. Lymphocytes detect antigens that elicit an immune response.

Granular Leukocytes

- Neutrophils
- Eosinophils
- Basophils

Lymphocytes

- **NK (natural killer)**
- **B lymphocytes**
  - Plasma cells
  - Memory B cells (B_M)
- **T lymphocytes**
  - Helper T cells (T_H)
  - Cytotoxic T cells (T_C)
  - Memory T cells (T_M)

Lymph nodes

- Occipital
- Submandibular
- Deep cervical
- Cubital
- Axillary
- Tracheobronchial
- Bronchial
- Thoracic duct
- Hepatic
- Gastric
- Mesenteric
- Common iliac
- Inguinal

Vessels of the lymphatic system

- Right lymphatic duct
- Thoracic duct
- Jugular trunk
- Subclavian trunk

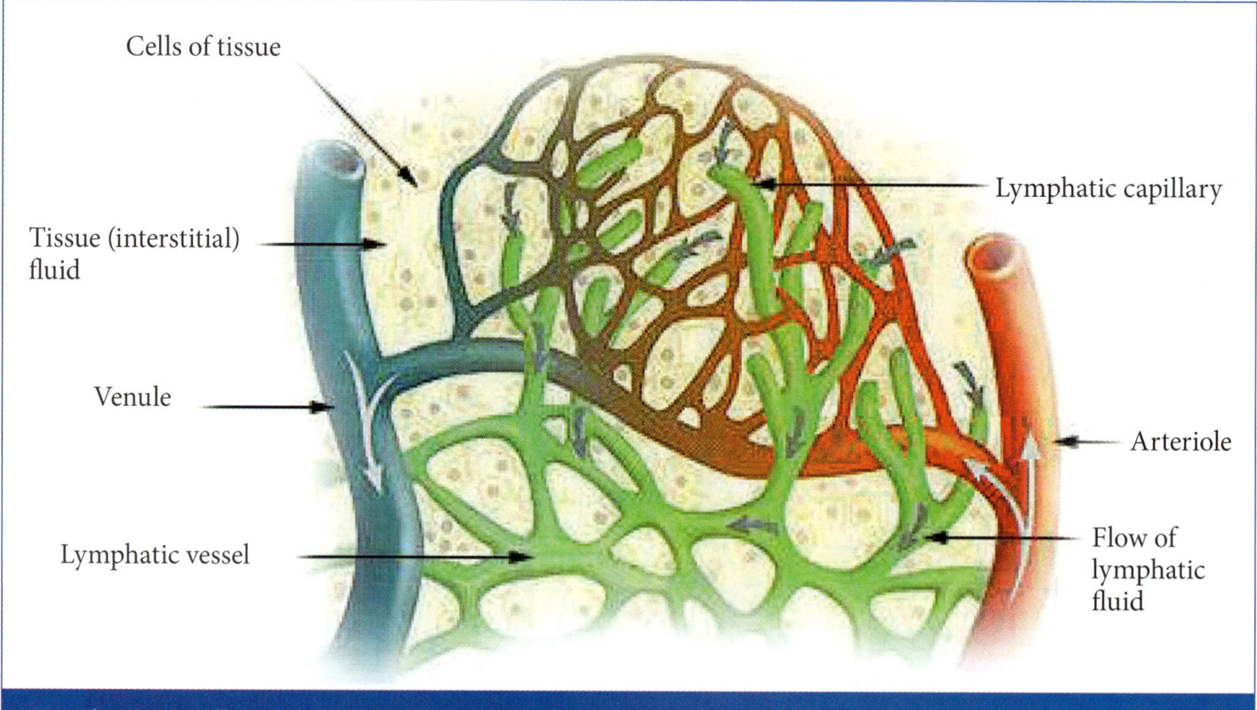

Circulatory capillary bed and lymphatic capillary network

There are three primary functions of the lymphatic system: fluid recovery, lipid absorption, and immunity.

Fluid recovery returns fluid that was not reabsorbed directly back into the bloodstream from tissues. It occurs after filtration, the process of blood plasma (without the large proteins) being filtered out of blood capillaries into the interstitial spaces by hydrostatic pressure. As blood moves through a capillary bed, the pressure drops. The decrease in blood pressure allows the filtered fluid (called interstitial fluid in the tissue) to be reabsorbed near the venule end of the capillary bed by colloid osmotic pressure. About 85% of filtered fluid is reabsorbed back into the bloodstream. The other 15% is taken up by lymphatic capillaries (fluid recovery) that carry this fluid (now called lymphatic fluid) in lymphatic vessels. Lymphatic fluid passes through lymph nodes which possess many lymphocytes ready to attack or tag foreign invaders. Lymph enters back into blood circulation near the heart. Any blockage of this flow can lead to edema, a swelling of tissues caused by excess fluid. Elephantiasis is an extreme example of edema as a result of impeded lymph flow.

Dietary fats (digested to free fatty acids) make their way into our bodies through columnar epithelial cells of the small intestines. In these cells, chylomicrons (droplets of triglycerides, cholesterol, phospholipids and protein) are produced and secreted on their basal side. Chylomicrons are too large to enter blood capillaries and enter a lacteal (lymphatic capillary with large openings). They travel through lymphatic vessels and nodes and make their way back into the bloodstream via the subclavian vein near the heart.

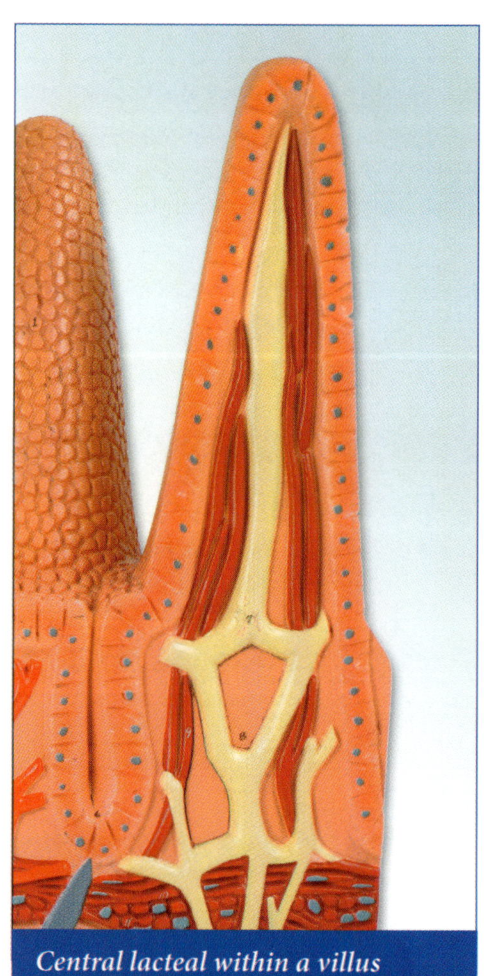

Central lacteal within a villus

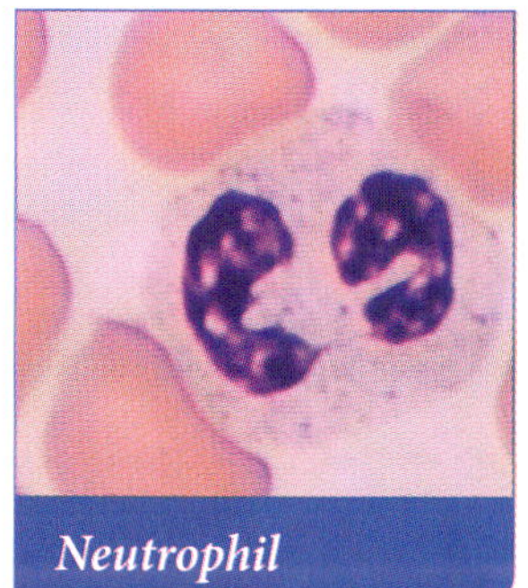

Neutrophil

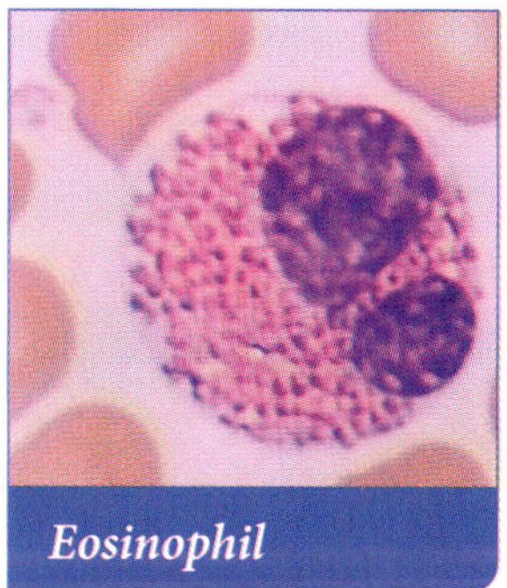

Eosinophil

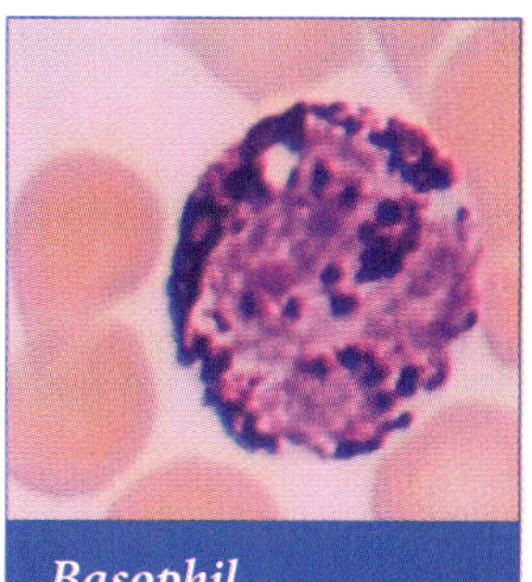

Basophil

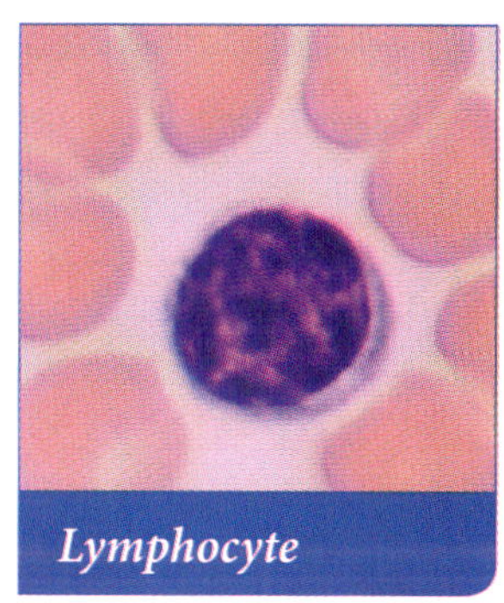

Lymphocyte

60–70% of all leukocytes	**2–4% of all leukocytes**	**0.4–1% of all leukocytes**	NK (natural killer)
Bacteria killers. Perform phagocytosis of bacteria and release hydrogen peroxide, hypochlorite, and superoxide into tissue, killing many bacteria and themselves.	Produce hydrogen peroxide and superoxide, phagocytize antibody tagged antigen, promote basophil action, and release enzymes that break down histamine.	Secrete histamine (vasodilator) and heparin (anticoagulant) to increase and maintain blood flow, and leukotrienes that attract and promote neutrophil and eosinophil activity. Mast cells, similar to basophils, can release histamine and heparin in response to exposure of an allergen. Tropomyosin of shellfish or peanuts can elicit mast cells to release their products which can lead to anaphylactic shock.	Are part of immune surveillance of non-specific resistance. They hunt and kill bacteria, cells with viruses, and cancer cells based on foreign proteins.

Specific immunity involves lymphocytes and macrophages derived from monocytes

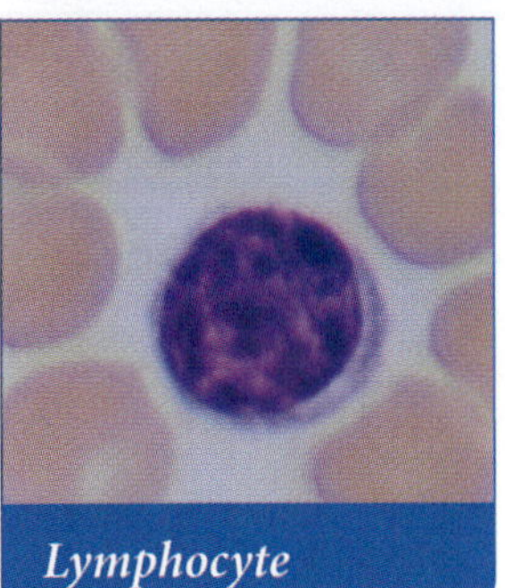

Lymphocyte

B lymphocytes
Plasma cells
Memory B cells (B_M)

T lymphocytes
Helper T cells (T_H)
Cytotoxic T cells (T_C)
Memory T cells (T_M)

SEM of a lymphocyte

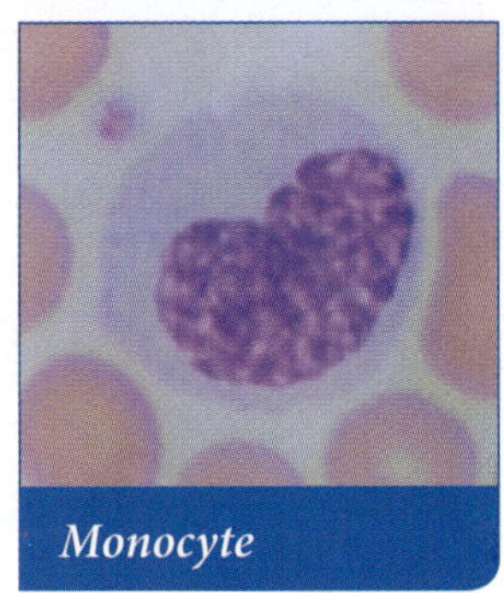

Monocyte

25–30% of all leukocytes
Lymphocytes are the major players in the immune system and perform a variety of functions. B, T and NK (natural killer) are the three primary types. T and B lymphocytes play the key roles in specific immunity. Helper T cells (T_H) secrete a variety of interleukins in response to non-self antigens and are necessary for most immune responses. B lymphocytes that happen to have receptors for a specific invading antigen will bind the antigen and bring it inside itself. Some B cells will actively divide and produce clones. Others need the aid of helper T cells (T_H) to divide. Most B cells transform into plasma cells that will produce and secrete antibodies that will bind to invading cells or toxins. Others will differentiate into memory B cells (B_M) in case of future infection. Cytotoxic T (T_C) cells must be stimulated by antigen presenting cells (APC) to divide rapidly and directly attack bacteria or infected cells (viral or cancerous). T_C cells release perforin and granzymes (creates holes in the target's cell membrane and introduces digestive enzymes, respectively) and interferon, a viral inhibitor.

3–8% of all leukocytes
Monocytes can leave the circulatory system and become wandering or fixed macrophages. Macrophages are phagocytic machines that live throughout the body. They also play a role in the immune system by acting as an antigen presenting cell (APC). They display digested epitopes (antigen region that elicits immune response). Macrophages are not classified as leukocytes.

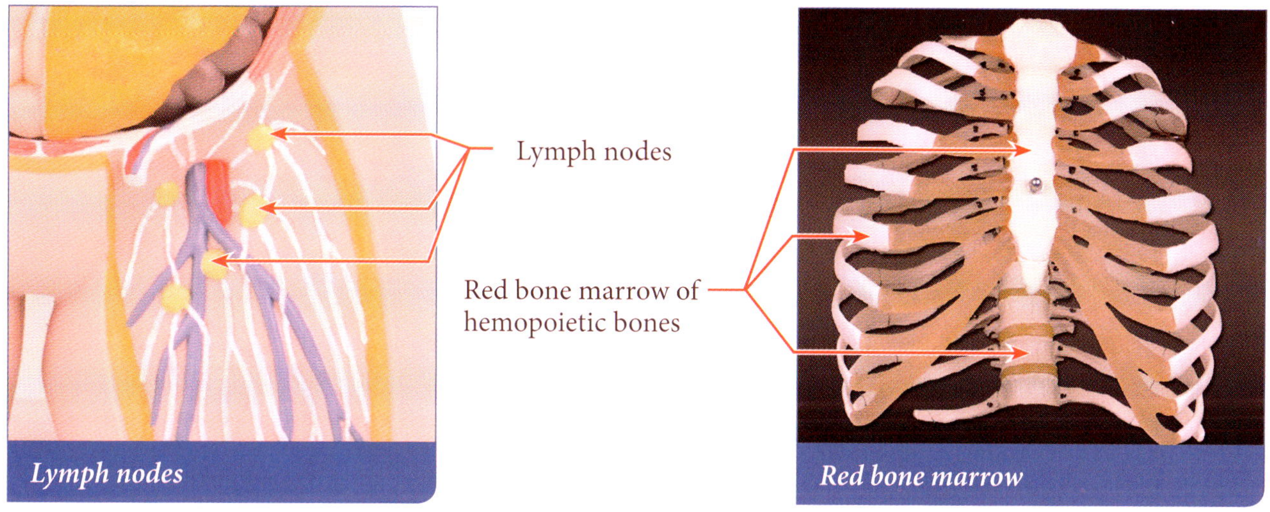

Lymph nodes

Red bone marrow

COMPONENTS OF THE LYMPHATIC SYSTEM

Lymph nodes – Filter lymphatic fluid returned from tissue (interstitial fluid) and contain lymphocytes, plasma cells, macrophages and reticular cells.

Red bone marrow – Bones of the thoracic cage and pelvis contain hemopoietic marrow that produce each of the formed elements of blood. Leukocytes are the active cells in resistance to pathogens.

Palatine tonsils – Located in the lateral region of the oropharynx, help resist airborne and ingested pathogens.

The spleen – The largest of the lymphatic organs, contains macrophages that engulf invaders and dying erythrocytes. Lymphocytes detect antigens that elicit an immune response.

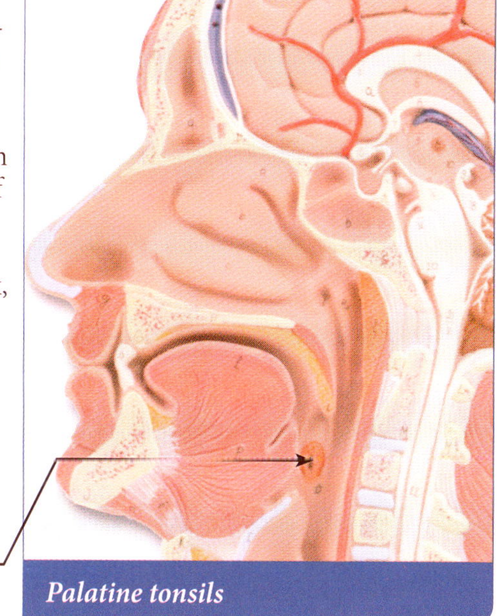

Palatine tonsils

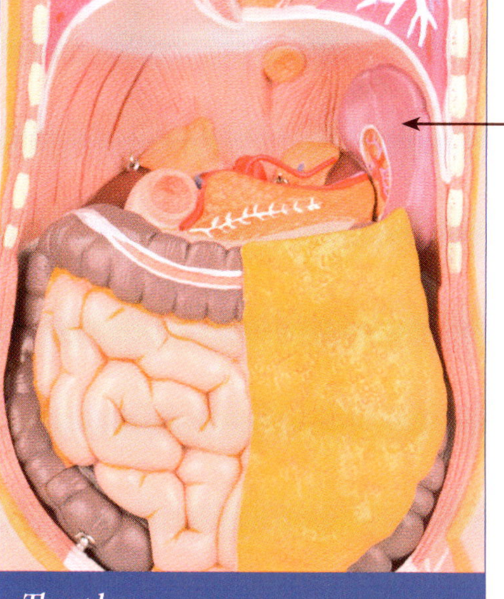

The spleen

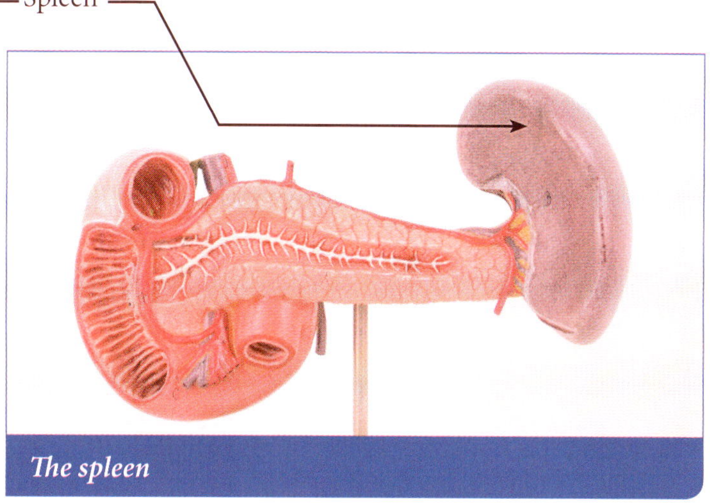

The spleen

LYMPH NODES

- Ⓐ Occipital
- Ⓑ Submandibular
- Ⓒ Deep cervical
- Ⓓ Cubital
- Ⓔ Axillary
- Ⓕ Tracheobronchial
- Ⓖ Bronchial
- Ⓗ Hepatic
- Ⓘ Gastric
- Ⓙ Mesenteric
- Ⓚ Common iliac
- Ⓛ Inguinal

LYMPHATIC VESSELS

- Ⓜ Jugular trunk
- Ⓝ Subclavian trunk
- Ⓞ Right lymphatic duct
- Ⓟ Thoracic duct

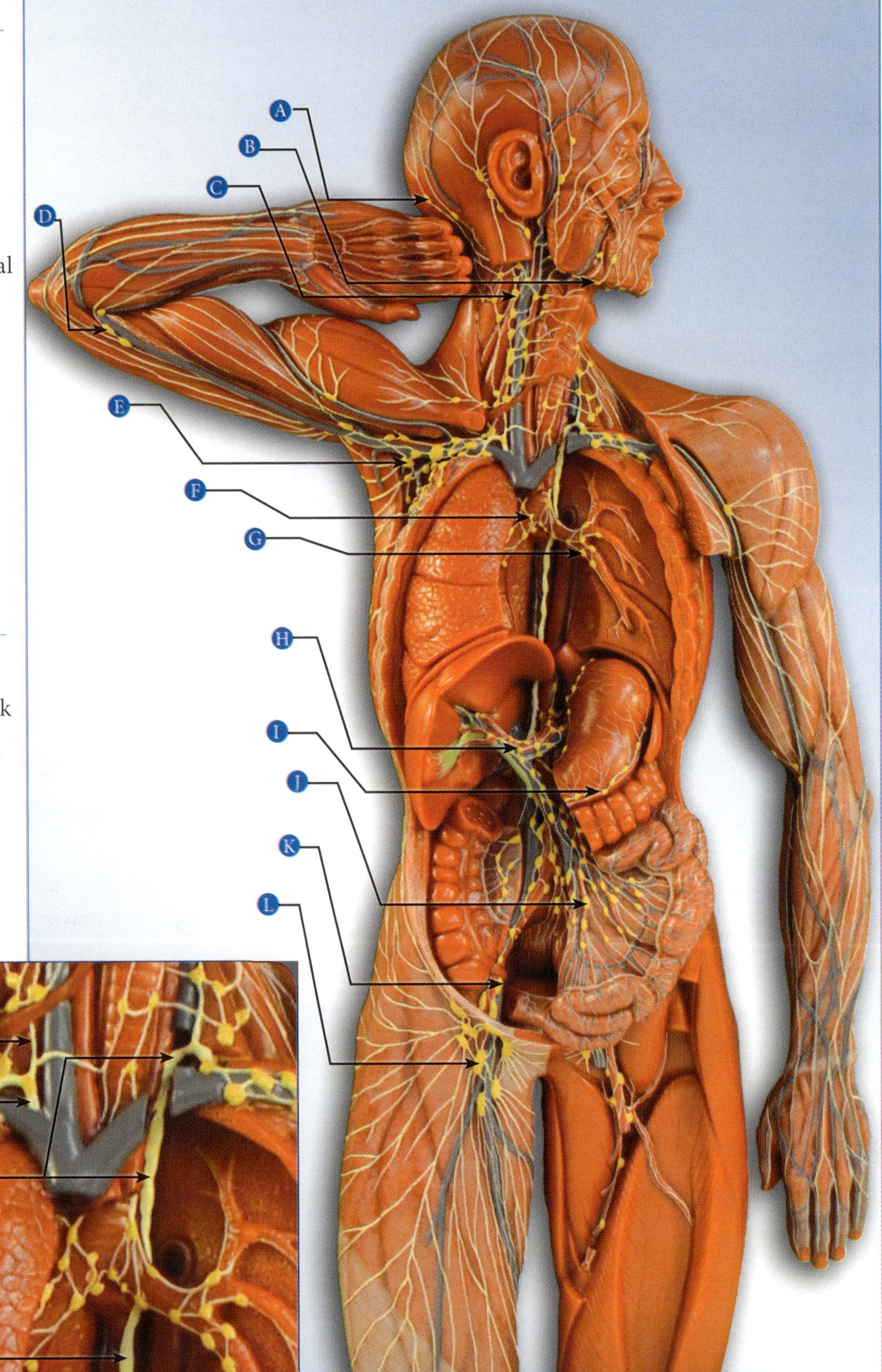

This model shows only a portion of approximately 500 lymph nodes in our bodies

Respiration

Airways and Gas Exchange

The respiratory system involves delivering atmospheric air to the lungs and allowing gas exchange between air and blood to occur in tiny air sacs known as alveoli. Our constant need to breathe is driven by the consumption of oxygen and the production of carbon dioxide by quadrillions of mitochondria that live in our cells. Although the need for oxygen is obvious, it is actually the level of CO_2, or more specifically the changes in pH of cerebrospinal fluid (CSF) caused by changes in CO_2 levels, that regulate our breathing rate.

The upper respiratory tract

- Vestibule
- Superior conchae
- Middle conchae
- Inferior conchae
- Superior meatus
- Middle meatus
- Inferior meatus
- Auditory tube
- Soft palate
- Uvula
- Nasopharynx
- Oropharynx
- Laryngopharynx
- Respiratory epithelium (pseudostratified columnar epithelium with cilia and goblet cells) – found in nasal cavity of upper respiratory tract and lower respiratory tract up to respiratory bronchioles

Breathing muscles

- External intercostals
- Scalenes
- Sternocleidomastoid
- Diaphragm
- Internal intercostals
- External abdominal obliques
- Rectus abdominis

The lower respiratory tract

- Hyoid bone
- Thyrohyoid ligament
- Thyroid cartilage
- Cricoid cartilage
- Tracheal cartilage
- Corniculate cartilage
- Arytenoid muscle
- Arytenoid cartilage
- Vestibular fold (false vocal cord)
- Vocal cord
- Glottis (opening into the trachea)
- Epiglottis
- Carina
- Main (primary) bronchus
- Lobar (secondary) bronchi
- Segmental (tertiary) bronchi
- Alveolar capillaries
- Alveolar lymphatic capillaries
- Alveoli
- Alveolar lymphatic capillaries
- Respiratory bronchiole
- Alveolar sacs (alveoli)
- Pulmonary venule
- Pulmonary arteriole
- Terminal bronchiole

Lung volumes

- Tidal Volume (TV)
- Inspiratory Reserve Volume (IRV)
- Expiratory Reserve Volume (ERV)
- Vital Capacity (VC)
- Forced Expiratory Volume (FEV_1)

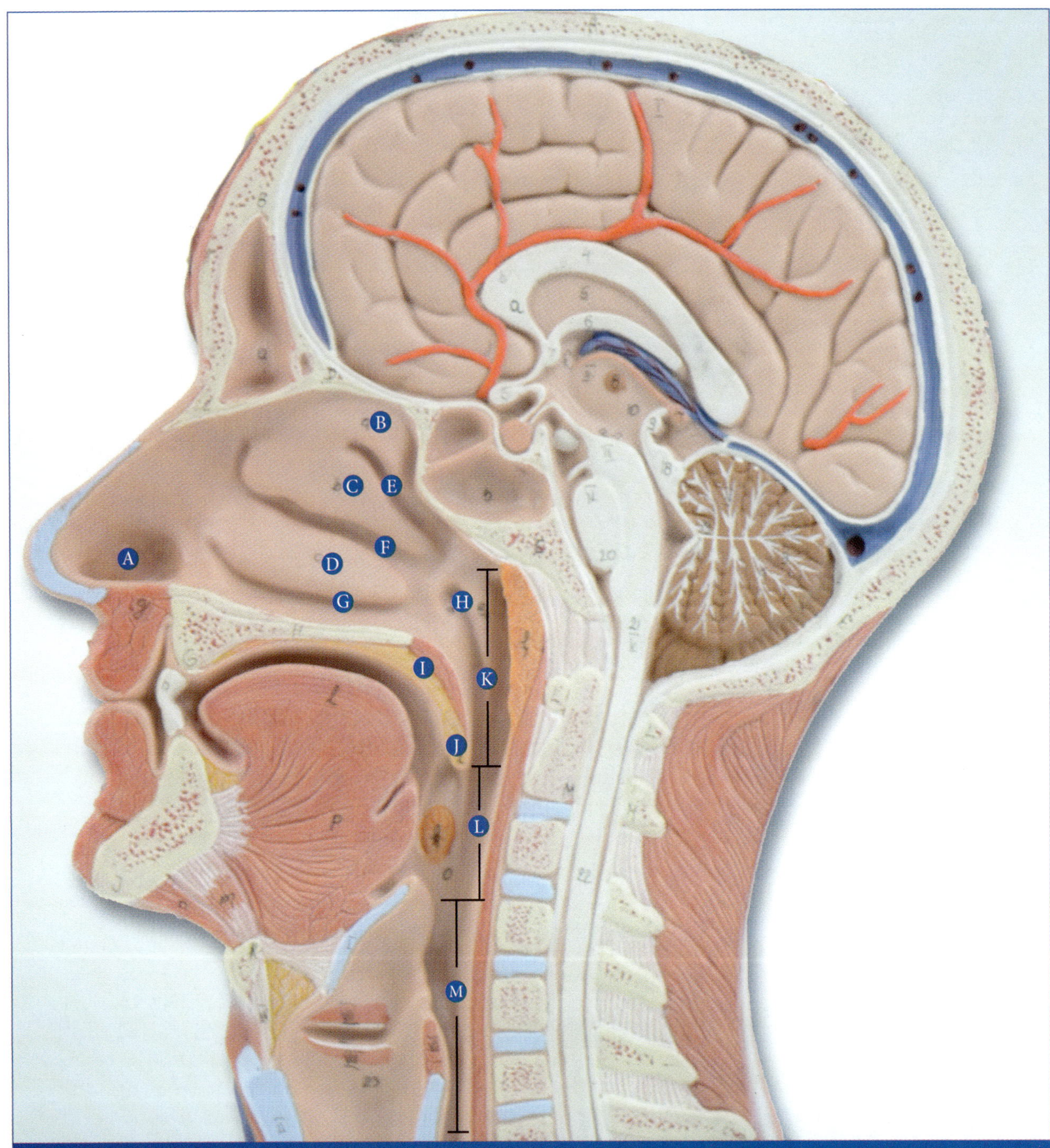

Midsagittal section of the head showing the upper respiratory tract

Ⓐ	Vestibule	Ⓕ	Middle meatus	Ⓚ	Nasopharynx
Ⓑ	Superior conchae	Ⓖ	Inferior meatus	Ⓛ	Oropharynx
Ⓒ	Middle conchae	Ⓗ	Auditory tube	Ⓜ	Laryngopharynx
Ⓓ	Inferior conchae	Ⓘ	Soft palate		
Ⓔ	Superior meatus	Ⓙ	Uvula		

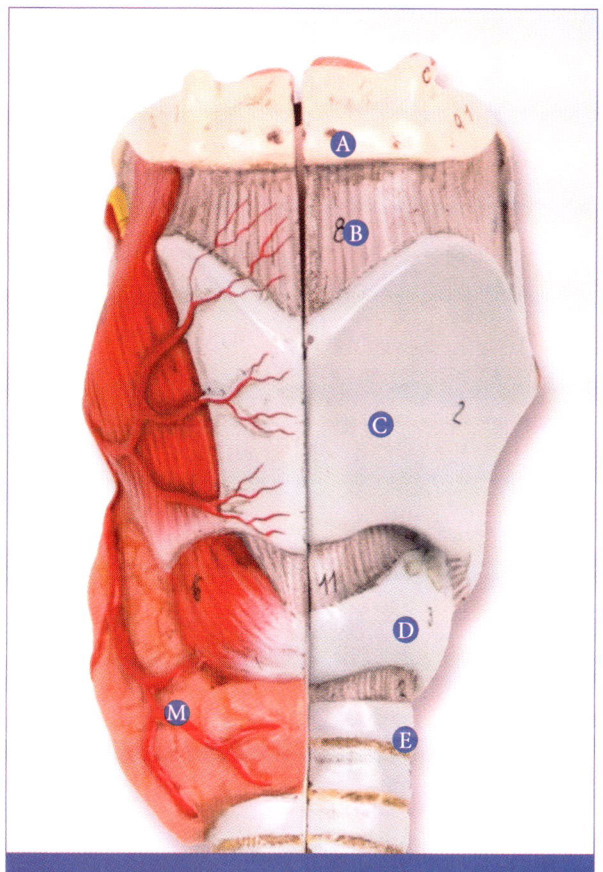

Anterior view of the larynx

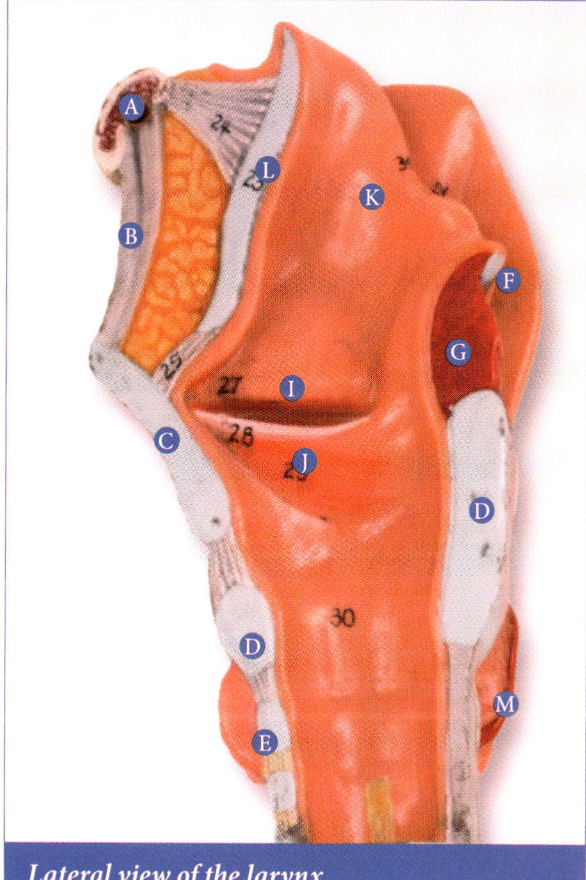

Lateral view of the larynx

Lateral view of the larynx

Ⓐ Hyoid bone

Ⓑ Thyrohyoid ligament

Ⓒ Thyroid cartilage

Ⓓ Cricoid cartilage

Ⓔ Tracheal cartilage

Ⓕ Corniculate cartilage

Ⓖ Arytenoid muscle

Ⓗ Arytenoid cartilage

Ⓘ Vestibular fold (false vocal cord)

Ⓙ Vocal cord

Ⓚ Glottis (opening into the trachea)

Ⓛ Epiglottis

Ⓜ Thyroid gland

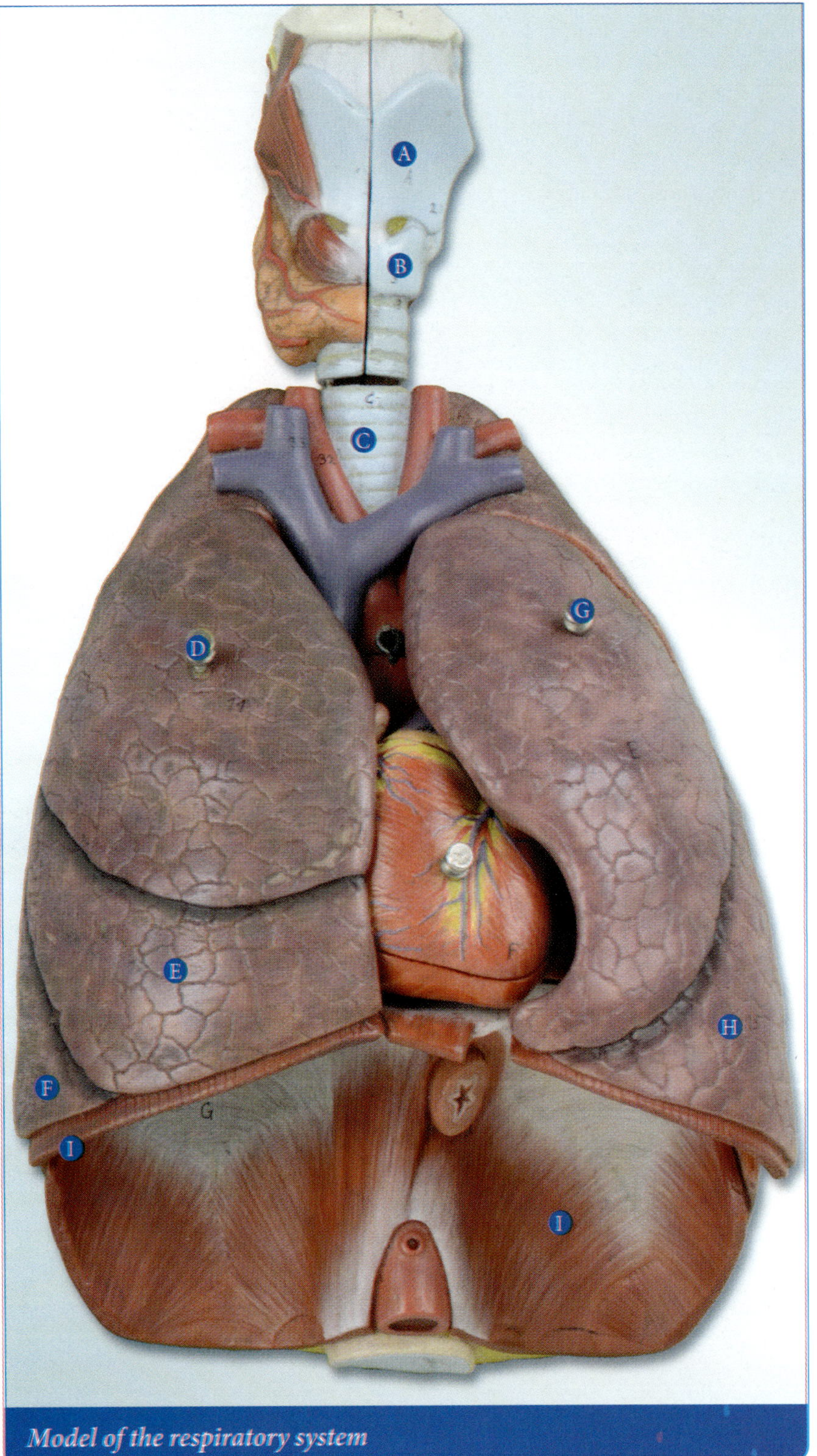

Model of the respiratory system

The respiratory system is structurally divided into the upper respiratory system (nose and pharynx) and the lower respiratory system (larynx, trachea, bronchi, and lungs). Functionally, the conducting division (nose, pharynx, larynx, trachea, bronchi, and bronchioles) delivers air into and out of the lungs. Portions of the nasal cavity, trachea, bronchi, and larger bronchioles are lined with respiratory epithelium (ciliated psuedostratified columnar epithelium with goblet cells) which help to humidify and filter inspired (inhaled) air. In the trachea, the mucociliary escalator helps keep the airways free of pathogens and excess mucous accumulation by the action of beating cilia that transport a thin film of mucous (with trapped debris) up to the pharynx, where it can be swallowed and broken down by the acid bath in the stomach. The trachea contains C-shaped cartilage rings, with the open end facing the esophagus. This allows the passage of food in the esophagus that runs along the posterior side. The trachealis muscle connects the open ends of the "C."

The respiratory division (respiratory bronchioles and alveoli) is the site of gas exchange between air and blood.

A Thyroid cartilage	**D** Superior lobe of right lung	**G** Superior lobe of left lung
B Cricoid cartilage	**E** Middle lobe of right lung	**H** Inferior lobe of left lung
C Trachea	**F** Inferior lobe of right lung	**I** Diaphragm

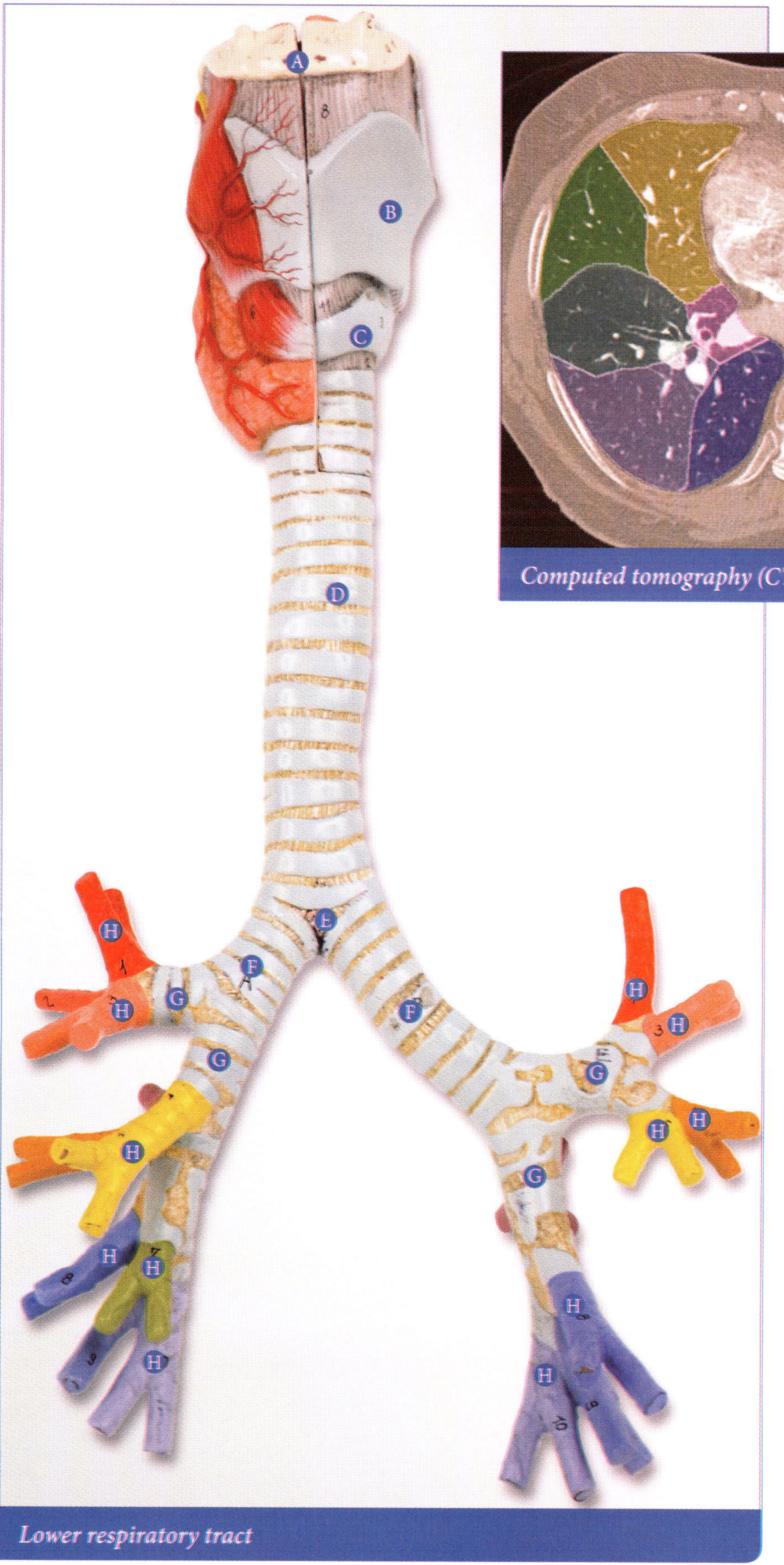

Computed tomography (CT) scan of the lungs

The above CT scan has five bronchopulmonary seg-ments falsely colored in the right lung and six broncho-pulmonary segments falsely colored in the left lung.

- **A** Hyoid bone
- **B** Thyroid cartilage
- **C** Cricoid cartilage
- **D** Tracheal cartilage
- **E** Carina
- **F** Main (primary) bronchus
- **G** Lobar (secondary) bronchi
- **H** Segmental (tertiary) bronchi

Each tertiary bronchus supplies an independent unit of the lung called a bronchopulmonary segment. The right lung contains ten of these and the left lung has eight.

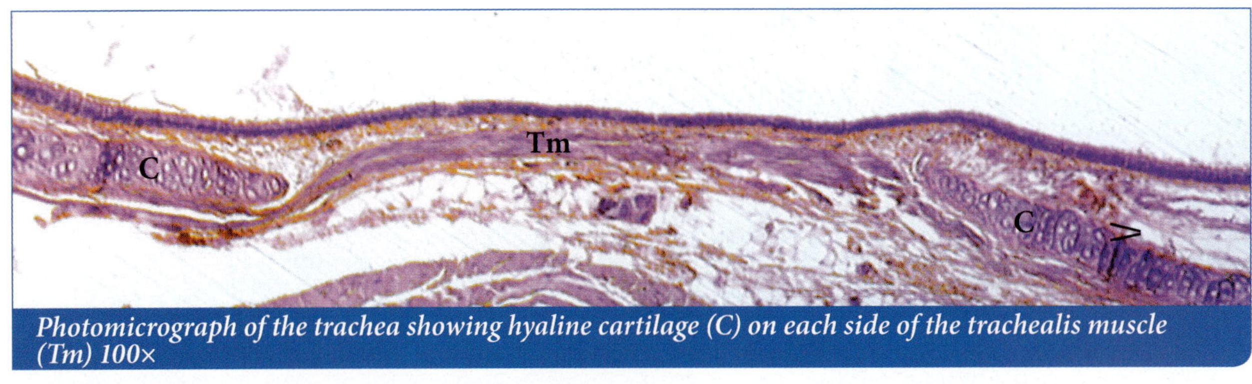

Photomicrograph of the trachea showing hyaline cartilage (C) on each side of the trachealis muscle (Tm) 100×

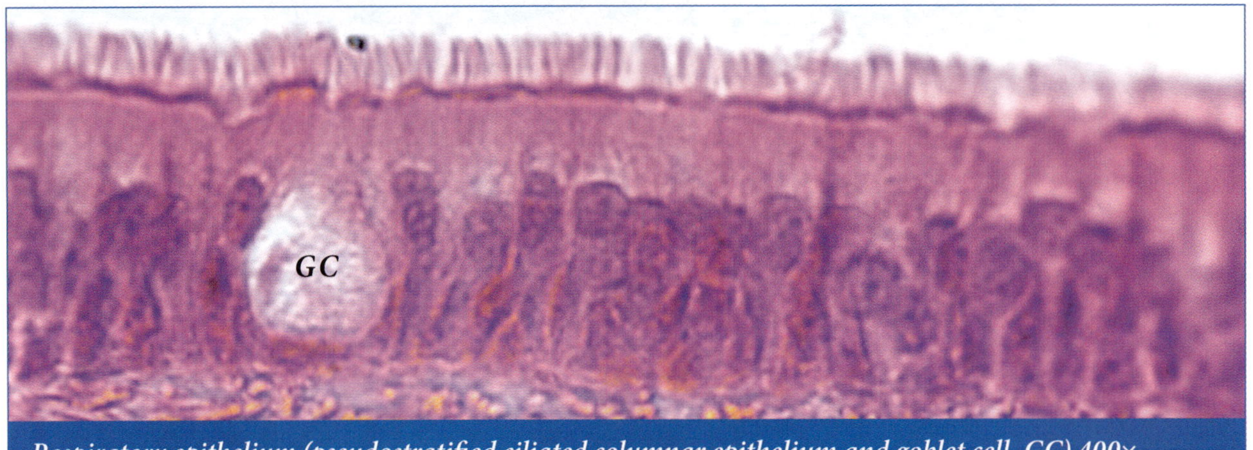

Respiratory epithelium (pseudostratified ciliated columnar epithelium and goblet cell, GC) 400×

SEM of respiratory epithelium (pseudostratified columnar epithelium with cilia) 2,650×

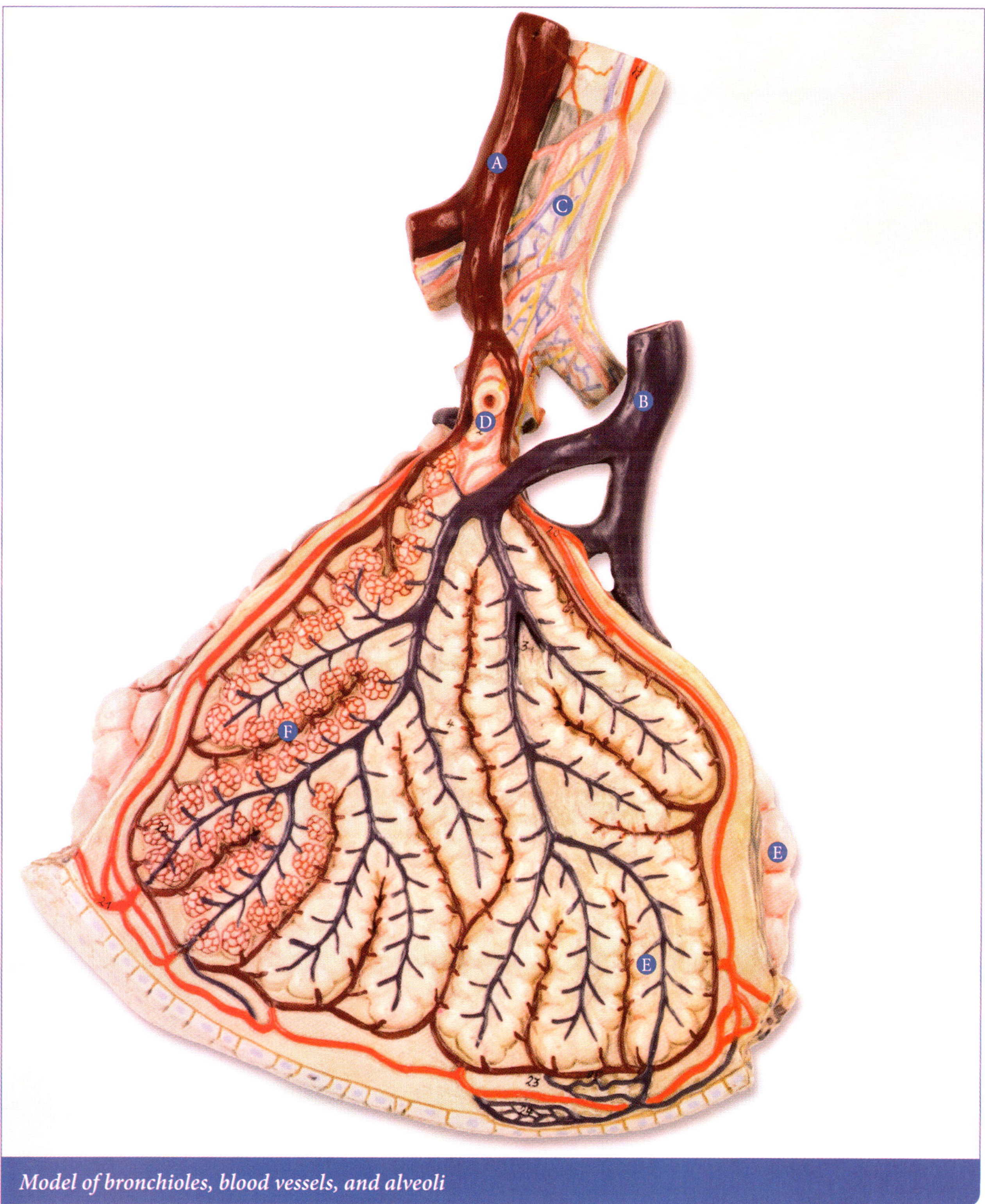

Model of bronchioles, blood vessels, and alveoli

Ⓐ	Pulmonary venule	Ⓓ	Respiratory bronchiole
Ⓑ	Pulmonary arteriole	Ⓔ	Alveolar sacs (alveoli)
Ⓒ	Terminal bronchial	Ⓕ	Alveolar capillaries

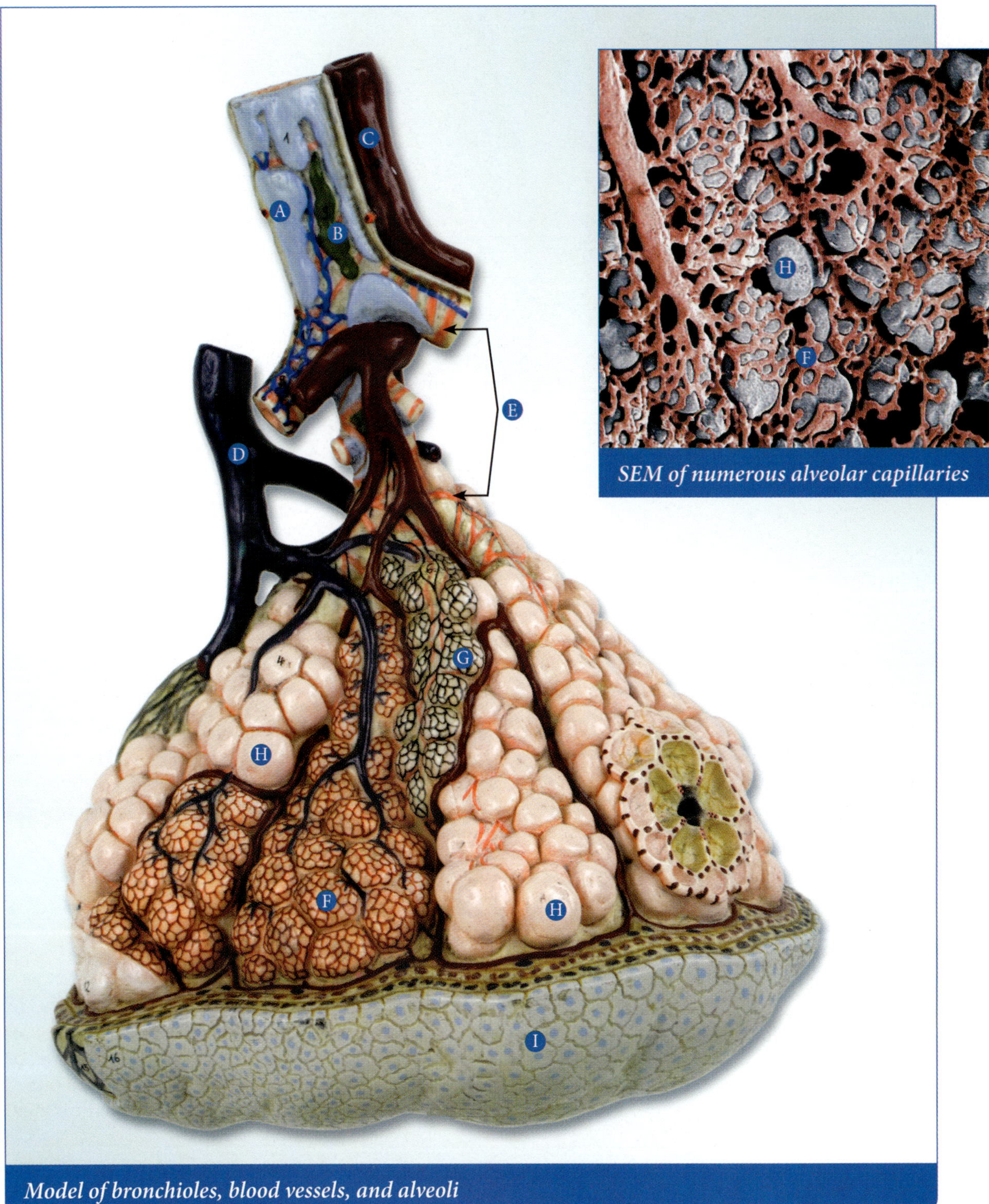

SEM of numerous alveolar capillaries

Model of bronchioles, blood vessels, and alveoli

A Cartilaginous plates

B Lymphatic vessel

C Pulmonary venule

D Pulmonary arteriole

E Smooth muscle cells

F Alveolar capillaries

G Alveolar lymphatic capillaries

H Alveolar sacs

I Visceral pleura

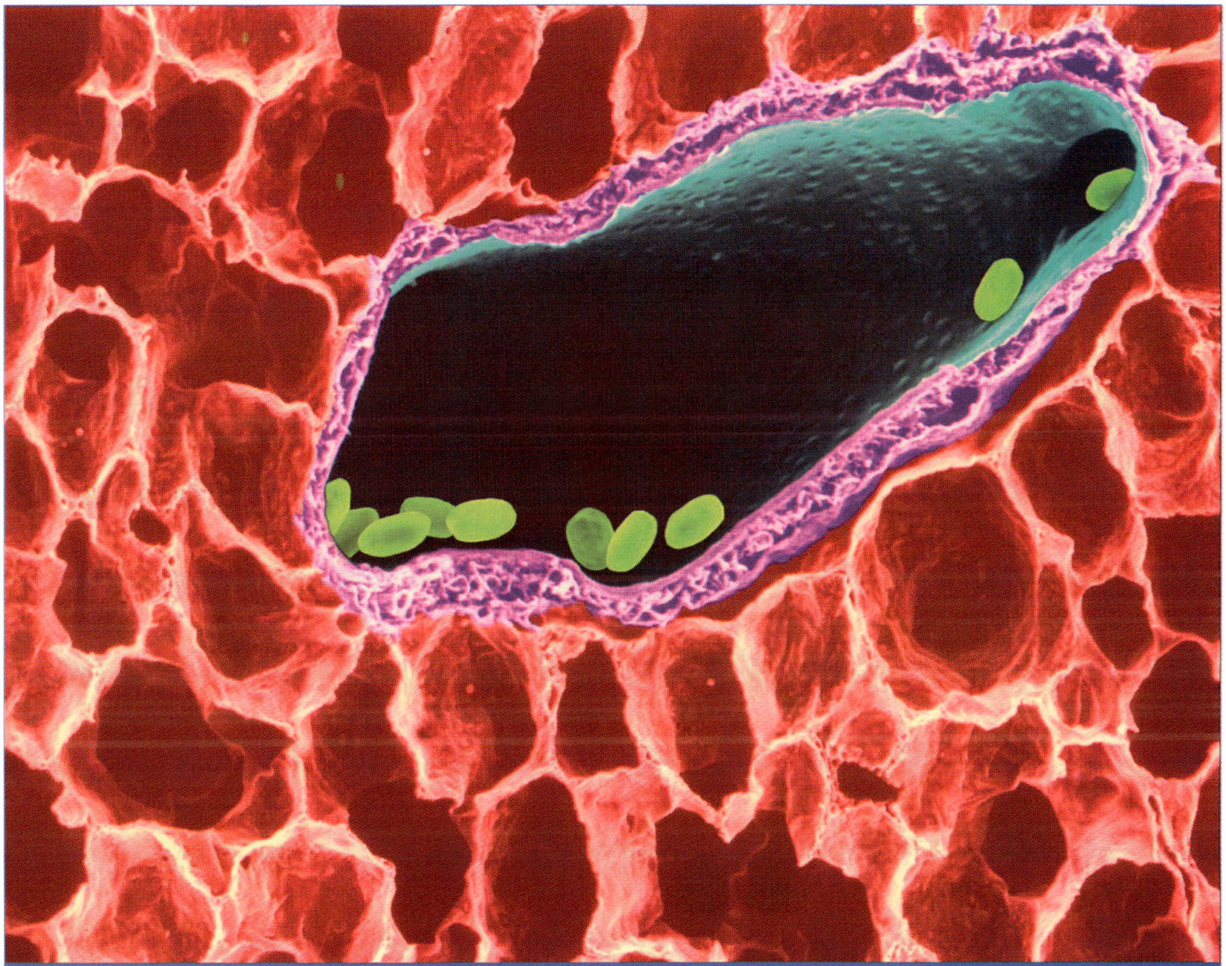

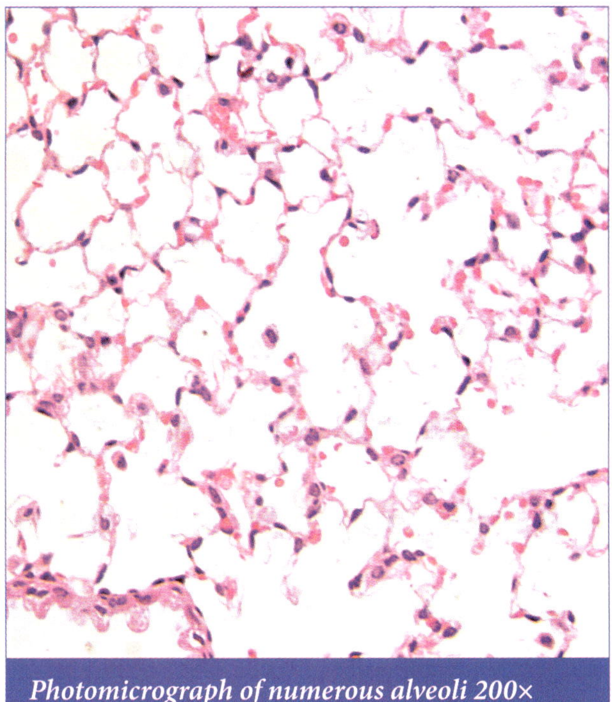

SEM of a bronchiole (with green bacterial spores inside) surrounded by alveoli 2,690×

Photomicrograph of numerous alveoli 200×

There are around 300 million alveoli in our lungs that create about 70 m² of surface area (the size of a racquetball court!). Alveoli allow gas exchange (the respiratory division) in the lungs to occur. The structures up to the alveolus are transport airways (the conducting division) that deliver atmospheric air to these microscopic sacs. It is here that oxygen diffuses into our blood stream, and carbon dioxide is eliminated into the alveolar air to be exhaled. The respiratory membrane is the barrier that oxygen and CO_2 must cross to get into and out of the bloodstream, respectively. From the alveolus, O_2 must travel through a surfactant fluid layer (produced by type II cells), the alveolar epithelium (type I cells), then through basement membranes, and finally the capillary endothelium where it binds to hemoglobin. The respiratory membrane is extremely thin, about 0.5 micrometers, which is vital in allowing rapid diffusion between alveolar air and blood.

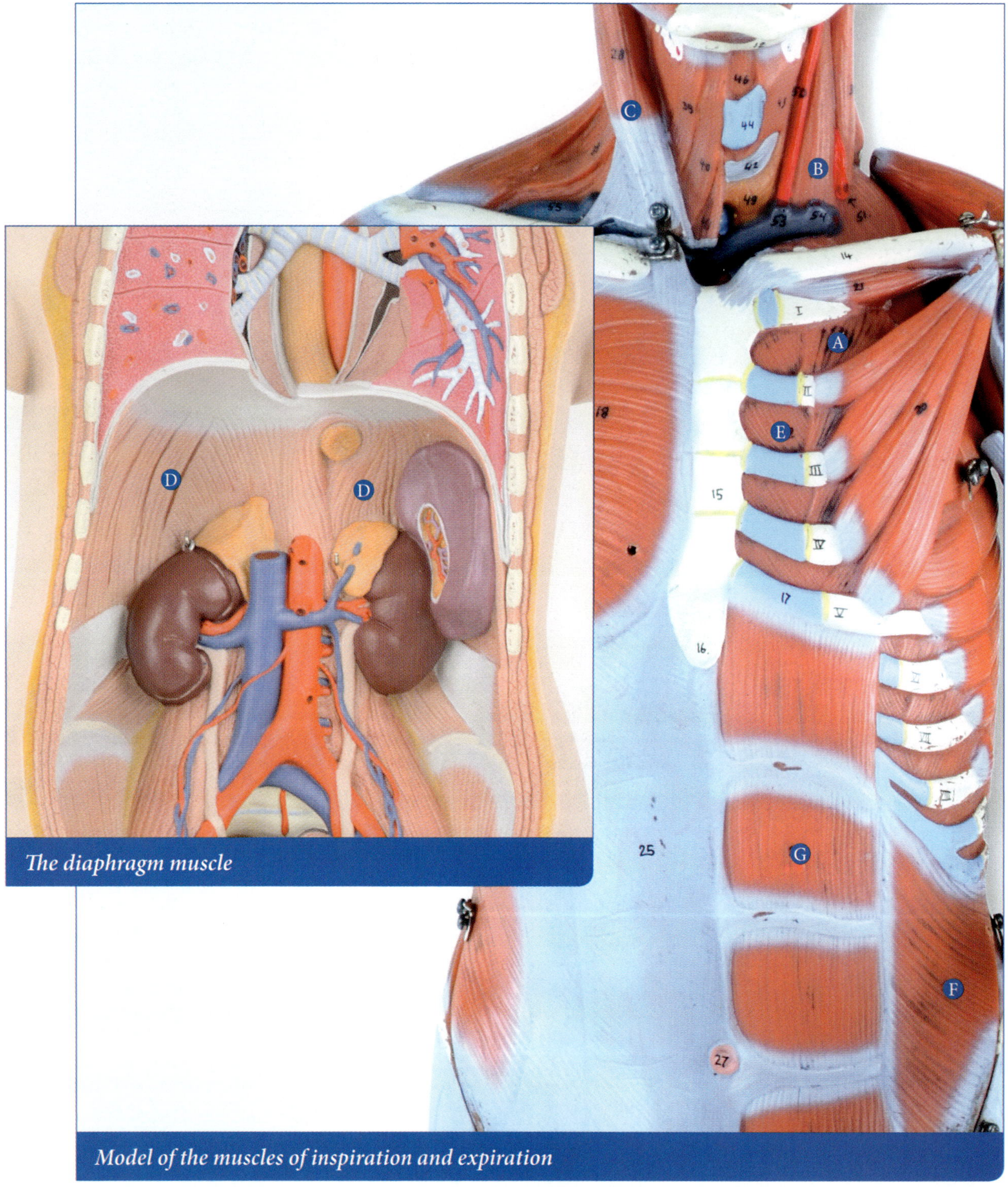

The diaphragm muscle

Model of the muscles of inspiration and expiration

MUSCLES OF INSPIRATION

- Ⓐ External intercostals
- Ⓑ Scalenes
- Ⓒ Sternocleidomastoid
- Ⓓ Diaphragm

MUSCLES OF EXPIRATION

- Ⓔ Internal intercostals
- Ⓕ External abdominal obliques
- Ⓖ Rectus abdominis

Pulmonary ventilation (breathing): Inspiration and expiration is accomplished by varying the volume of the thoracic cavity. Boyle's Law states that air pressure is inversely related to the volume in a sealed container, i.e., as volume increases, pressure decreases, and as volume decreases, air pressure increases. At rest, the muscle of inspiration and passive expiration is the diaphragm. As this inspiratory breathing muscle contracts, the volume of the thoracic cavity increases, the pressure within the lungs drops below atmospheric pressure, and air rushes into the lungs. When the diaphragm relaxes, elastic recoil and surface tension within air spaces return the muscle to its relaxed dome shape position. Thoracic volume decreases, causing an increase in air pressure within the lungs, and air moves out of the lungs.

LUNG VOLUMES

Tidal Volume (TV) is the amount of air moved into and out of the lungs at rest. 500 ml

Inspiratory Reserve Volume (IRV) begins at peak of inspiration of TV and ends at maximum inflation of the lungs. 3,000 ml

Expiratory Reserve Volume (ERV) begins at expiration of TV and ends when the lungs cannot expel any more air. 1,200 ml

Vital Capacity (VC) is the total amount of air that can be moved into and out of the lungs in one breath. 4,800 ml

Forced Expiratory Volume at 1 Second (FEV$_1$) is used as a tool to detect obstructive disorders (limiting airflow from reduced diameter or blockage of airways). FEV$_1$ is the percentage of vital capacity exhaled in one second. The normal range is 75–85% of tidal volume exhaled in one second.

Functional residual capacity (FRC) is the volume of air that always occupies the lungs unless atelectasis occurs (collapsed lung), i.e., the volume that exists after complete exhalation. 1,300 ml

Total lung capacity (TLC) is the total volume of air the lungs can hold. 6,100 ml

As we rest, we breathe about 12 times per minute. If each breath at rest is 500 ml, then we are moving about 6 L of air per minute in and out of our lungs. This may increase up to 20 times during exercise. During intense exercise, breathing rate may increase to 40 times per minute and the volume up to 3,000 ml per breath. That is 120 L of air per minute!

Air flow and lung volumes of normal and forced breathing of a subject at rest

Spirometry is the measurement of air volume that is moved into and out of the lungs during breathing. A spirometer is a device used to record these values. Spirometers may directly measure the air moving in and out of the lungs by a mechanical device or indirectly using a tube containing a screen-like membrane that partially obstructs flow, creating pressure differentials that allow estimation of pressure and flow. Spirometry is an important diagnosis tool in assessing chronic obstructive pulmonary diseases (COPD) asthma, bronchitis, and emphysema.

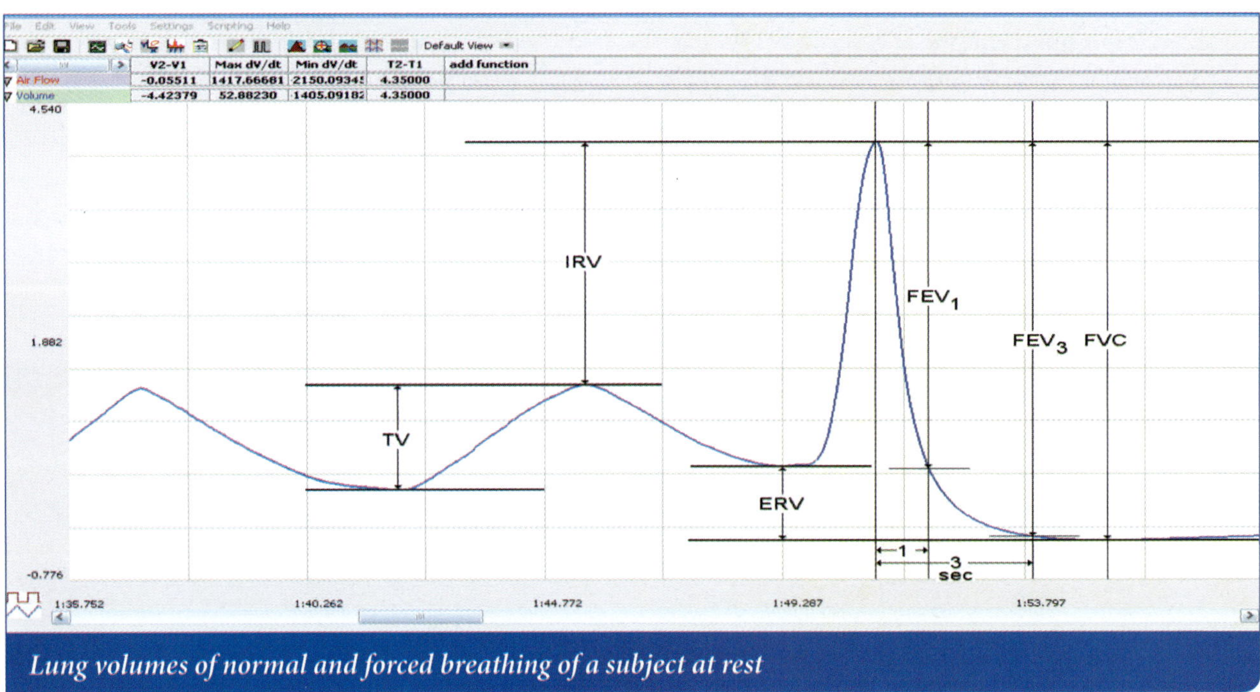

Lung volumes of normal and forced breathing of a subject at rest

Above is a recording of normal and forced lung volumes taken from a subject at rest and displayed on the Volume channel in the Analysis window. The normal breathing cycles are to the left of the forced inspiration and expiration. Lines and labels were added to the above figure to indicate the following volumes: Tidal Volume (TV) about 500 ml, Inspiratory Reserve Volume (IRV) about 3,000 ml, Expiratory Reserve Volume (ERV) about 1,200 ml, Vital Capacity (VC) about 4,800 ml, and Forced Expiratory Volume at 1 Second (FEV$_1$), the percentage of vital capacity exhaled in one second.

9 CHAPTER

Urinary

Kidneys, Renal Corpuscles, and Nephrons

The urinary system eliminates metabolic wastes from the blood stream while maintaining salt, water, and acid-base balance. The kidneys receive about 20% of cardiac output at rest (about 1 liter per minute). Each kidney contains over 1 million nephrons, the functional unit, which filter and alter our blood chemistry. A little over half of our blood volume (approximately 2½ liters) is fluid or plasma. Our kidneys filter this fluid, producing around 180 liters of filtrate (fluid forced out of blood) every day with approximately 99% being reabsorbed along the nephron. Our kidneys can produce dilute or concentrated urine depending on hydration levels. Urea, sodium, chloride, and potassium ions in the medulla of the kidneys produce an osmotic gradient that is about 4 times greater than most body fluids. This ocean like environment of the medulla allows the kidneys to eliminate wastes, concentrate filtrate, and conserve water, all at the same time.

Urinary System

- Kidney
- Renal cortex
- Renal medulla
- Renal pyramid
- Minor calyx
- Major calyx
- Renal pelvis
- Ureter
- Urinary bladder
- Descending aorta
- Inferior vena cava
- Renal artery
- Renal vein
- Arcuate artery
- Arcuate vein
- Interlobar artery
- Interlobar vein
- Interlobular arteries

Types of Nephrons

- Juxtamedullary nephrons
- Cortical nephrons

Nephrons and Associated Structures

- Efferent arteriole
- Afferent arteriole
- Glomerulus
- Podocyte nucleus
- Filtration slits
- Glomerular (Bowman's) capsule
- Proximal convoluted tubule
- Descending limb of loop (thick segment)
- Descending limb of loop (thin segment)
- Ascending limb of loop (thin segment)
- Ascending limb of loop (thick segment)
- Distal convoluted tubule
- Collecting duct
- Macula densa
- Juxtaglomerular cells
- Vasa recta
- Renin-angiotensinogen-angiotensin II
- Counter current multiplier
- Counter current exchanger

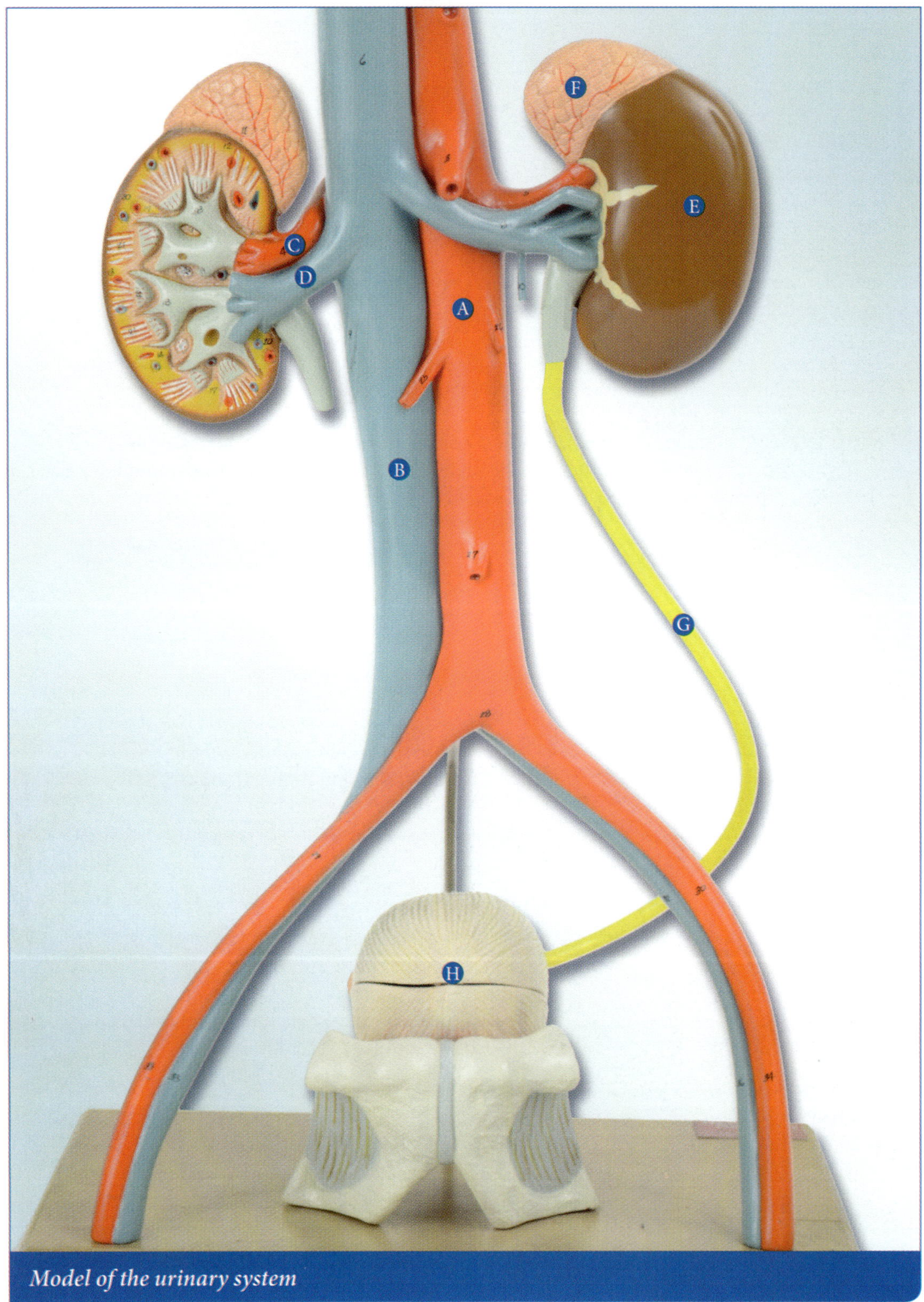

Model of the urinary system

A Descending aorta	**D** Renal vein	**G** Ureter
B Inferior vena cava	**E** Kidney	**H** Urinary bladder
C Renal artery	**F** Adrenal gland	

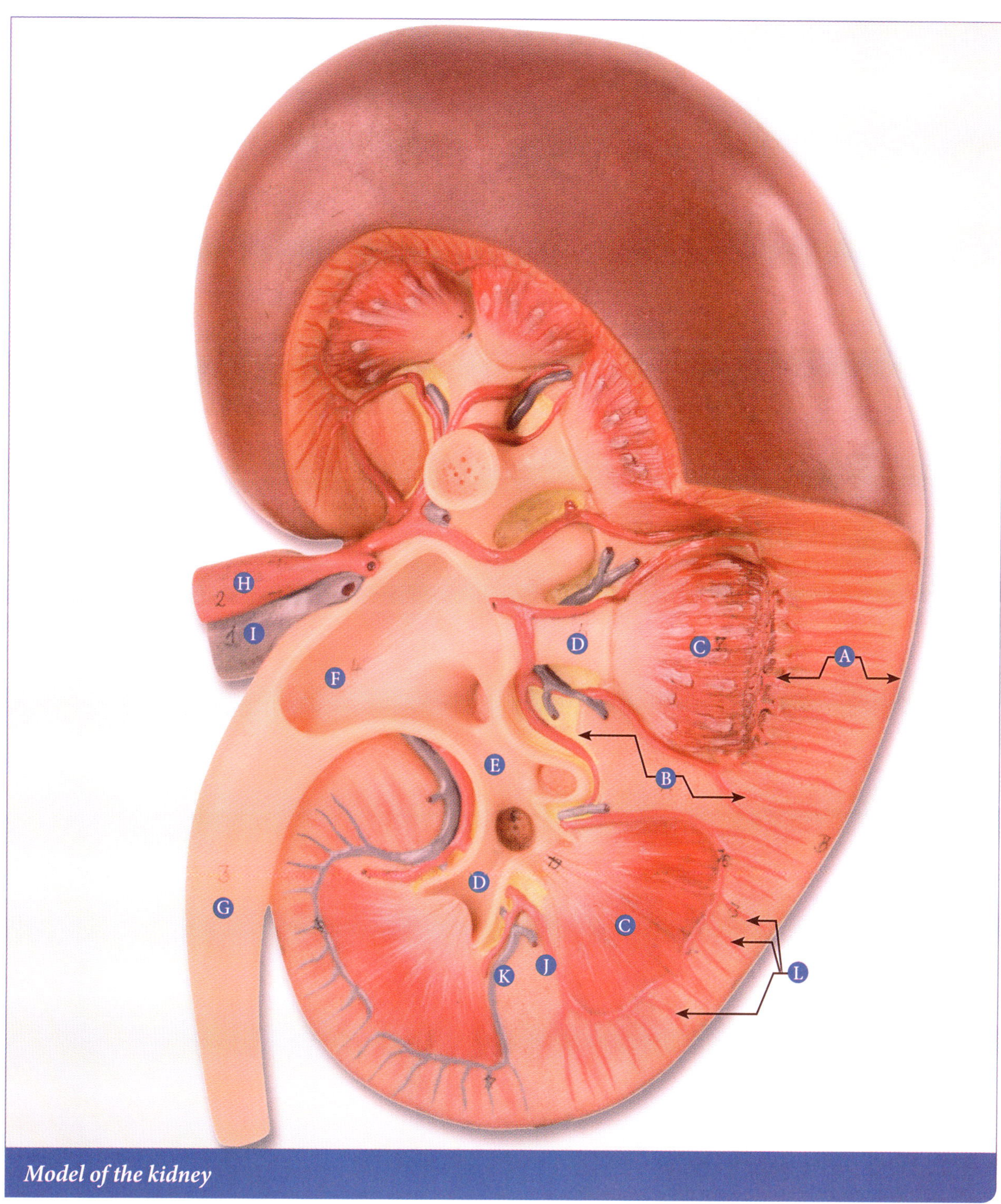

Model of the kidney

A	Renal cortex	E	Major calyx	I	Renal vein
B	Renal medulla	F	Renal pelvis	J	Interlobar artery
C	Renal pyramid	G	Ureter	K	Interlobar vein
D	Minor calyx	H	Renal artery	L	Interlobular arteries

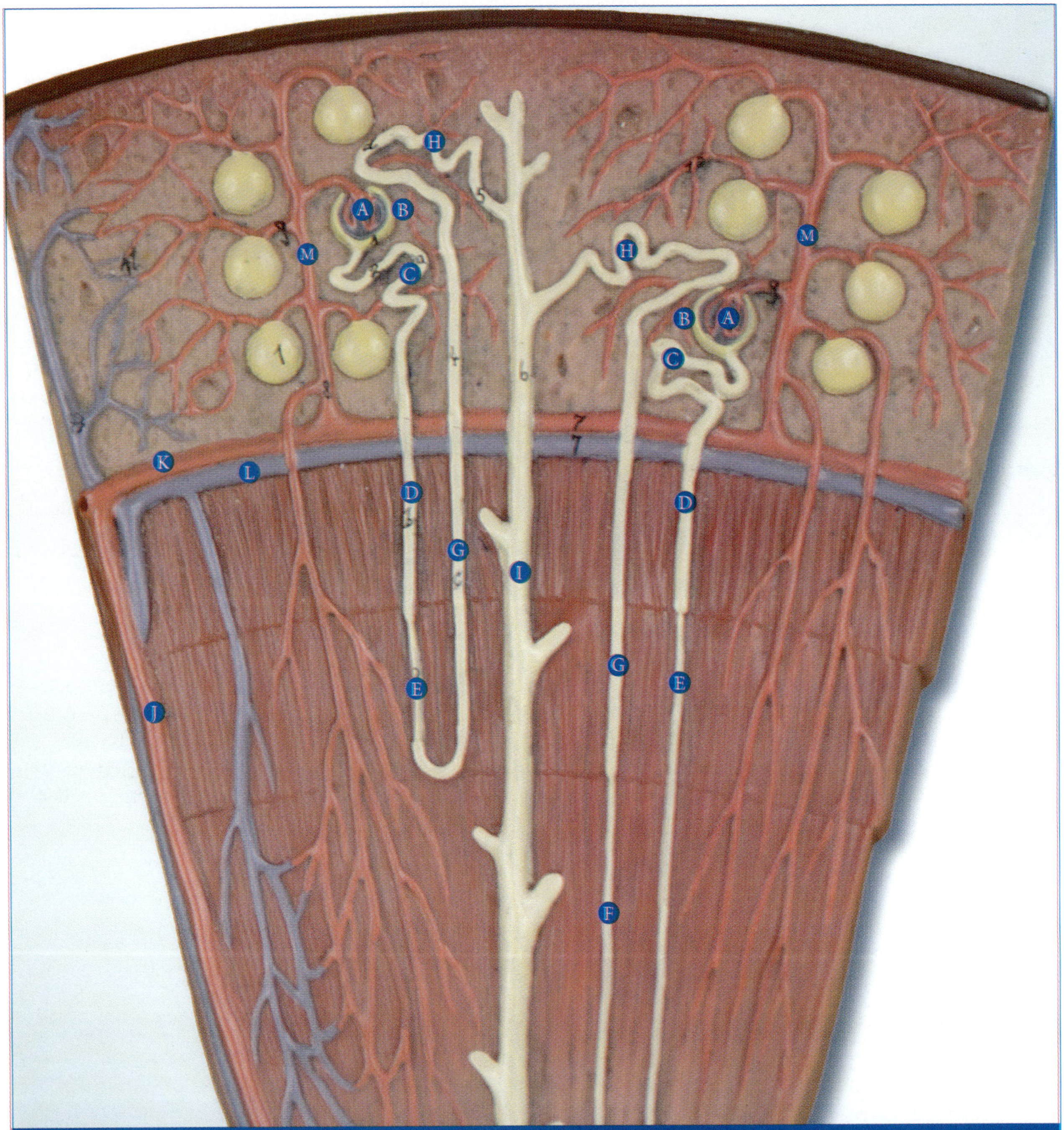

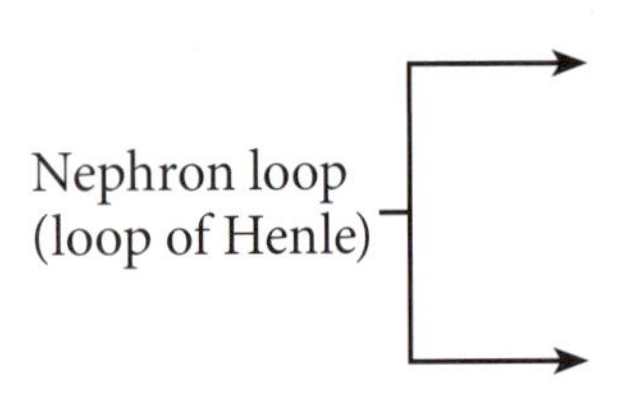

A Glomerulus		**H** Distal convoluted tubule
B Glomerular (Bowman's) capsule		**I** Collecting duct
C Proximal convoluted tubule		**J** Interlobar artery
D Descending limb (thick segment)		**K** Arcuate artery
E Descending limb (thin segment)		**L** Arcuate vein
F Ascending limb (thin segment)		**M** Interlobular artery
G Ascending limb (thick segment)		

Nephron loop (loop of Henle)

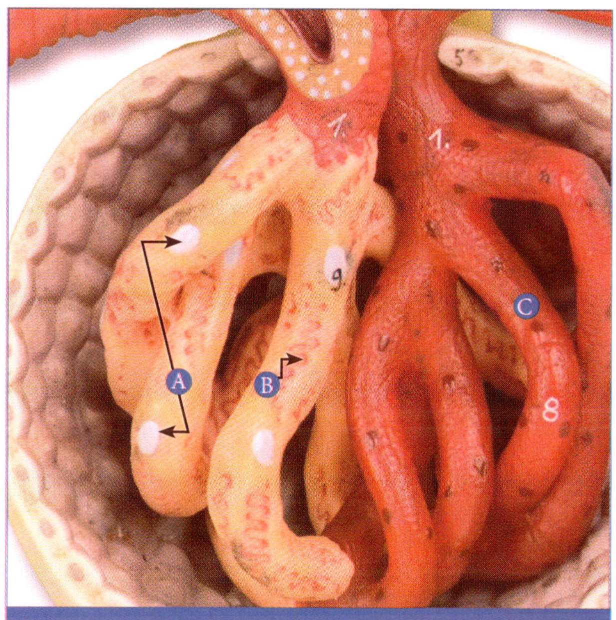

Model of the glomerulus covered with podocytes

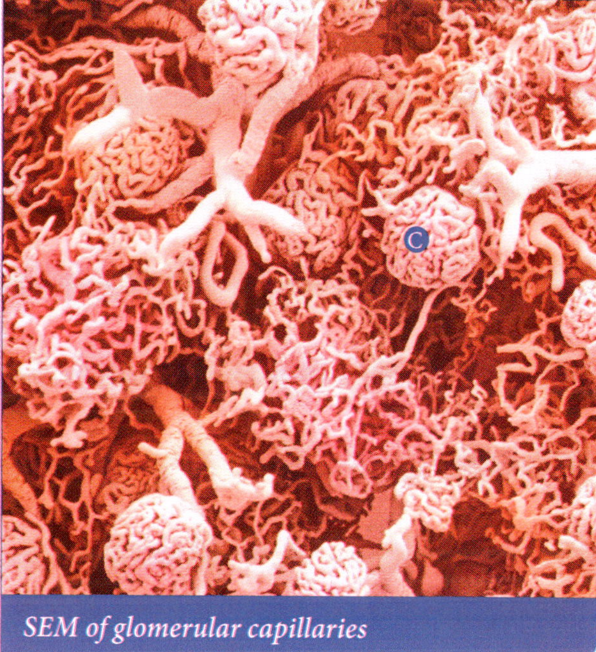

SEM of glomerular capillaries

Filtration of the blood begins in the glomerulus, a fenestrated (holes) capillary bed C. The glomerulus has the highest blood pressure of any capillary bed, about 60 mmHg (typical capillaries have about 10 mmHg). Wrapped around the glomerular capillaries are cells called podocytes that form filtration slits about 30 nm wide. They allow water, urea, electrolytes, glucose, amino acids, fatty acids, uric acid, creatinine, and vitamins to exit the blood stream, while preventing proteins and blood cells from exiting. As blood plasma and the small, dissolved substances are squeezed out of the glomerulus, the fluid (called filtrate at this point until it exits the collecting duct) is captured by the glomerular (Bowman's) capsule and enters the proximal convoluted tubule. The glomerulus and the glomerular capsule collectively are known as the renal corpuscle.

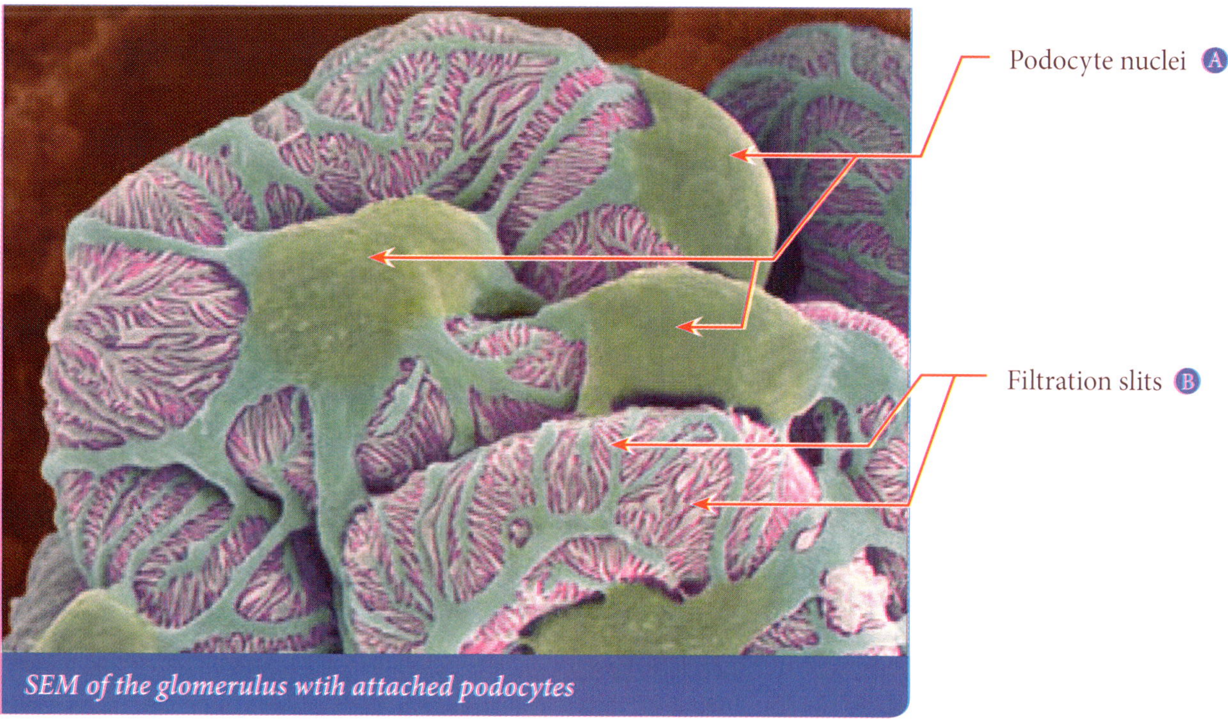

SEM of the glomerulus wtih attached podocytes

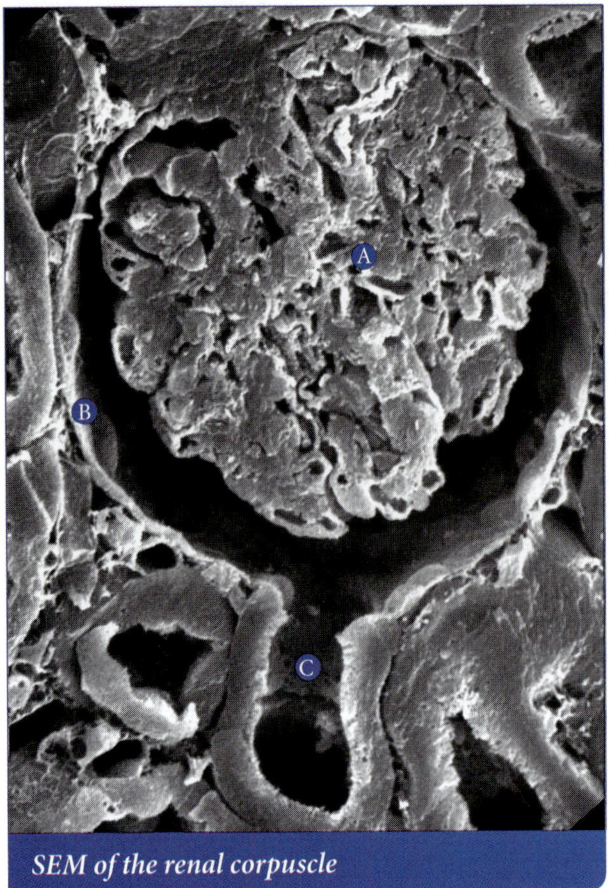

SEM of the renal corpuscle

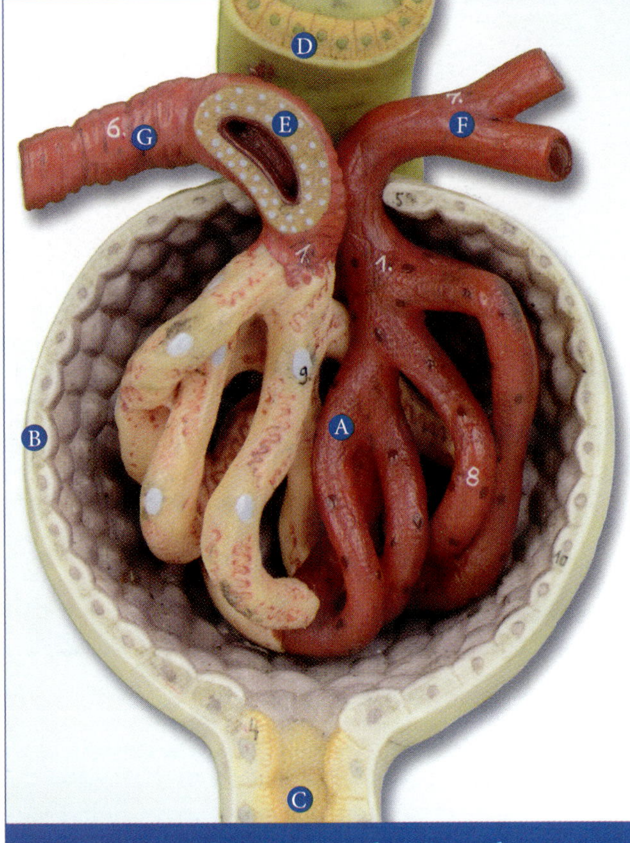

Model of a renal corpuscle and associated structures

The above EM and below photomicrograph clearly show the glomerular capsule **B** and the specialized capillary bed, the glomerulus **A** .

A Glomerulus

B Glomerular (Bowman's) capsule

Renal corpuscle

C Proximal convoluted tubule

D Macula densa of the end of the nephron loop

E Juxtaglomerular cells of the afferent arteriole

F Efferent arteriole (carries blood exiting the glomerulus)

G Afferent arteriole (carries blood entering the glomerulus)

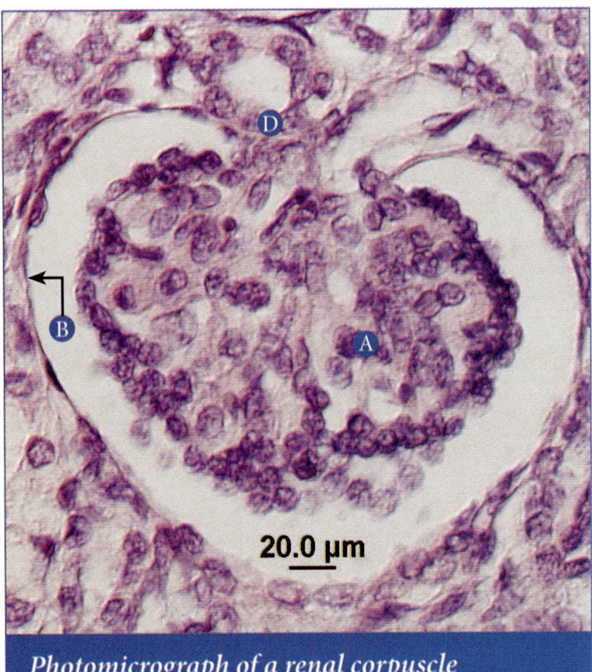

Photomicrograph of a renal corpuscle

The Juxtaglomerular cells, or JG cells (smooth muscle cells of the afferent arteriole) are in contact with, and are influenced by, the macula densa (patch of epithelial cells of distal convoluted tubule that monitor filtrate). When the macula densa detects an increase in filtrate flow, it sends paracrine molecules that stimulate JG cells to contract, constricting the afferent arteriole and reducing glomerular blood pressure and glomerular filtration rate (GFR).

If a drop in blood pressure and GFR occurs, other paracrine messengers cause relaxation of JG cells that dilate afferent artioles and increase GFR. JG cells also secrete renin, an enzyme that converts angiotensinogen to angiotensin I, which is converted to angiotensin II by angiotensin converting enzyme (ACE) in the lungs. Angiotensin II causes an increase in systemic blood pressure by vasoconstriction, increased thirst, and aldosterone secretion which promotes sodium and water retention.

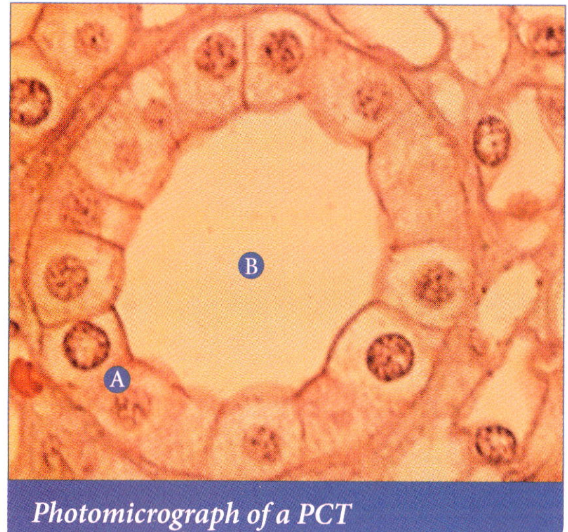

Photomicrograph of a PCT

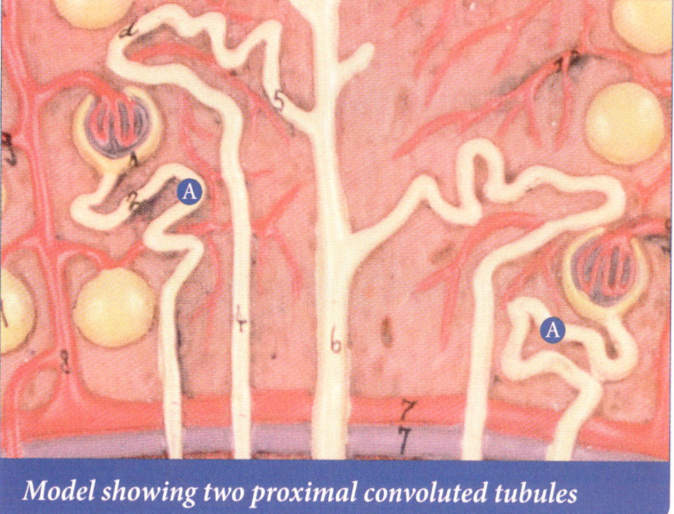

Model showing two proximal convoluted tubules

The proximal convoluted (twisted) tubule (PCT) **A** is the longest section of the nephron. Simple cuboidal epithelial cells that line the lumen **B** contain long, abundant microvilli (see SEM below) that enable rapid reabsorption of filtrate back into the blood stream. About 60% of filtrate formed by glomeruli is reabsorbed here.

This prevents the loss of valuable substances such as water, electrolytes, glucose, amino acids, and fatty acids. Sodium chloride is the most important salt of reabsorption. As NaCl is transported into the simple cuboidal epithelial cells, other molecules like glucose are co-transported. This increases osmotic movement of water back into the bloodstream.

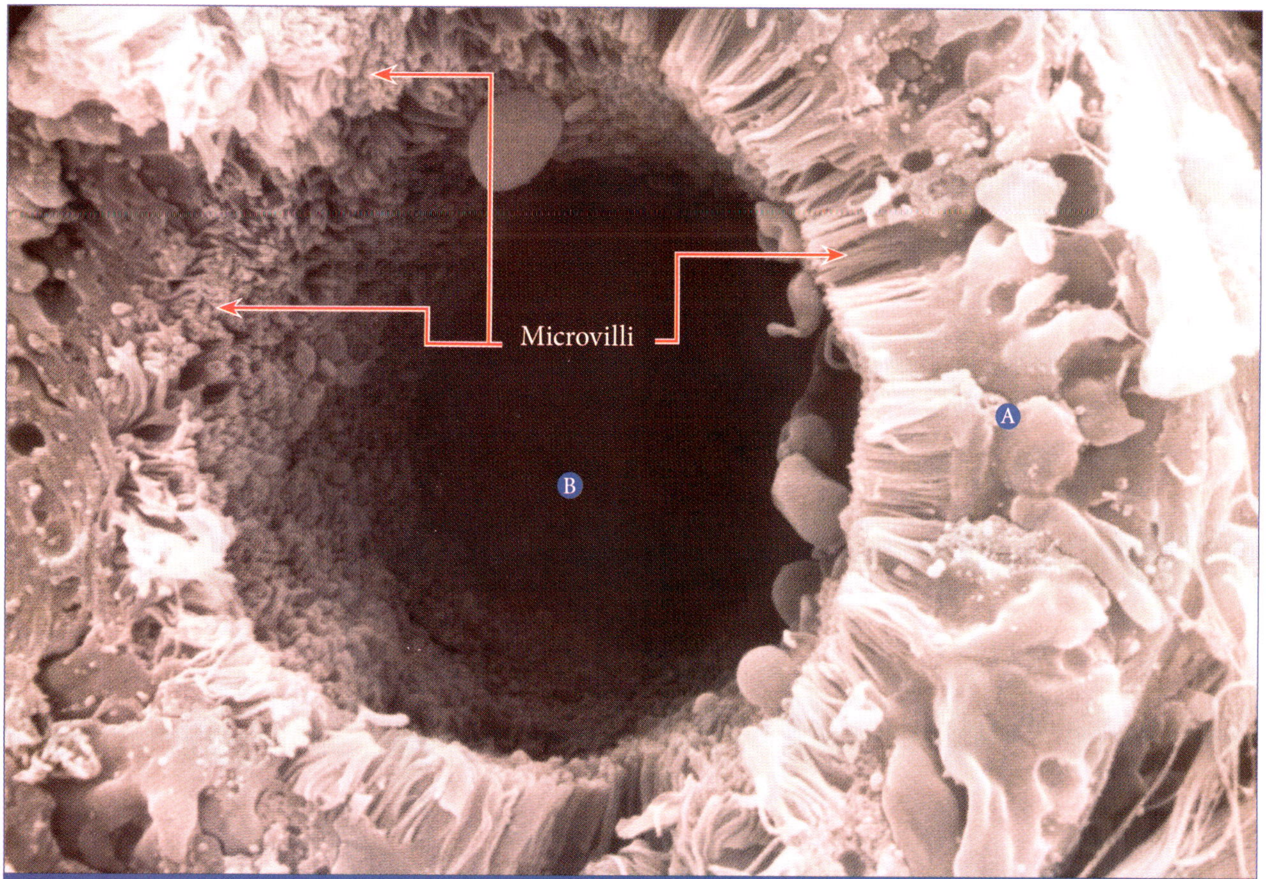

EM of proximal convoluted tubule with microvilli

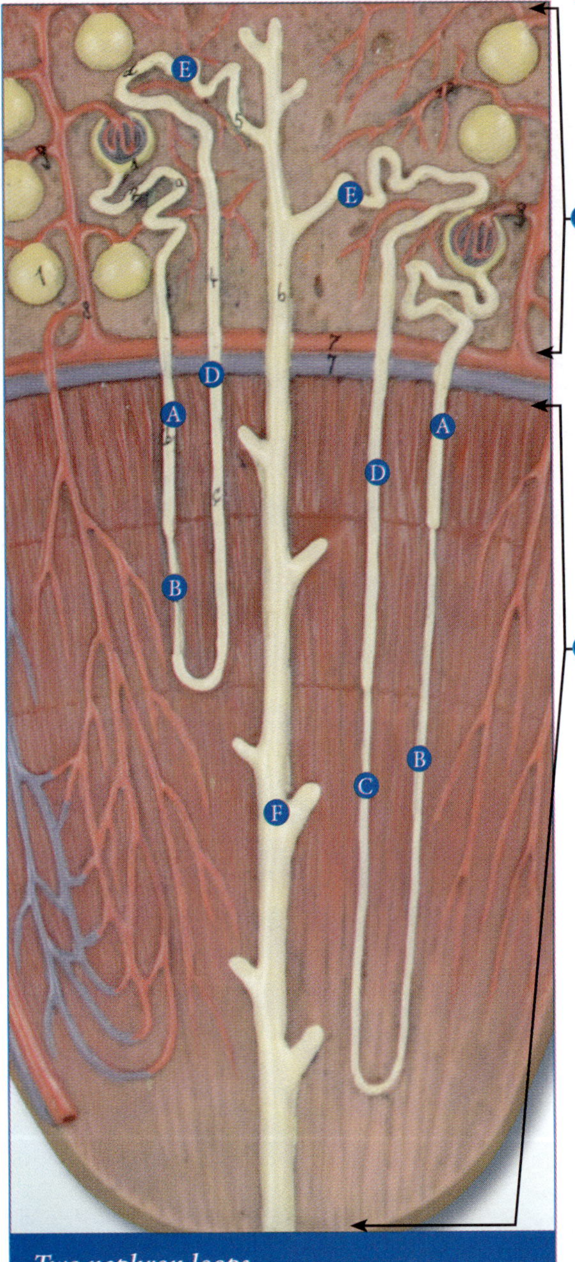

Two nephron loops

The cortex of the kidneys contains glomeruli, PCTs, and distal convoluted tubules (DCT). The nephron loop and collecting duct penetrate into the salty medulla where concentration of urine can occur. Urea accounts for about 40% of the salts in the medulla from a constant recycling from collecting duct back into the nephron loop.

The nephron loop consists of the descending limb (into the salty medulla) and ascending limb (out of the medulla and back into the cortex). The loop contains thin and thick segments that perform different functions. The thin segment is permeable to water, so water leaves the tubule osmotically into the medulla (where it immediately enters blood vessels, the vasa recta). The thick segment is impermeable to water and has numerous sodium, potassium, and chloride pumps that actively transport these ions into the medulla (along with urea), maintaining a high salinity gradient that will allow the collecting duct to concentrate urine. Nephron loops can penetrate deep into the salty medulla (juxtamedullary nephrons) or relatively superficially into the medulla (cortical nephrons). Approximately 85% are cortical nephrons and as their name implies, have shorter loops and lie closer to the cortex. Juxtamedullary nephrons (15%) have loops that run deep into the medulla and are responsible for maintaining high osmolarity and the ability to concentrate urine (counter current exchange and multiplier).

The DCT is responsible for modifying and monitoring the filtrate. Secretions of K^+ and H^+ from the bloodstream into the DCT are important functions of water and acid-base balance. Aldosterone stimulates cells in the DCT to reabsorb sodium (water piggybacks sodium) into blood and secrete potassium into filtrate in response to decrease blood pressure or blood sodium levels. About 20% of the glomerular filtrate is reabsorbed here. Over half of the osmotic gradient in the tubular fluid at this point is produced by urea.

A	Descending limb (thick segment)
B	Descending limb (thin segment)
C	Ascending limb (thin segment)
D	Ascending limb (thick segment)
E	Distal convoluted tubule
F	Collecting duct
G	Cortex
H	Medulla

Nephron loop (loop of Henle): A, B, C, D

The collecting duct makes the final alterations to the filtrate. As the filtrate enters the collecting duct, it is isotonic (equal osmotic pressure with surrounding tissue, about 300 milliosmole). In extreme states of dehydration, the filtrate flowing through collecting ducts can become 4 times as concentrated (around 1,200 milliosmole—about as salty as ocean water) as other bodily fluids. This is possible because of the high osmolarity of the medulla from urea recycling and counter current exchange and mutiplier. In mild states of dehydration (this happens every night while we sleep), ADH from the posterior pituitary causes protein "holes" called aquaporins to be incorporated into the collecting duct. As the filtrate passes along through the increasingly salty medulla, water exits through aquaporin channels of the collecting duct and re-enters the bloodstream. At the end of the collecting duct, there are no further modifications to the filtrate, and it is now called urine.

Urine enters the minor then major calyces, into the renal pelvis, then down the ureters to the bladder. As the bladder fills and stretches, receptors initiate spinal reflexes that relax the internal urethral sphincter (involuntary). The external urethral sphincter remains closed until appropriate timing of urination (micturition) can occur.

 Collecting duct

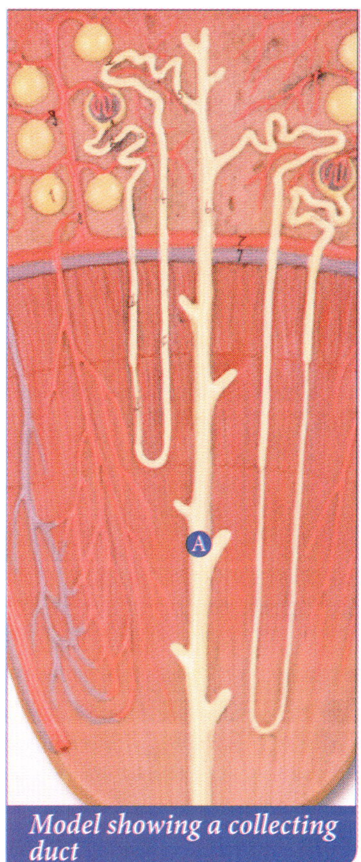
Model showing a collecting duct

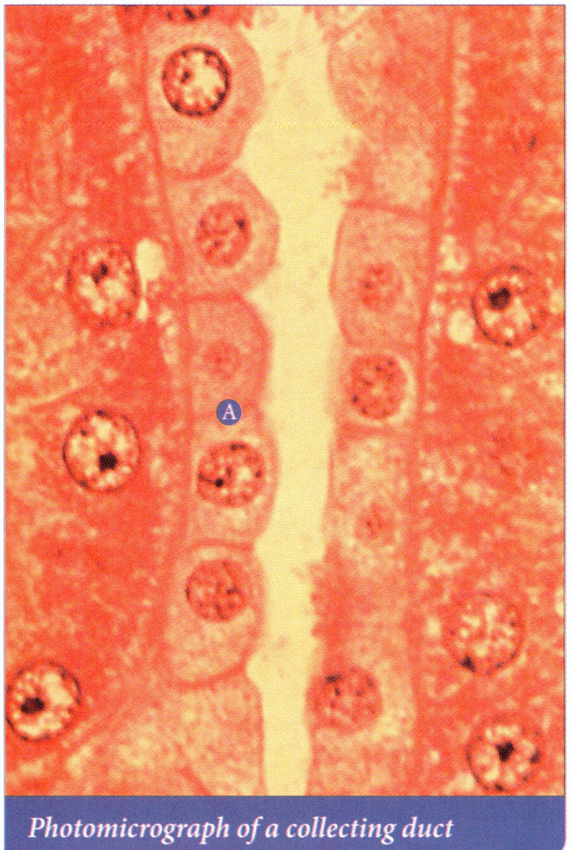
Photomicrograph of a collecting duct

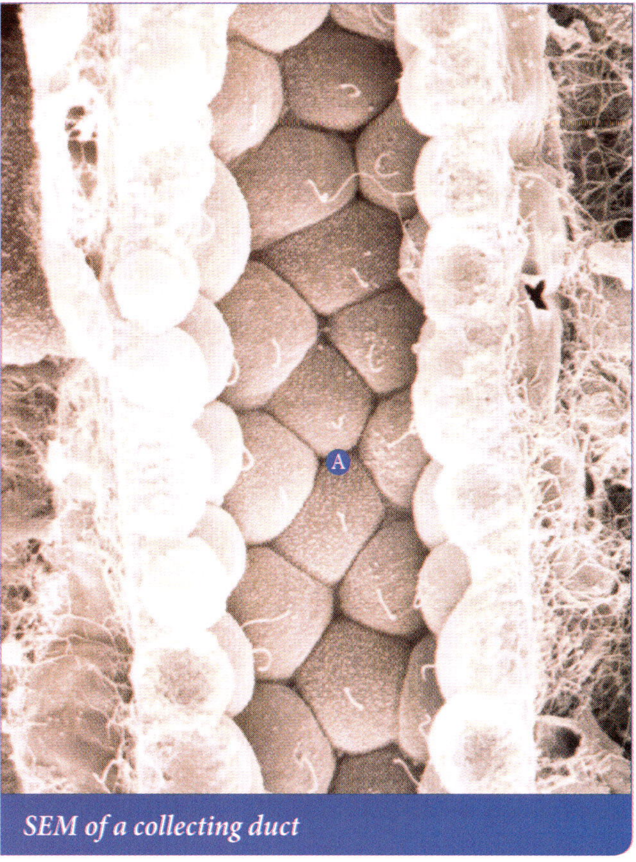
SEM of a collecting duct

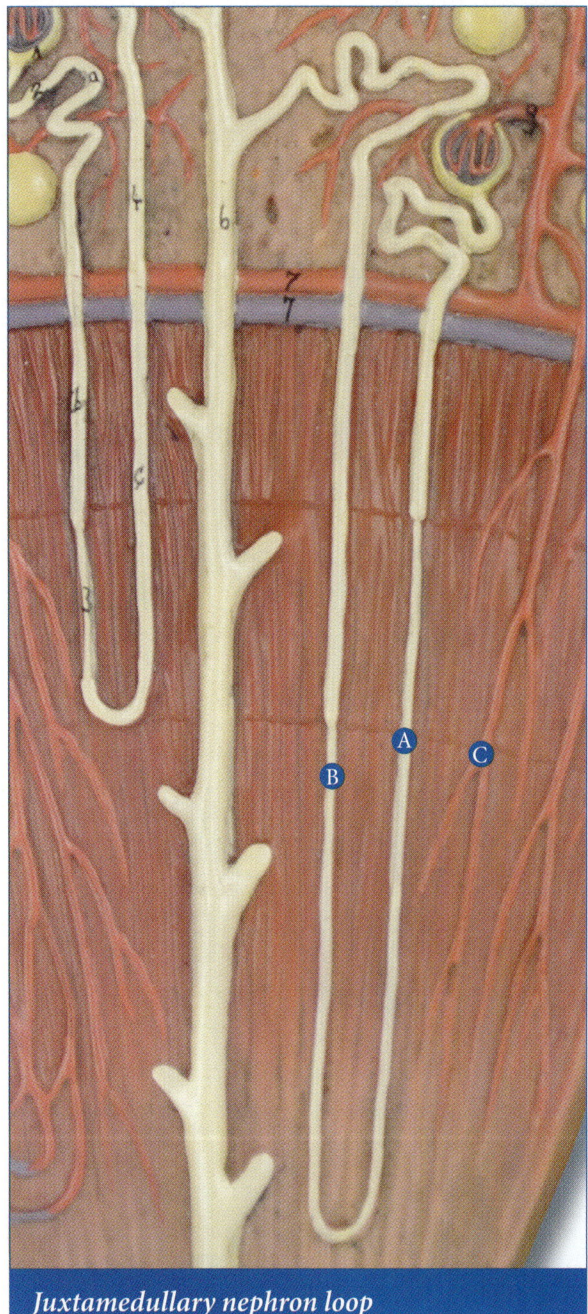

Juxtamedullary nephron loop

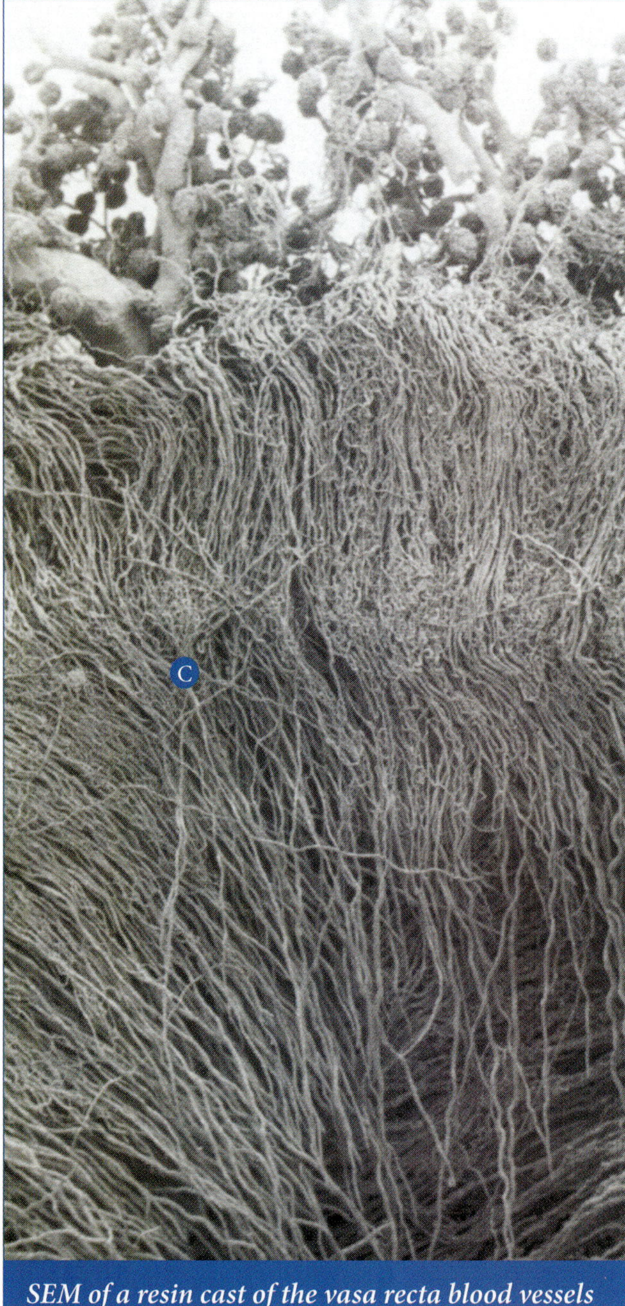

SEM of a resin cast of the vasa recta blood vessels

The **counter current multiplier** maintains the high salinity of the renal medulla and involves juxtamedullary nephrons that have long loops that run deep in the medulla. The counter current portion involves filtrate moving toward the medulla in the descending limb of the nephron loop **A** and moving back towards the cortex in the adjacent ascending limb **B**. This flow in opposite directions in two adjacent tubes allows efficient exchange. The multiplier portion refers to the deposition of salts in the medulla. Tubular fluid entering the loop has a concentration of approximately 300 milliosmole where fluid leaving the loop is only 100 milliosmole.

The vasa recta **C** are blood vessels that supply the medulla's metabolic needs while maintaining the high salinity of that region. The **countercurrent exchange** system ensures that arterial and venous vasa recta vessels do not reduce the saltiness of the medulla. The vasa recta vessels are formed from the efferent arterioles of glomeruli and travel along the nephron loops. Blood flow occurs in the opposite direction than fluid flowing in the nephron loop and adjacent capillaries that readily exchange water and salt. This allows the vasa recta to carry more water out of the medulla than it delivers, further contributing to the high osmolarity of the medulla.

CHAPTER 10

Digestion

Gastrointestinal Tract and Accessory Structures

The gastrointestinal (GI) tract consists of a tube that extends from mouth to anus. The contents of this tube are surprisingly not part of our bodies until they are digested and absorbed into our blood stream to be used for cellular metabolism and growth. Mechanical digestion begins in the mouth (buccal cavity) by mastication (chewing). This increases surface area to volume ratio and moistens food with secretions from salivary glands. Once swallowing is initiated, one direction movement is accomplished by rhythmic, muscular contractions (peristalsis) of the muscularis externa. Chemical digestion is accomplished by enzymes from the mouth, stomach, and most importantly, pancreatic secretions into the small intestines. Our digestive tract breaks down ingested food to macromolecules such as glucose, amino acids, fatty acids, and nucleic acids using enzymes during the process of digestion. Absorption of water and nutrients occurs through the intestinal lining. The small intestine absorbs most nutrients, vitamins, and water. The large intestine hosts a plethora of microbes, completes water reabsorption, and compacts feces.

Mastication and the buccal cavity

- Teeth
  - Incisors – 8
  - Canines – 4
  - Premolars – 8
  - Molars – 8
  - Wisdom teeth (3rd molars) – 4
- Tongue
- Parotid salivary gland

Layers of the alimentary canal

- Mucosa
- Submucosa
- Longitudinal muscularis externa
- Circular muscularis externa
- Serosa or Adventitia

From mouth to stomach

- Esophagus
- Stomach
  - Cardiac region
  - Fundic region
  - Circular muscle layer
  - Oblique muscle layer
  - Antrum
  - Pylorus
  - Pyloric sphincter
  - Gastric rugae
  - Gastric pit

Organs of digestion

- Pancreas
- Liver
- Gallbladder

Small intestines

- Duodenum
  - Major duodenal papillae
  - Duodenal glands in the submucosa
- Jejunum
- Ileum
  - Lymphatic nodules (Peyer's patches)

Small intestine specializations

- Mucosa
  - Plicae circulares
  - Villi
  - Central lacteal
  - Simple columnar epithelial cells
  - Microvilli
  - Goblet cell
  - Muscularis mucosae
  - Lymphatic nodules (Peyer's patches)
  - Submucosa
    - Duodenal glands in the submucosa

Large intestines

- Ileocecal valve
- Cecum
- Appendix
- Ascending colon
- Transverse colon
- Greater omentum
- Descending colon
- Sigmoid colon
- Rectum
- Anus

A — Teeth
B — Esophagus
C — Stomach
D — Small intestines
E — Ascending colon
F — Transverse colon
G — Descending colon
H — Greater omentum
I — Liver
J — Gallbladder

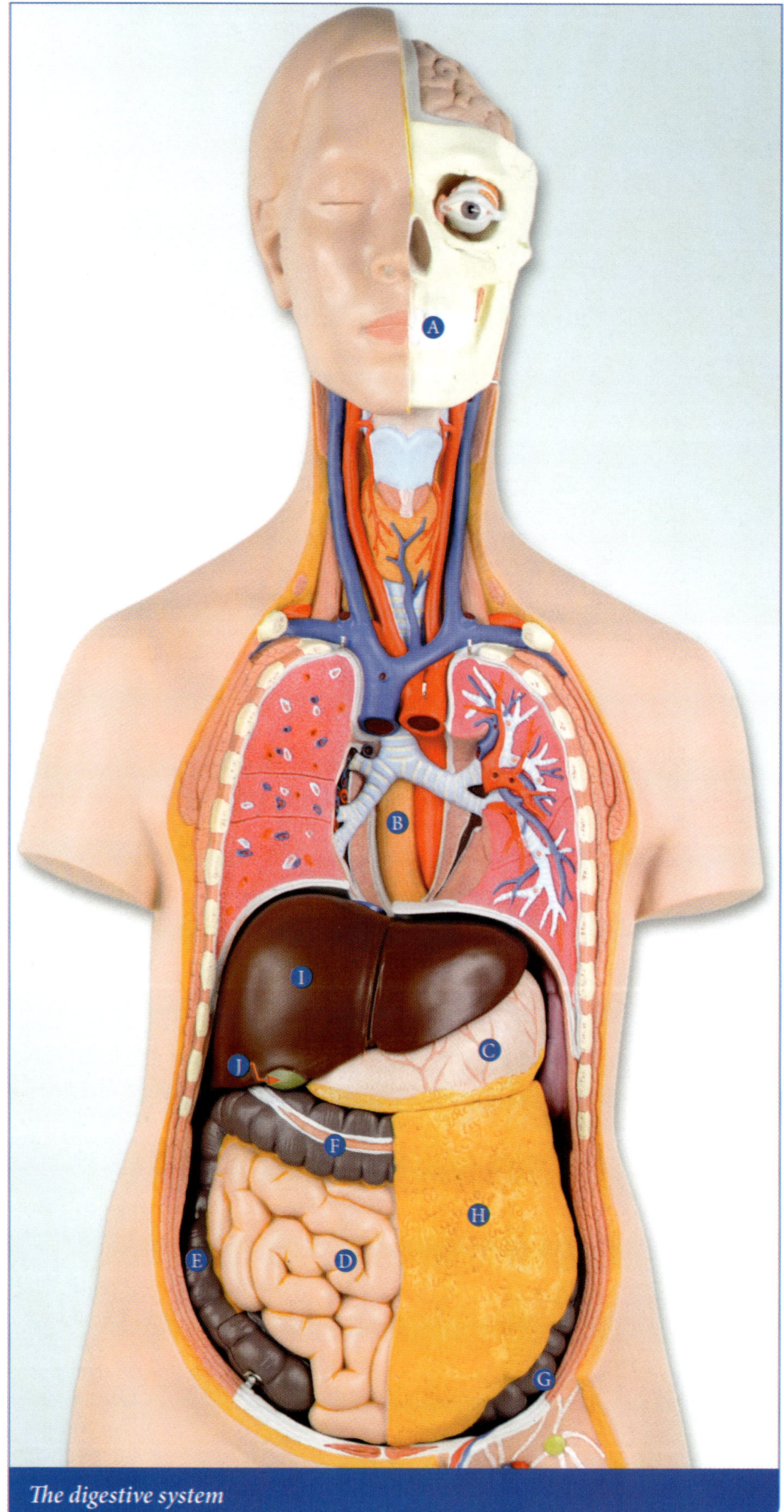

The digestive system

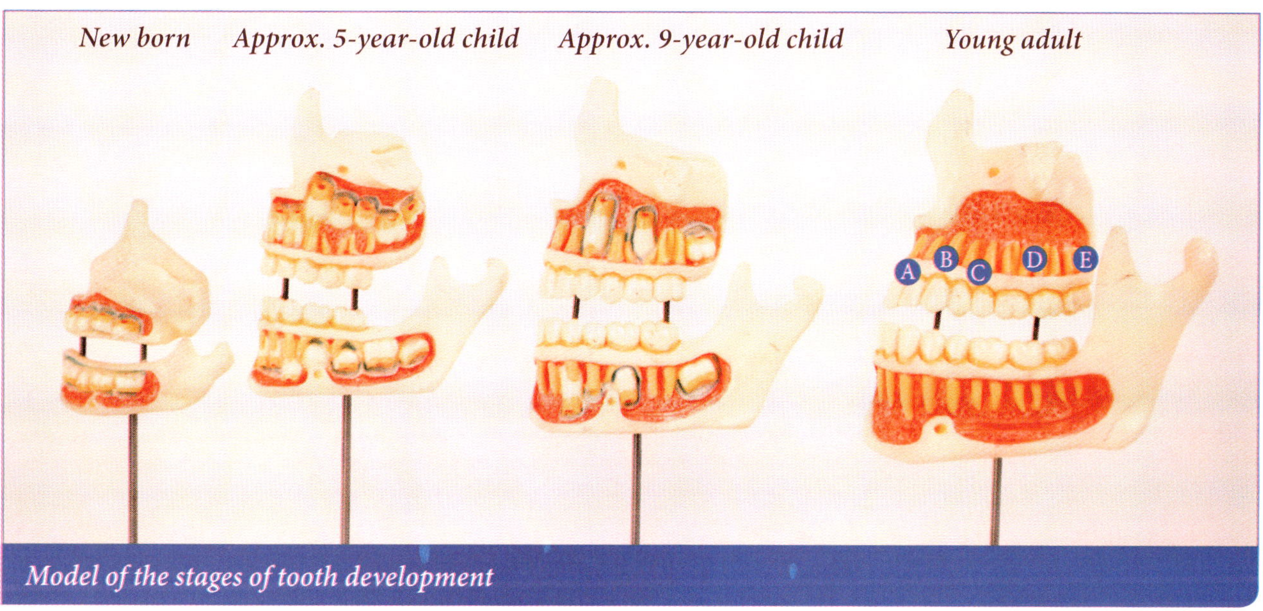

Model of the stages of tooth development

Ⓐ Incisor(s) – 8, 4 top and 4 bottom center of smile

Ⓑ Canine – 4, 2 top and 2 bottom, made for ripping flesh

Ⓒ Premolar(s) – 8, 4 top and 4 bottom

Ⓓ Molar(s) – 8, 4 top and 4 bottom

Ⓔ Wisdom teeth (3rd molars) – 4, 2 top and 2 bottom, sometimes do not descend or have to be pulled to make room

Ⓕ Parotid salivary gland – produces the majority of saliva, which helps moisten masticated food (bolus) through the esophagus

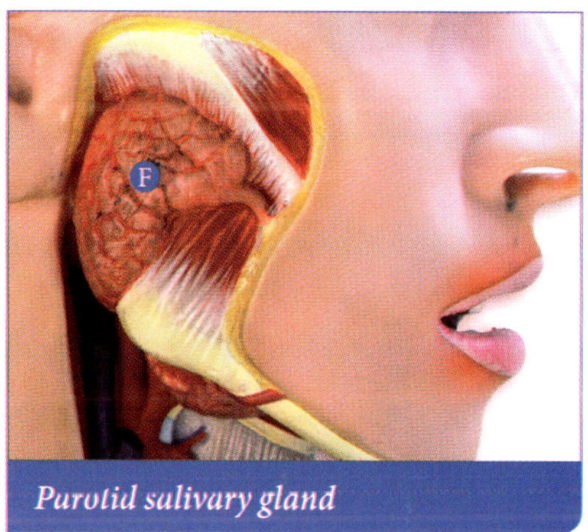

Parotid salivary gland

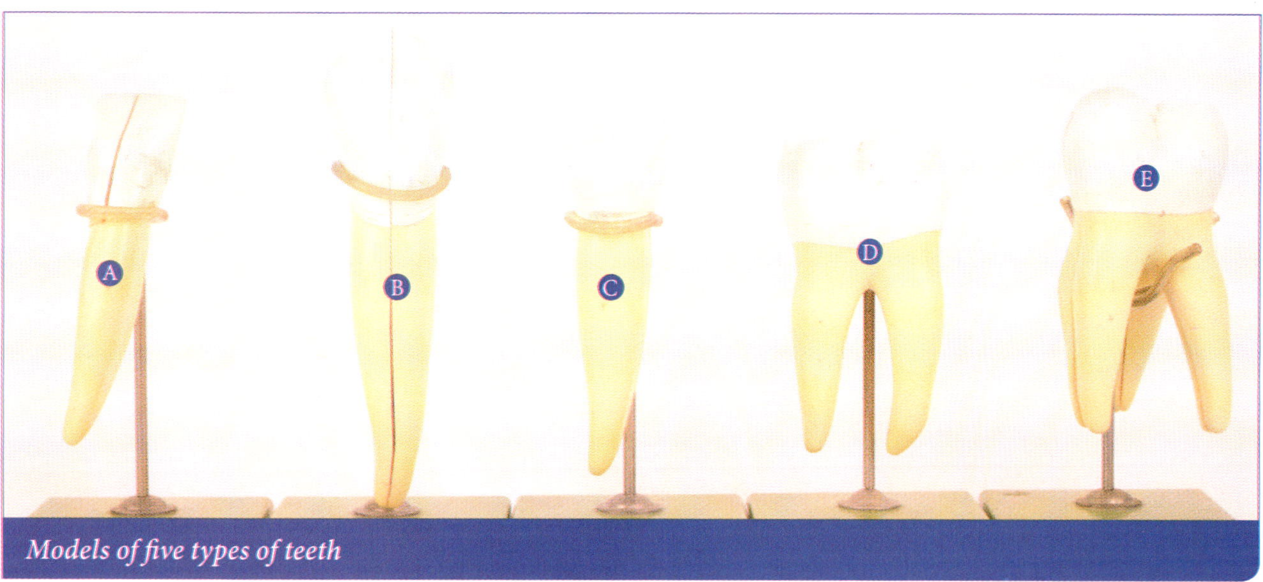

Models of five types of teeth

Mastication (chewing) is mechanical digestion that increases surface-area-to-volume ratio of ingested food. Salivary glands help moisten and lubricate chewed food for swallowing.

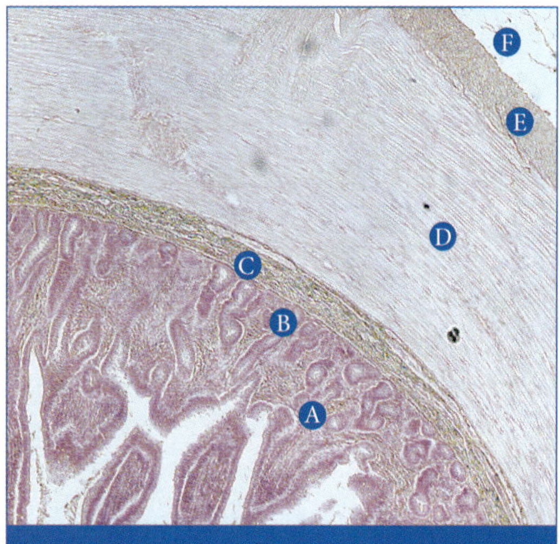

Photomicrograph of the small intestine

- **A** Mucosa
- **B** Muscularis mucosae
- **C** Submucosa
- **D** Circular muscularis externa
- **E** Longitudinal muscularis externa
- **F** Serosa

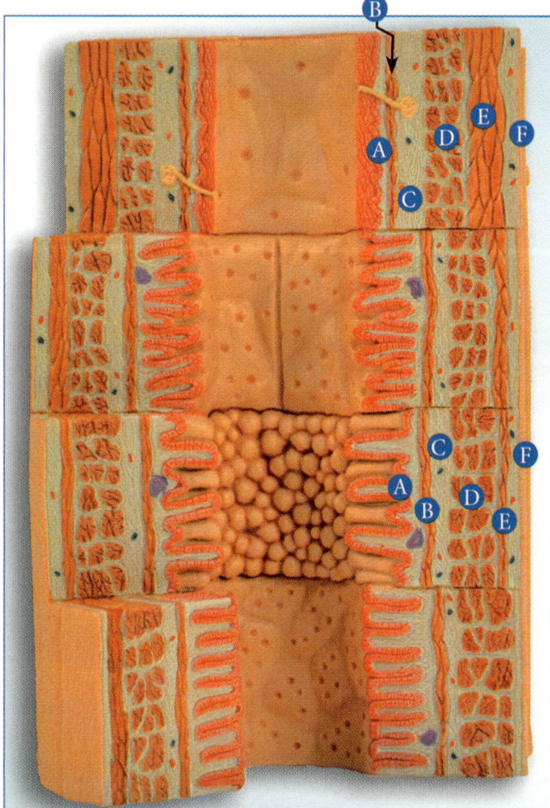

The digestive tract layers

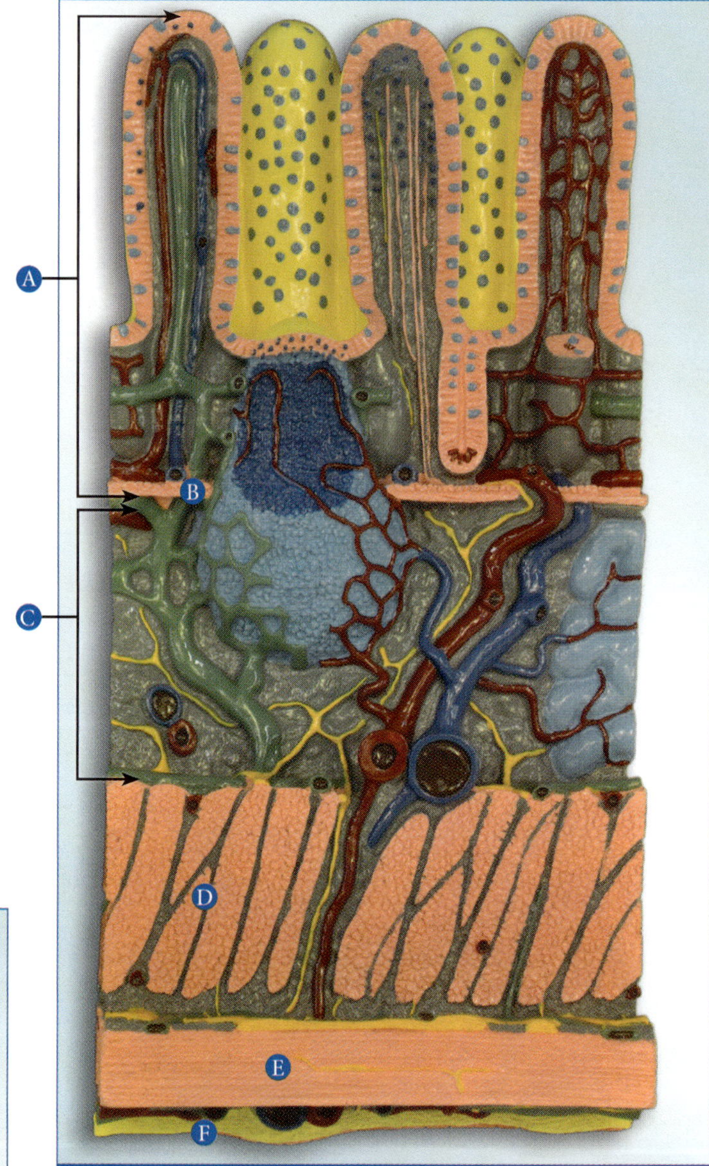

Model showing layers of the GI tract

The digestive tract, alimentary canal, or GI (gastrointestinal) tract, consists of a muscular tube that travels from the mouth to the anus. The entire length is made of four layers: the mucosa, submucosa, muscularis externa, and serosa (outermost thin layer of the last portion of esophagus to the sigmoid colon), or adventitia (outermost layer of the pharynx and most of esophagus and rectum). Both the serosa and adventitia serve to anchor the tube to surrounding tissue. Regional specializations occur in each major division of the GI tract. Modifications of the mucosa and submucosa layers dictate the type of secretion introduced into the lumen of the tube. The secretions from the entrire GI tract averages 7–8 liters/day. This large amount of body water must be reabsorbed, mainly by the intestines or dehydration can occur.

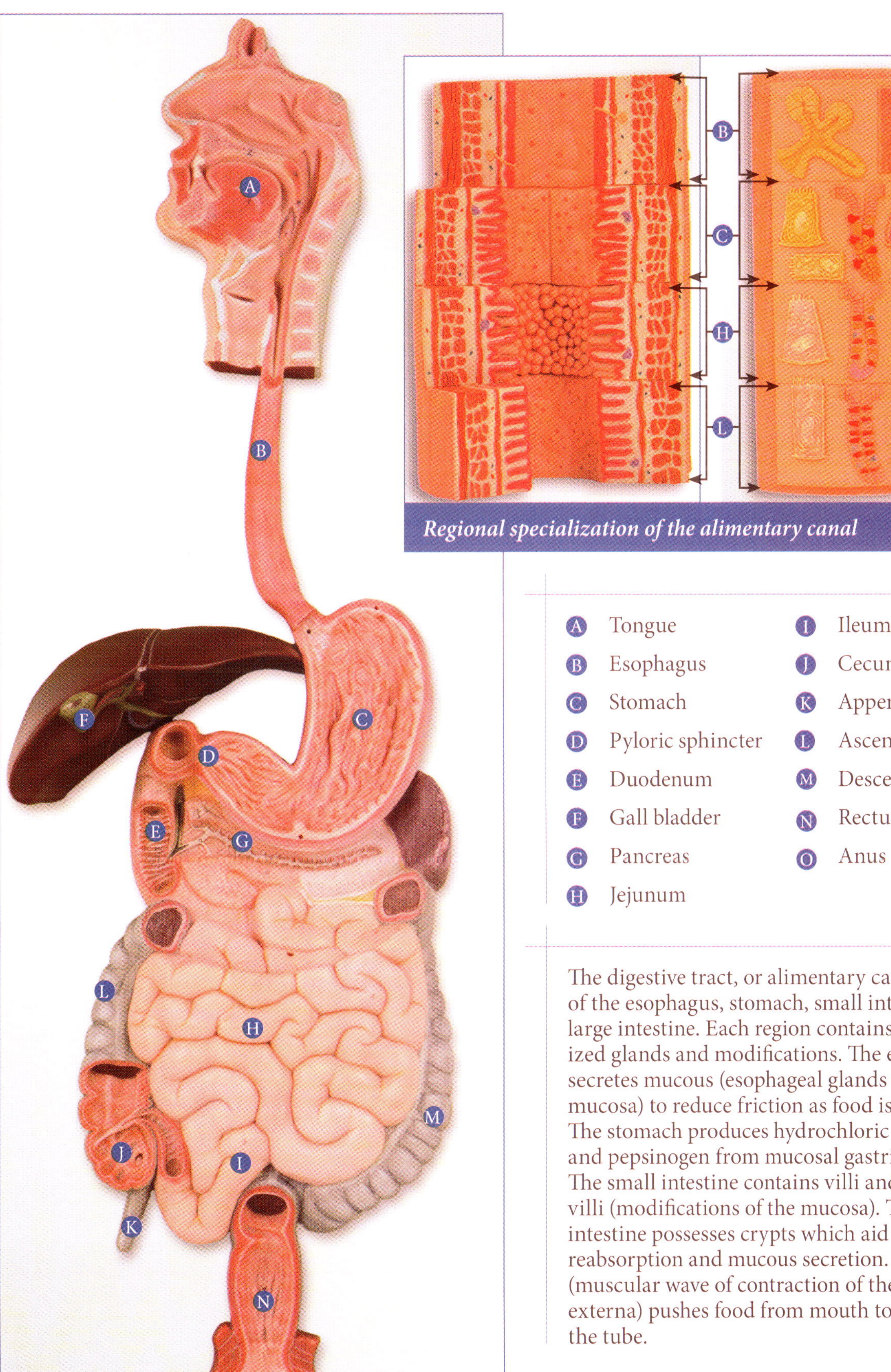

Regional specialization of the alimentary canal

The digestive system

Ⓐ	Tongue	Ⓘ	Ileum
Ⓑ	Esophagus	Ⓙ	Cecum
Ⓒ	Stomach	Ⓚ	Appendix
Ⓓ	Pyloric sphincter	Ⓛ	Ascending colon
Ⓔ	Duodenum	Ⓜ	Descending colon
Ⓕ	Gall bladder	Ⓝ	Rectum
Ⓖ	Pancreas	Ⓞ	Anus
Ⓗ	Jejunum		

The digestive tract, or alimentary canal, consists of the esophagus, stomach, small intestines, and large intestine. Each region contains specialized glands and modifications. The esophagus secretes mucous (esophageal glands in the submucosa) to reduce friction as food is swallowed. The stomach produces hydrochloric acid (HCl) and pepsinogen from mucosal gastric glands. The small intestine contains villi and microvilli (modifications of the mucosa). The large intestine possesses crypts which aid in water reabsorption and mucous secretion. Peristalsis (muscular wave of contraction of the muscularis externa) pushes food from mouth to anus along the tube.

Ⓐ Cardiac region

Ⓑ Fundic region

Ⓒ Circular muscle layer

Ⓓ Oblique muscle layer

Ⓔ Antrum

Ⓕ Pylorus

Ⓖ Pyloric sphincter

Ⓗ Gastric rugae

Ⓘ Gastric pit

As food fills the stomach (now called chyme), it stretches. The gastric rugae (wrinkles) allow this to occur. The muscular wall contractions mix the ingested contents with secretions from mucous cells (mucous), parietal cells (HCl), chief cells (pepsinogen, an inactive form of the protein digesting enzyme, pepsin), and enteroendocrine hormones that regulate digestion. HCl (hydrochloric acid) is a strong acid that helps kill ingested pathogens and activates pepsinogen into pepsin.

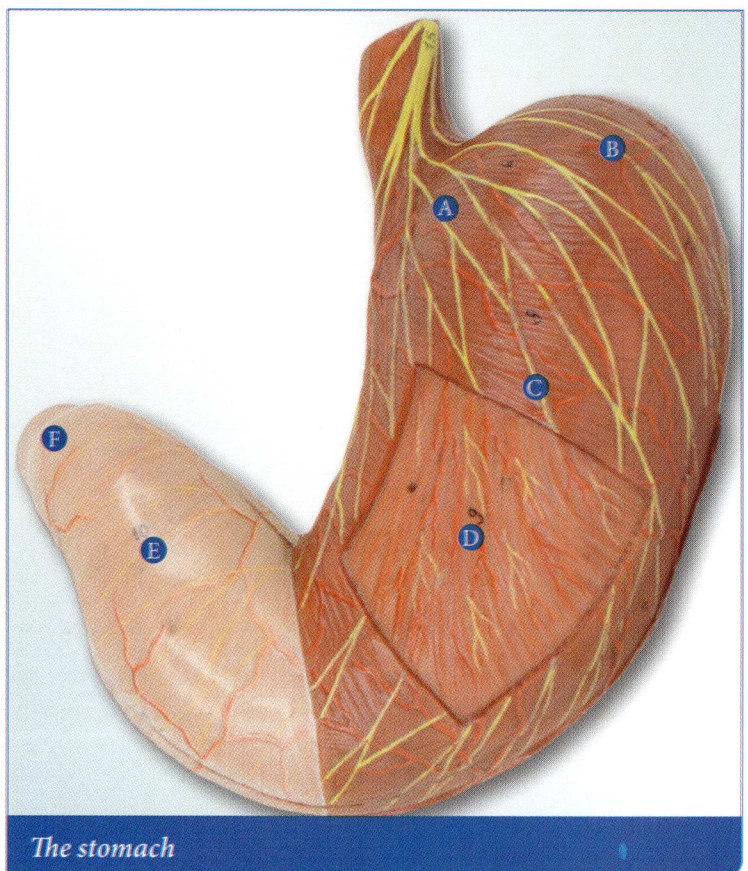

The stomach

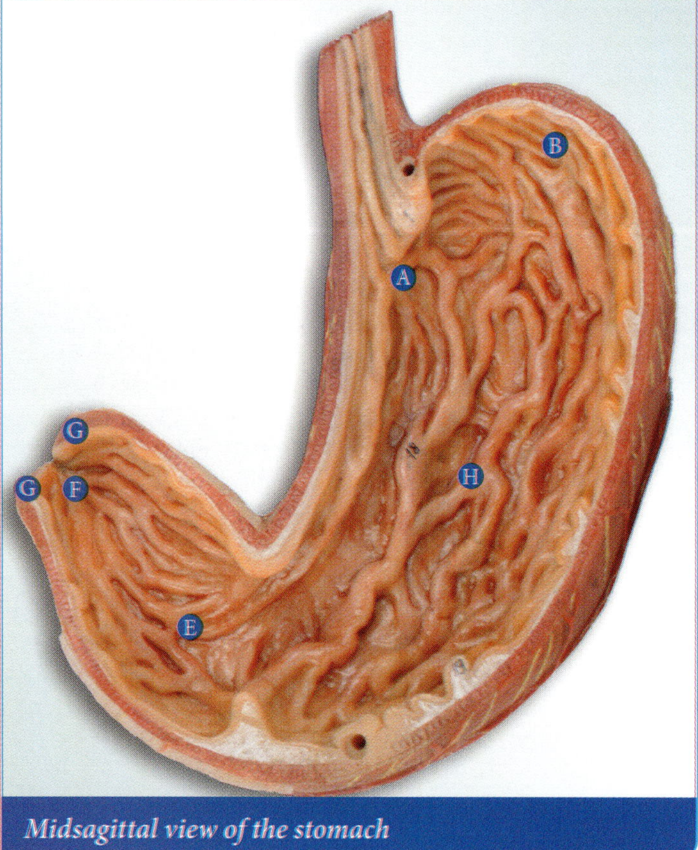

Midsagittal view of the stomach

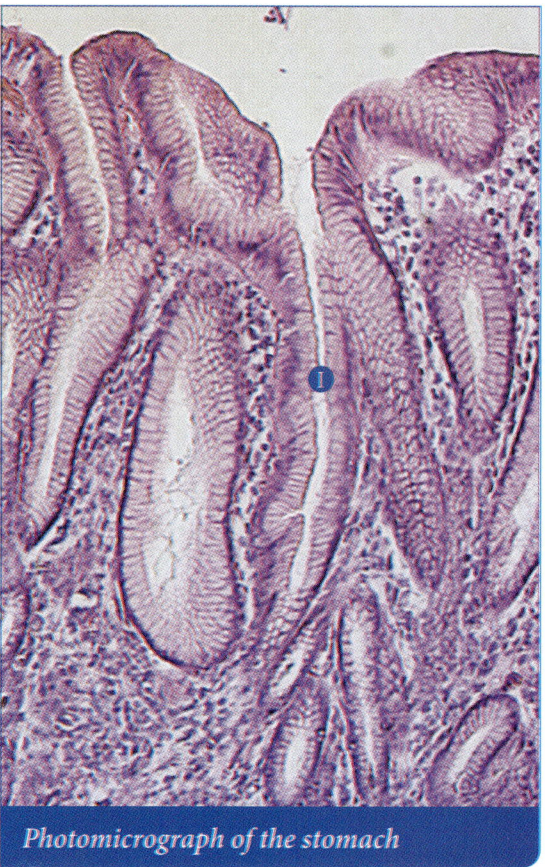

Photomicrograph of the stomach

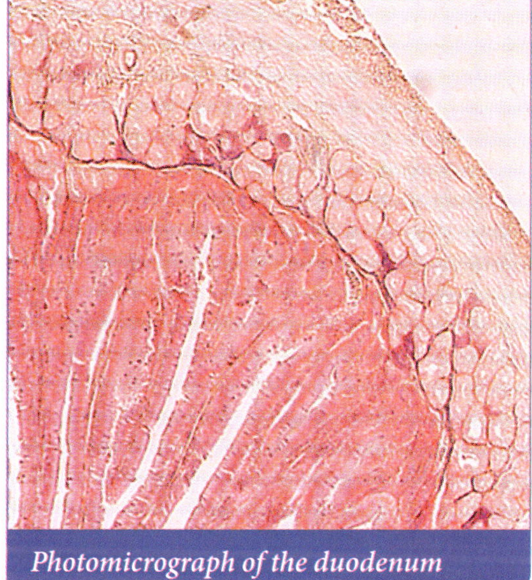

Photomicrograph of the duodenum

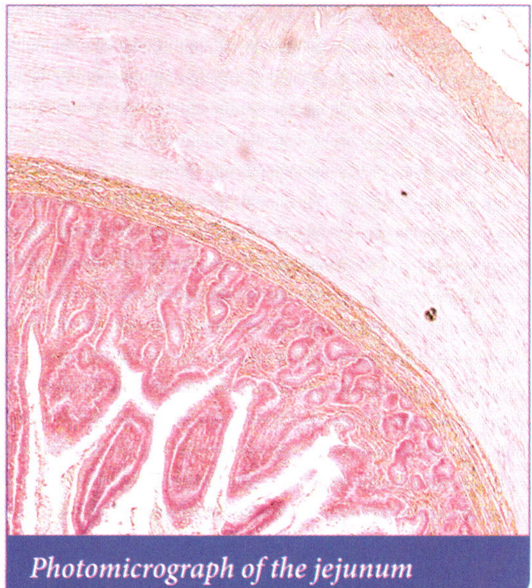

Photomicrograph of the jejunum

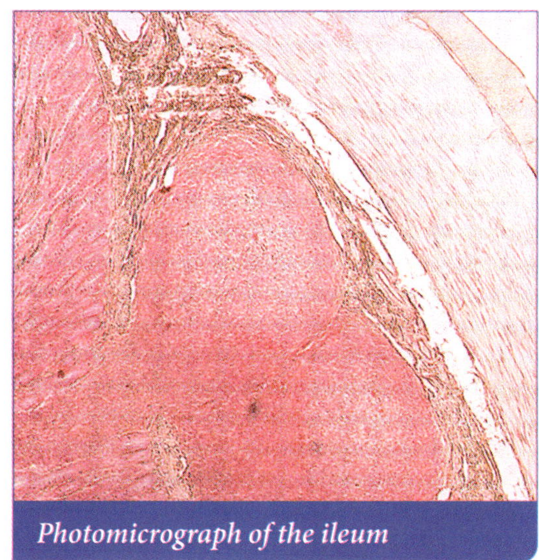

Photomicrograph of the ileum

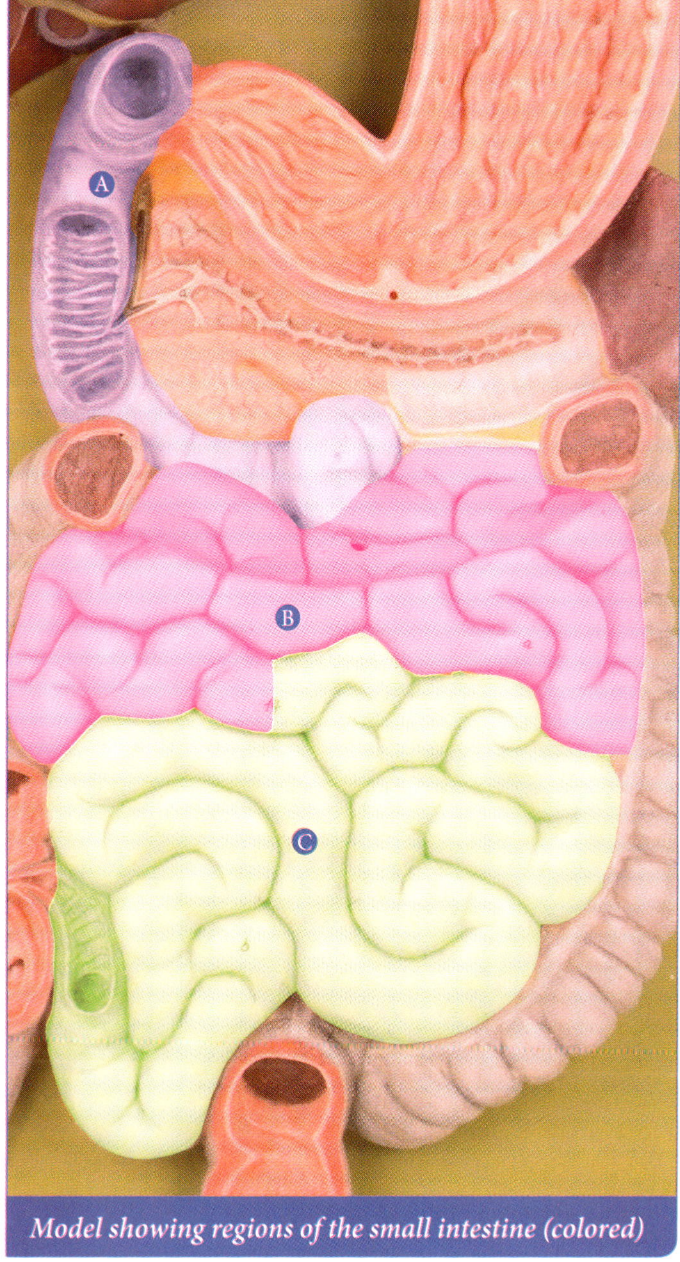

Model showing regions of the small intestine (colored)

The small intestine is divided into three main regions:

A Duodenum

B Jejunum

C Ileum

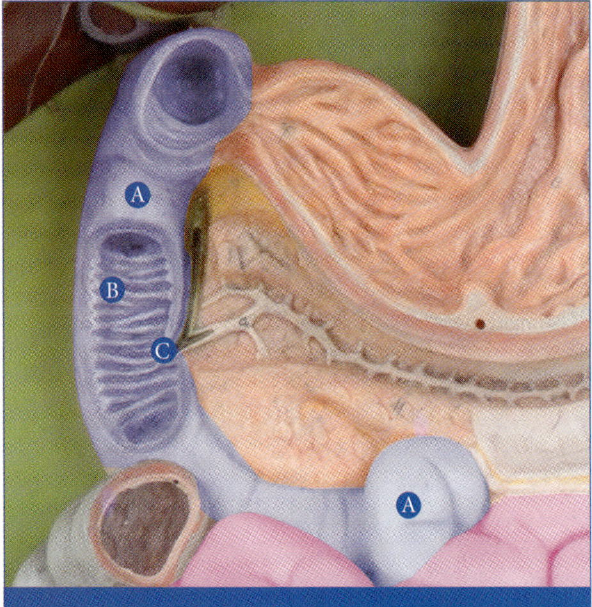

The duodenum (purple)

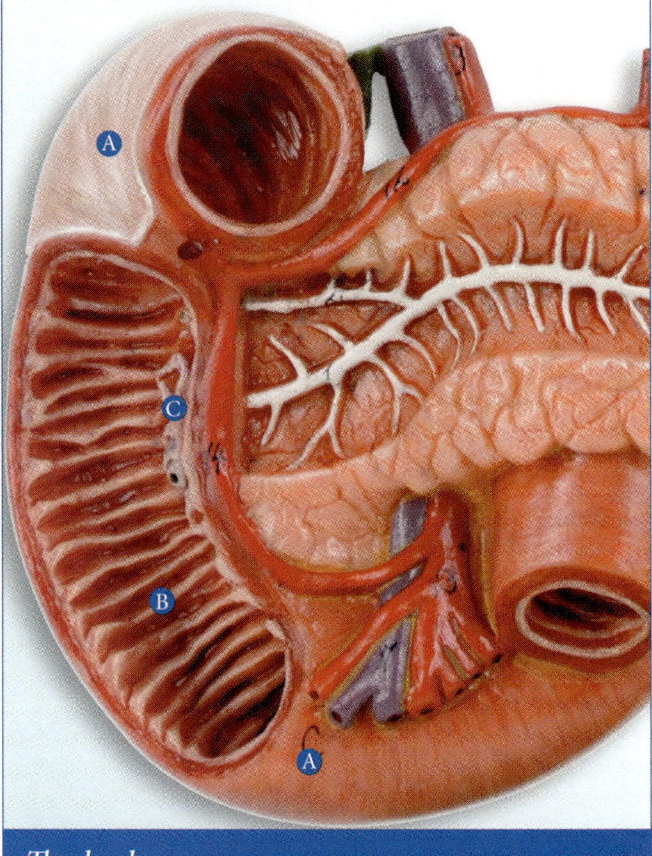

The duodenum

- (A) Duodenum
- (B) Plicae circulares (folds)
- (C) Major duodenal papillae
- (D) Duodenal glands in the submucosa

The duodenum is the first segment of the small intestines and is about 25 centimeters or 10 inches in length. As chyme exits the pylorus of the stomach, in small amounts controlled by the pyloric sphincter into the duodenum, the acidic contents and stretching of the duodenal wall trigger the release of enzymes, sodium bicarbonate, and zymogens from the pancreas. Bile salts from the gallbladder help emulsify (break large lipids into tiny, suspended droplets) fats. Alkaline mucus secretions of duodenal glands and brush border enzyme activity of the duodenal epithelium help neutralize stomach acid and activate pancreatic enzymes, respectively. The circular folds (plicae circulares) help mix the chyme with enzymes, mucus, and bile salts.

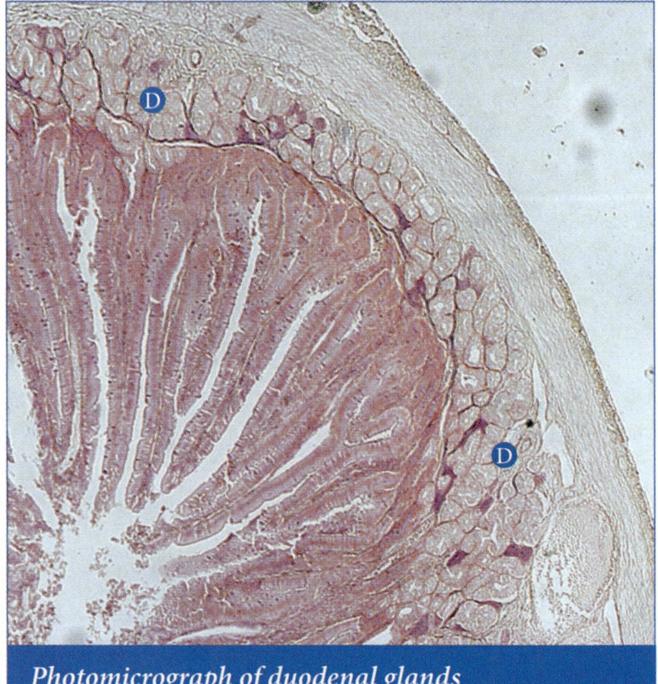

Photomicrograph of duodenal glands

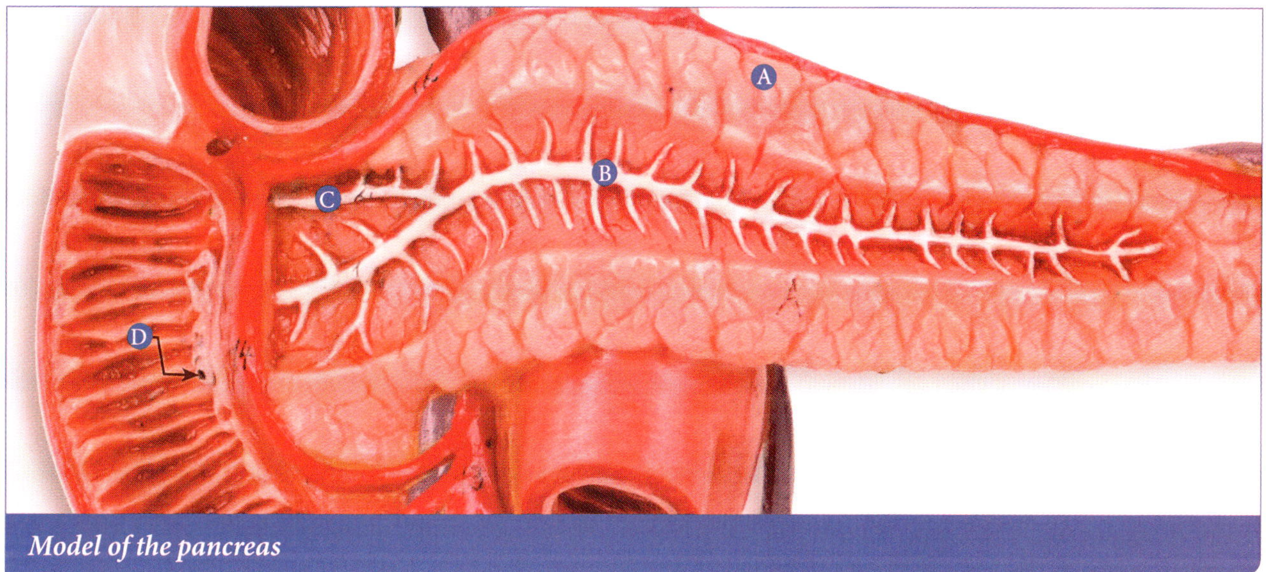

Model of the pancreas

- Ⓐ Pancreas
- Ⓑ Pancreatic duct
- Ⓒ Accessory pancreatic duct
- Ⓓ Major duodenal papillae
- Ⓔ Gallbladder

All enzymes of digestion are produced by the pancreas. Starch, fat, protein, DNA, and RNA are all broken down to the major macromolecules glucose, fatty acids, amino acids, and nucleotides, respectively. The pancreas produces and secretes enzymes, zymogens (inactive form of protein digesting enzymes), and sodium bicarbonate (buffers the acidic chyme from the stomach) into the duodenum.

The liver produces thousands of unique proteins for various functions. Bile formation has the only role in digestion. Bile is composed of bile salts, minerals, and pigments. Pigments are produced by the breakdown of the heme portion of hemoglobin into bilirubin. Only bile salts contribute to digestion by emulsifying and preparing fats for absorption. Bile salts are derived by the addition of glycine or taurine to cholesterol. This creates ampipathic molecules that separate fat globules into lipid droplets (emulsify). Bile salts then combine with fatty acids and monoglycerides and form tiny micelles, about 4–7 nanometers in diameter that can be absorbed into columnar epithelial cells of the small intestines.

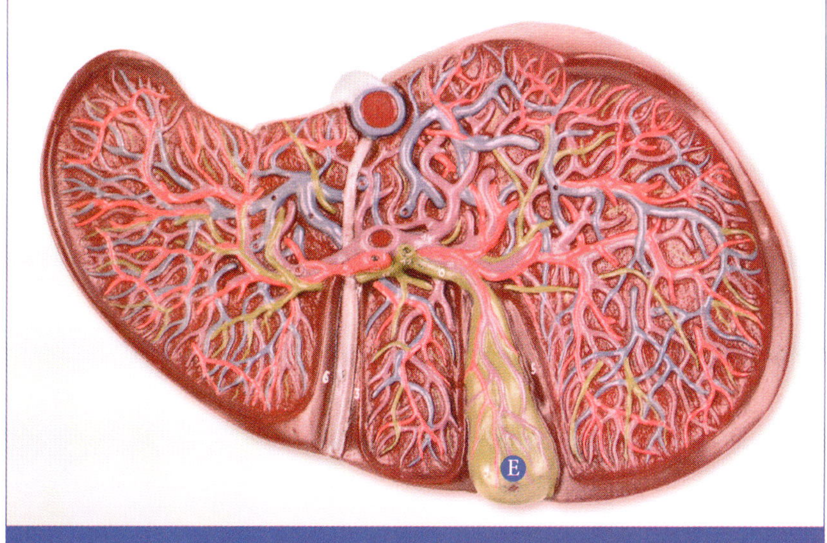

Model of the liver showing the gallbladder

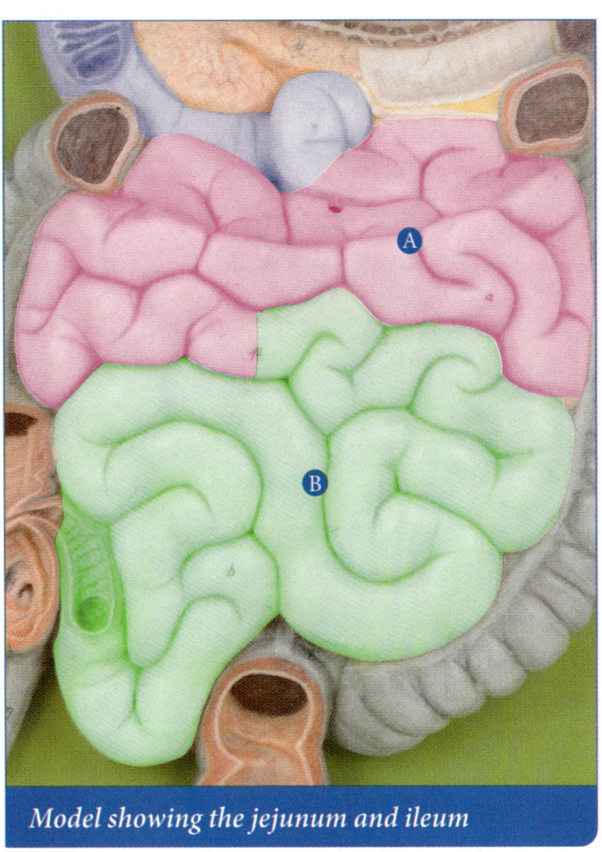

Model showing the jejunum and ileum

(A) Jejunum

(B) Ileum

(C) Peyer's patches

The jejunum is where most chemical digestion and absorption of nutrients takes place. It's about a meter and a half long and is highly vascular. Pancreatic enzymes continue to actively digest introduced chyme while villi and microvilli greatly increase the surface area of the jejunum's epithelium, where membrane transport pumps move nutrients into the circulatory system or lymphatic (fats) system. About the last half of the small intestines is the ileum. More digestion and absorption of nutrients takes place here. The ileum transforms in structure as it approaches the cecum of the large intestine. Clusters of lymphatic tissue (Peyer's patches) become more prevalent, housing numerous lymphocytes to resist infection.

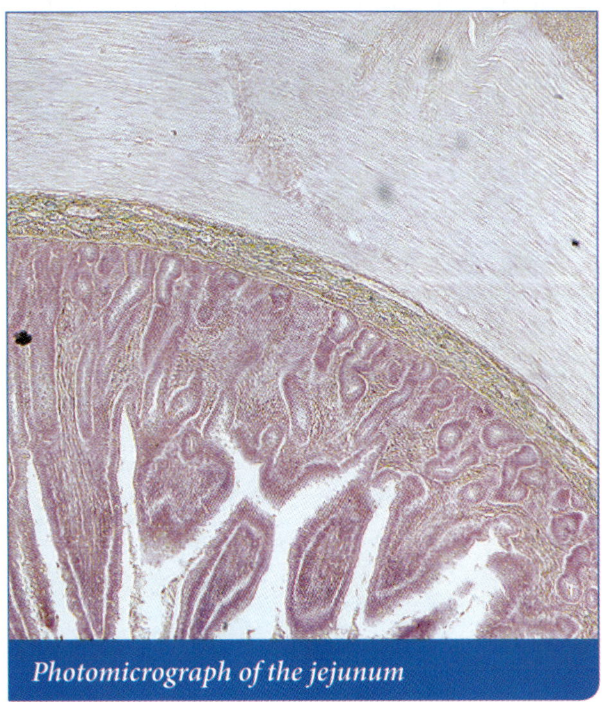

Photomicrograph of the jejunum

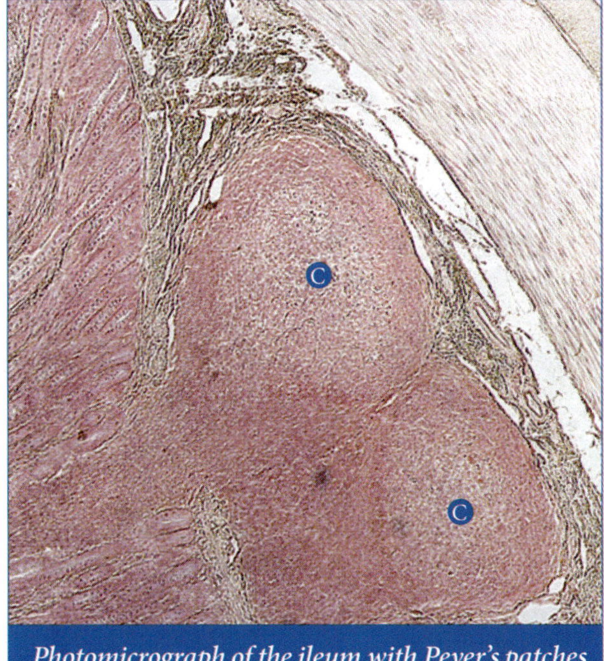

Photomicrograph of the ileum with Peyer's patches

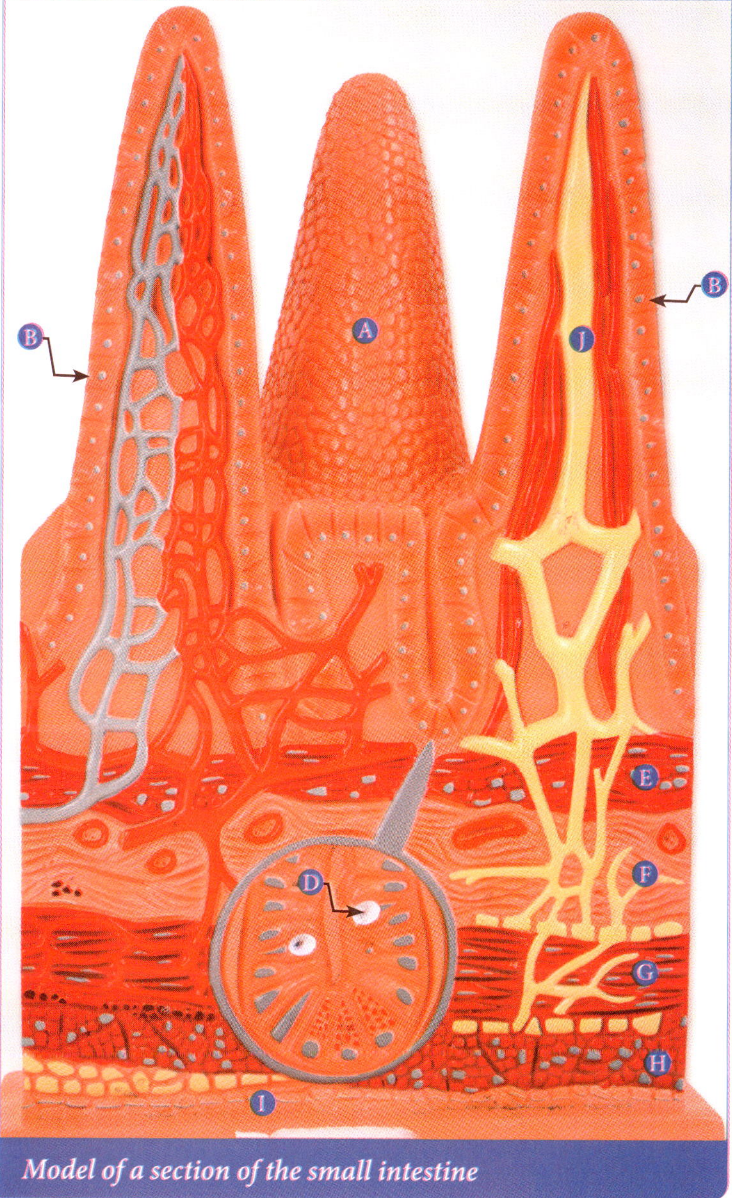

Model of a section of the small intestine

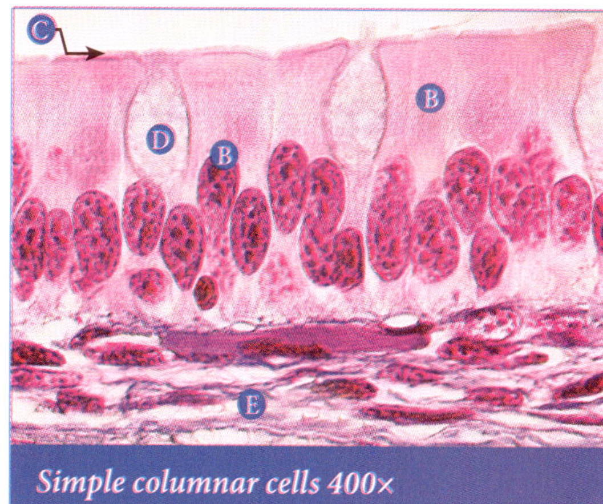

Photomicrograph showing villi 40×

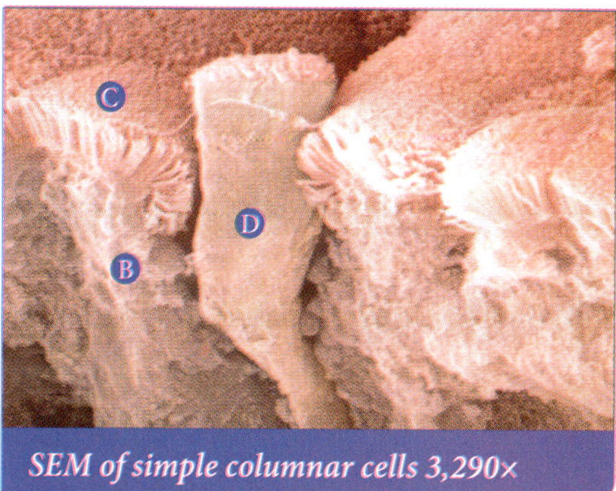

Simple columnar cells 400×

SEM of simple columnar cells 3,290×

Ⓐ Villus	Ⓖ Circular muscularis externa
Ⓑ Simple columnar epithelial cells	Ⓗ Longitudinal muscularis externa
Ⓒ Microvilli	Ⓘ Serosa
Ⓓ Goblet cell	Ⓙ Central lacteal
Ⓔ Muscularis mucosae	
Ⓕ Submucosa	

Villi and microvilli greatly increase the surface area of the small intestines (about 200 times). Without these structures, our intestines would need to be about 1,200 feet long to provide the same surface area. All nutrient absorption occurs through the simple columnar epithelial cells of the mucosa.

Microvilli model 80,000× actual size

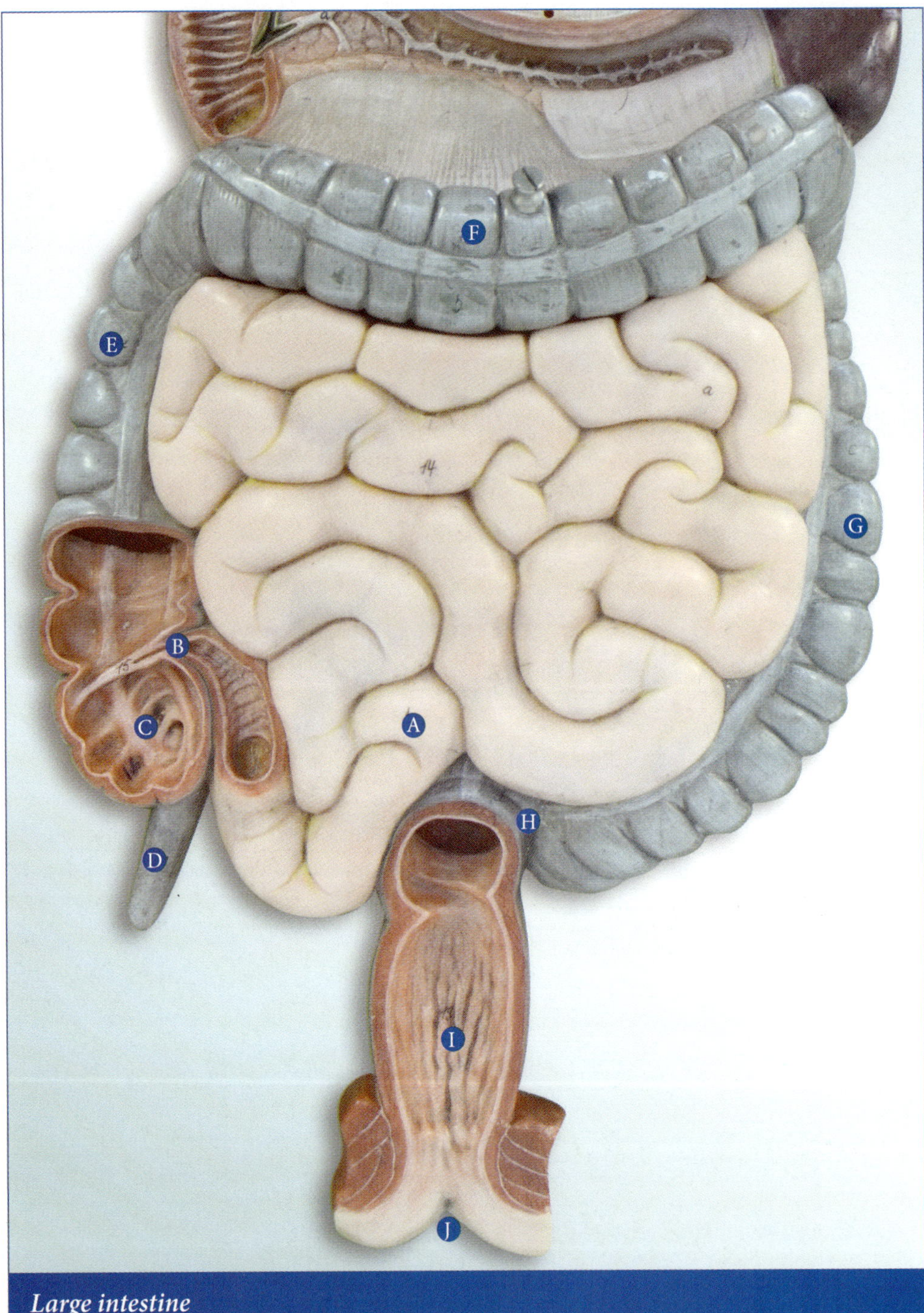

Large intestine

The large intestine is about 1.5 meters long and extends from the cecum to the rectum. The mucosa of the large intestine is simple columnar epithelium that's specialized for absorption. Motility is slower in the large intestine allowing most of the water of chyme that enters to be reabsorbed. This is important in preventing dehydration, which is a serious concern during extended bouts of diarrhea. The colon is home to about 800 species of bacteria that number in the billions. They further digest cellulose (we do not make enzymes that can break it down) and release glucose. Some bacteria produce sulphur containing gas that gives rise to the foul odor of flatus. About 60% of the dry mass of feces is bacteria. The epithelium makes a transition from simple columnar in the rectum to stratified squamous in the anal canal.

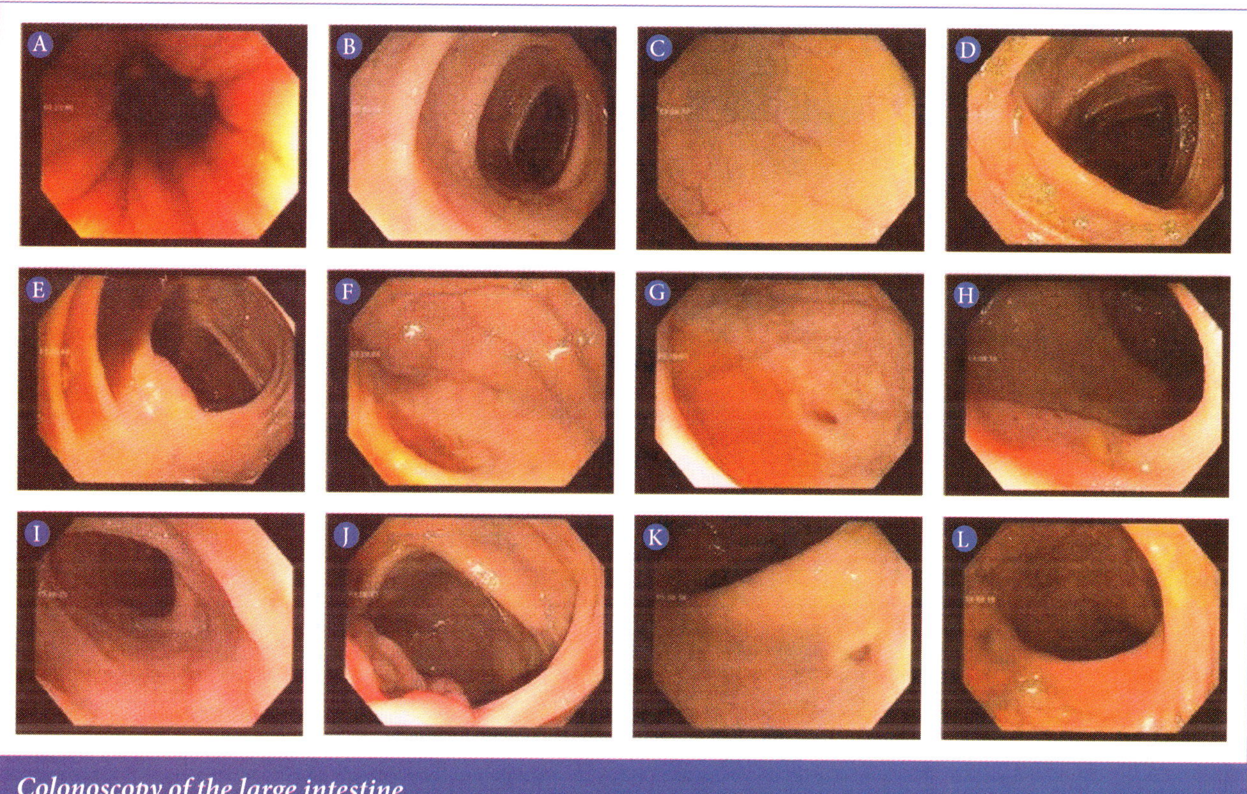

Colonoscopy of the large intestine

A   Anal canal – lined with stratified squamous epithelium. All other structures listed here contain simple columnar epitheium.

B   Mid-descending colon

C   Splenic flexure – the spleen can be seen through the colon wall at the sharp bend between the transverse and the descending colon

D   Mid-transverse colon

E   Cecum – beginning of the large intestine

F   Appendiceal orifice – opening into the appendix

G   Appendiceal orifice – opening into the appendix

H   Terminal ileum – last region of the small intestines

I   Terminal ileum – last region of the small intestines

J   Ileo-cecal valve – prevents material in the colon from traveling back into the small intestine

K   Diverticulum in the ascending colon – diverticula are pouches that form in the mucosa and submucosa causing a bulge

L   Rectum – about 12 cm long. As material fills, the rectal walls expand and stimulate stretch receptors involved in defecation.

11 CHAPTER *Reproduction*

Female and Male Hormones and Reproductive Systems

Without reproduction, a species is doomed to extinction. Mammals (humans being one of thousands) perform internal fertilization and development. In order to form offspring with 23 pairs of chromosomes, meiotic division in the gonads (ovaries and testes) produce haploid gametes, each containing 23 chromosomes. Females ovulate an egg about every month, and men produce sperm by the hundreds of millions per day. Sperm are transferred to the vagina by the penis and swim through the cervical opening to the uterus and finally to the uterine tube containing an egg, where fertilization takes place.

Female reproductive structures

- Labium majus
- Labium minus
- Clitoris
- Vagina
- Cervix
- Uterus
- Uterine (Fallopian) tube
- Fimbrae of the infundibulum
- Ovary
- Round ligament
- Ovarian ligament
- Urinary bladder
- Urethra

Ovaries, egg development & ovulation

- Primordial follicles
- Primary follicle
- Secondary follicle
- Rupturing follicle releasing its oocyte (egg)
- Oocytes
- Post ovulatory follicle
- Corpus luteum
- Corpus albicans

Pituitary & ovarian hormones

- Follicle stimulating hormone (FSH)
- Luteinizing hormone (LH)
- Estradiol
- Progesterone

Male reproductive structures

- Testis
- Epididymis
- Pampiniform plexus
- Ductus deferens
- Ampulla of ductus deferens
- Seminal vesicle
- Prostate gland
- Urethra
- Corpus cavernosum
- Corpus spongiosum
- Glans of penis
- Urinary bladder
- Ureter
- Urethra

Testes, sperm production

- Seminiferous tubule
- Spermatogonium
- Spermatocyte
- Spermatid
- Spermatozoa

Pituitary & testicular hormones

- Testosterone
- Follicle stimulating hormone (FSH)
- Luteinizing hormone (LH)

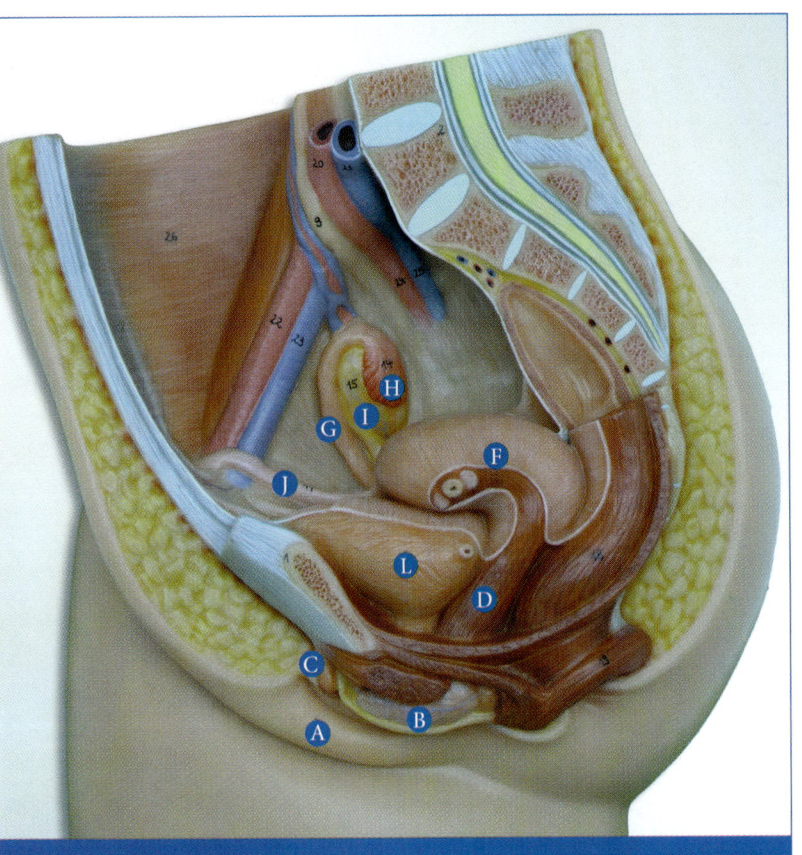

A — Labium majus
B — Labium minus
C — Clitoris
D — Vagina
E — Cervix
F — Uterus
G — Uterine (Fallopian) tube
H — Fimbrae of the infundibulum
I — Ovary
J — Round ligament
K — Ovarian ligament
L — Urinary bladder
M — Urethra

Model of the female reproductive system

The female reproductive cycle involves pituitary hormones (LH and FSH) that promote the development of gametes and ovarian hormones (estrogens and estradiol being the most abundant, as well as progesterone). These hormones prepare the uterine lining for implantation and pregnancy. Just prior to the menstrual flow, the pituitary gland increases secretion of follicle stimulating hormone (FSH). This stimulates about two dozen recruited follicles to grow and produce estradiol. As these select secondary follicles mature (each producing estradiol), one follicle, the dominant follicle, grows faster and larger. Approximately two weeks after menstruation, FSH levels decline and a sharp increase in luteinizing hormone (LH) triggers ovulation (release of oocyte, or egg). The ovulated follicle continues to be stimulated by LH and transforms into the corpus luteum, producing progesterone and estrogens. Both progesterone and estradiol stimulate the uterine lining (endometrium) to increase about ten times the thickness as the post menstrual endometrium. If pregnancy does not occur, the functionalis layer of the endometrium sloughs off and becomes part of the menstrual flow.

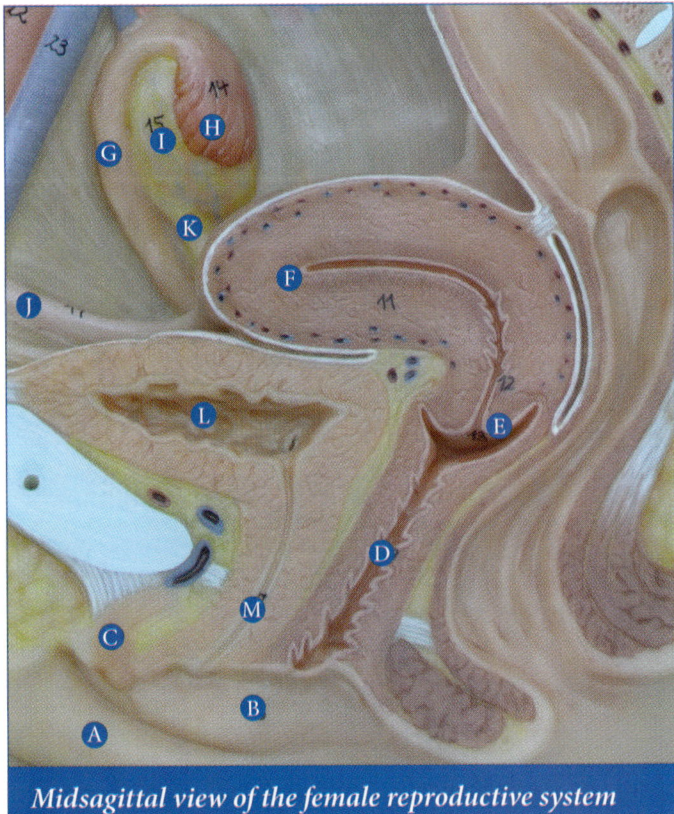

Midsagittal view of the female reproductive system

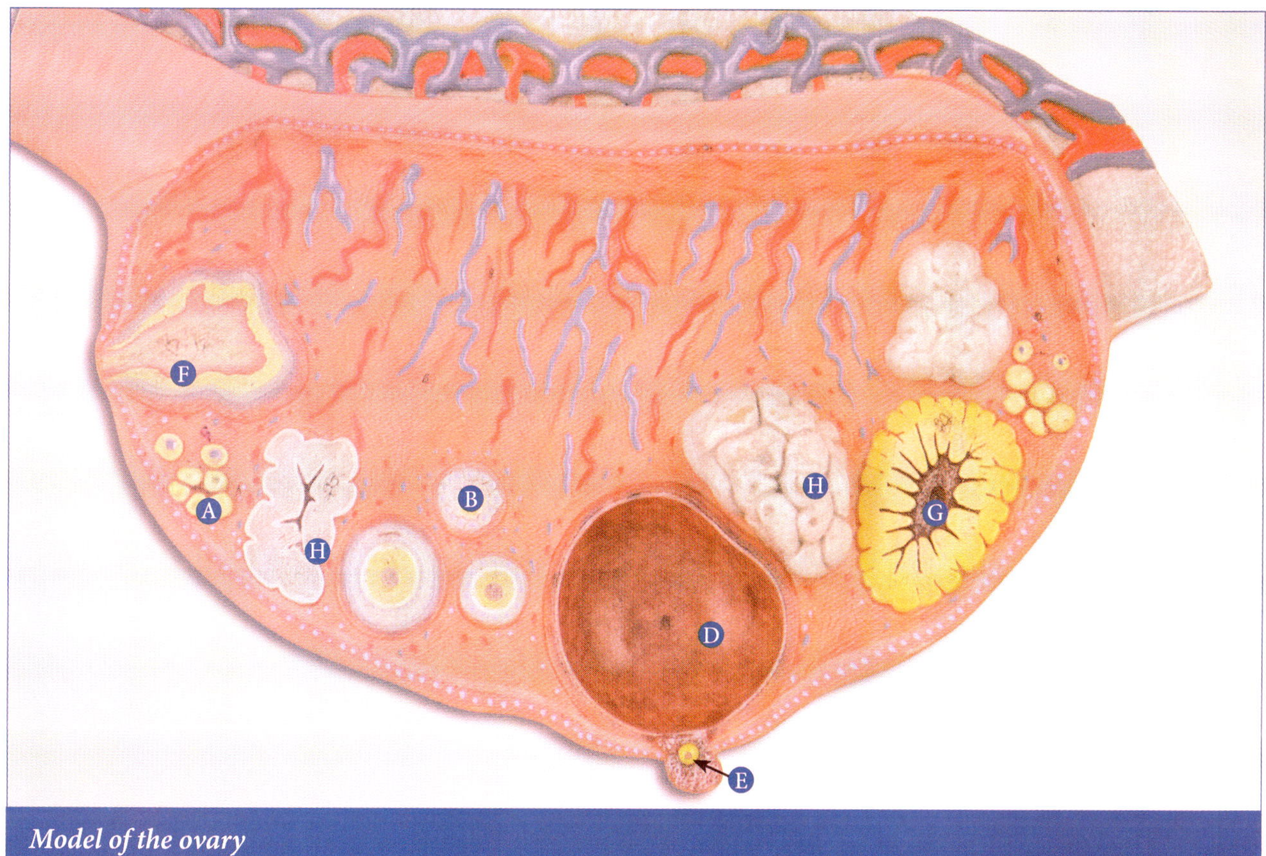

Model of the ovary

Ⓐ Primordial follicles

Ⓑ Primary follicle

Ⓒ Secondary follicles

Ⓓ Rupturing follicle releasing its oocyte (egg)

Ⓔ Oocyte

Ⓕ Post ovulatory follicle

Ⓖ Corpus luteum

Ⓗ Corpus albicans

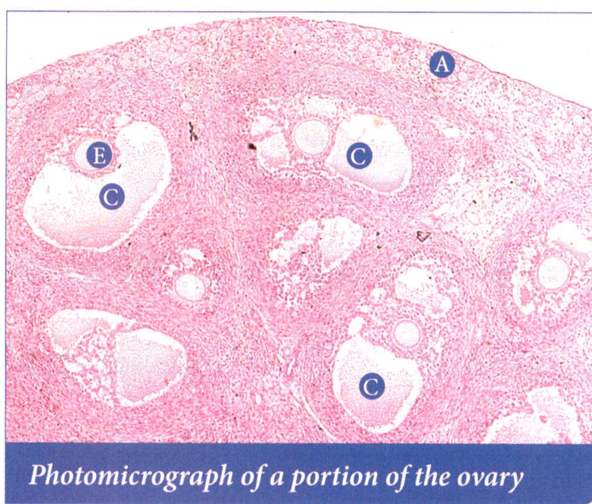

Photomicrograph of a portion of the ovary

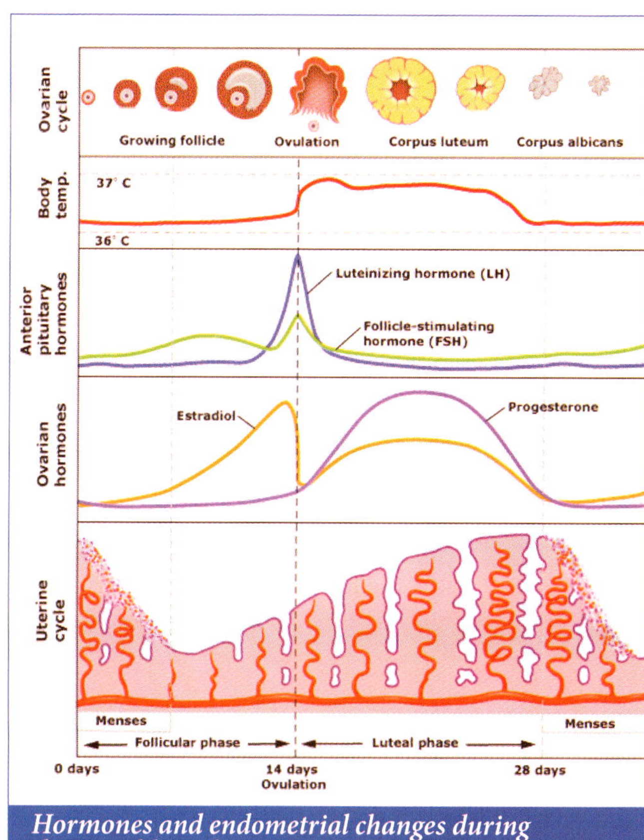

Hormones and endometrial changes during the monthly cycle

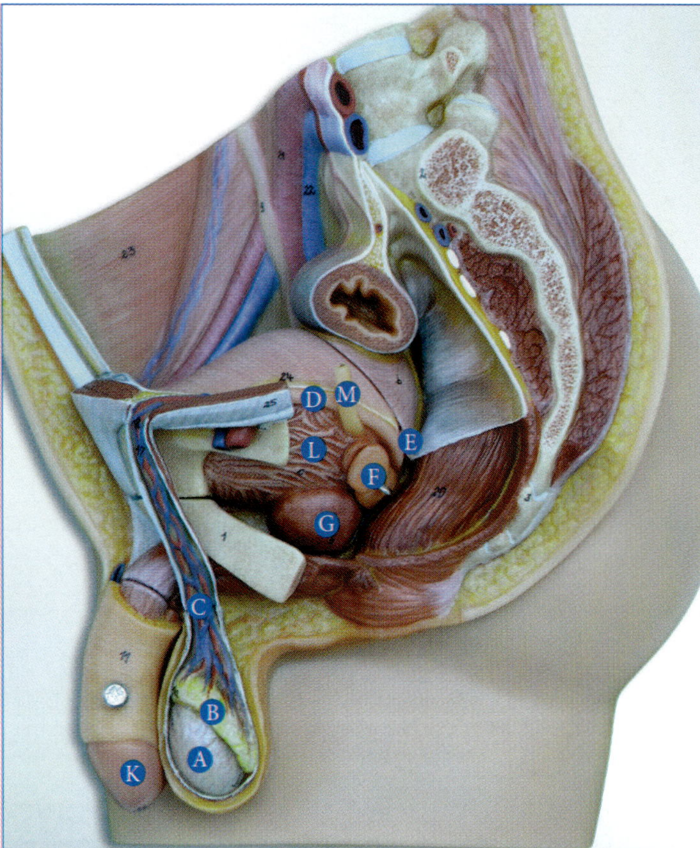

Model of the male reproductive system

- **A** Testis
- **B** Epididymis
- **C** Pampiniform plexus
- **D** Ductus deferens
- **E** Ampulla of ductus deferens
- **F** Seminal vesicle
- **G** Prostate gland
- **H** Urethra
- **I** Corpus cavernosum
- **J** Corpus spongiosum
- **K** Glans of penis
- **L** Urinary bladder
- **M** Ureter

The testis produces spermatozoa and androgens. The testis begins making sperm cells (spermatogenesis) and testosterone at the onset of puberty (pituitary release of gonadotropins, FSH and LH) and continues throughout a male's life. The majority of the testis consists of seminiferous tubules where spermatozoa are produced, stimulated by FSH. Spermatogenesis begins as spermatogonia cells in the periphery of the seminiferous tubules give rise to spermatocytes. Spermatocytes are the only cell type in the male body to undergo meiosis, producing four unique haploid cells (spermatids) from one diploid cell, and must be protected from the immune system by the blood-testis barrier (BTB). Spermatids continue the migration towards the lumen of the tubule and undergo a transformation (spermiogenesis) into an immature spermatozoon. Spermatozoa enter the lumen and flow to the epididymis where they mature into fully functional cells. This entire process takes about three days. A man produces about 300,000 spermatozoa per minute or 400,000,000 a day.

Testosterone is produced by interstitial, or Leydig cells. LH is the hormone that stimulates these cells to produce androgens, mainly testosterone.

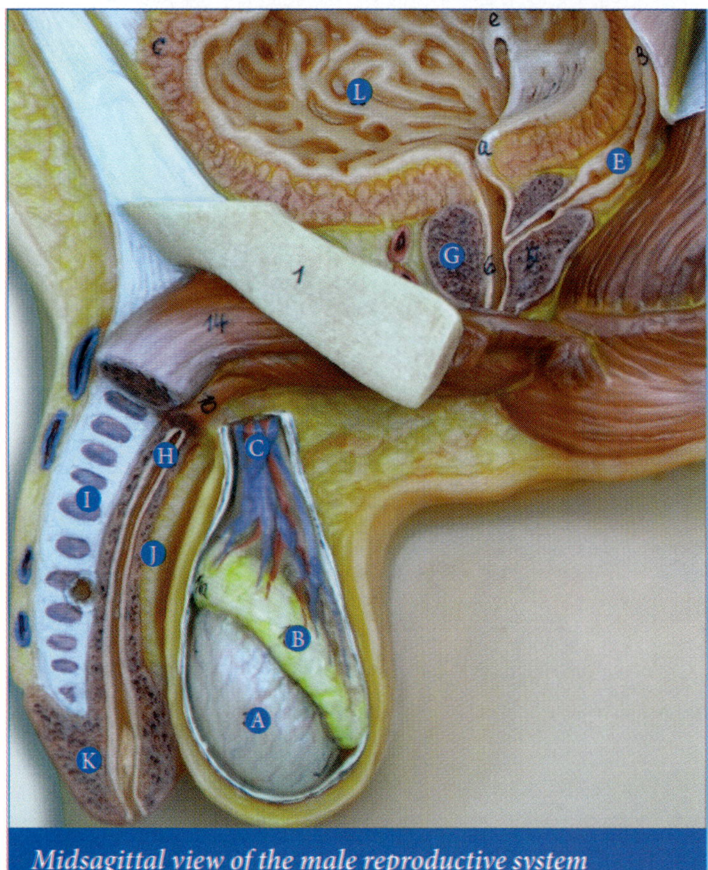

Midsagittal view of the male reproductive system

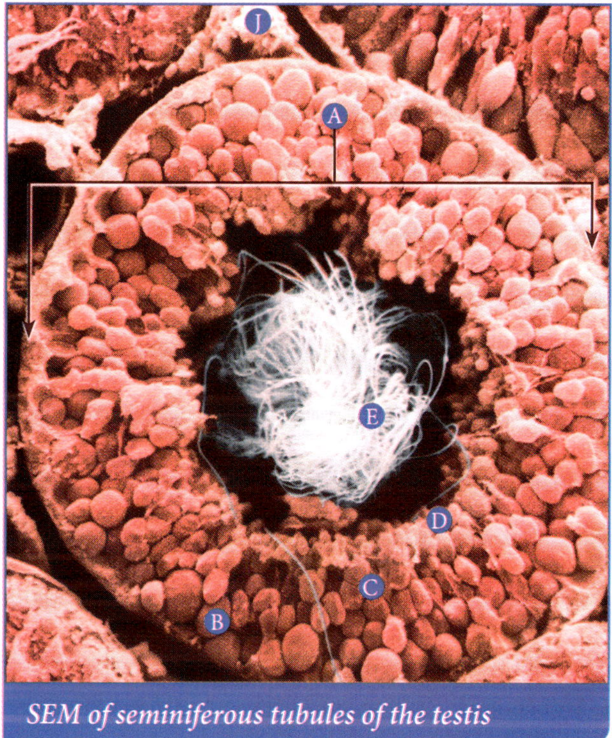

SEM of seminiferous tubules of the testis

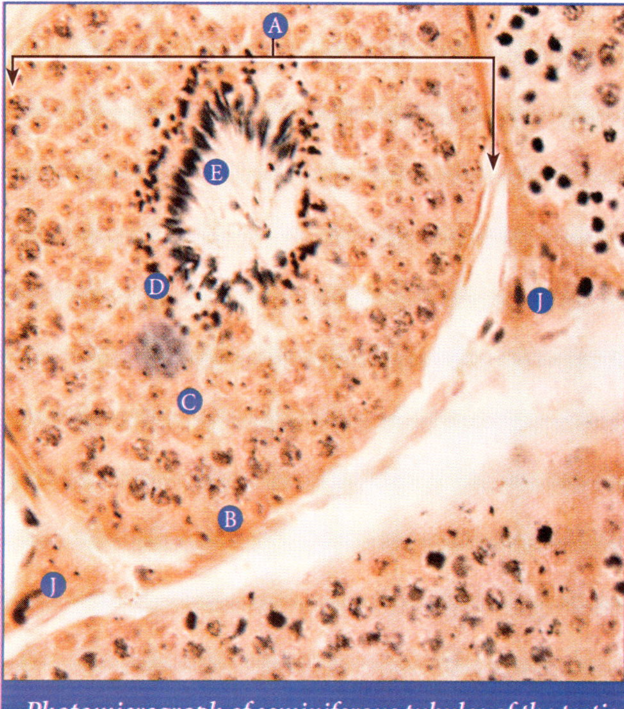

Photomicrograph of seminiferous tubules of the testis

A	Seminiferous tubule	F	Spermatozoon
B	Spermatogonia	G	Head
C	Spermatocytes	H	Midpiece
D	Spermatids	I	Tail
E	Immature spermatozoa	J	Interstitial (Leydig) cells

Meiotic divisions in the seminiferous tubules produce haploid spermatids which undergo spermiogenesis, changing into spermatozoa and enter the lumen.

Interspersed between tubules are clusters of Leydig cells (triangular shaped clusters above). The production of testosterone by interstitial, or Leydig cells, is stimulated by LH from the pituitary gland.

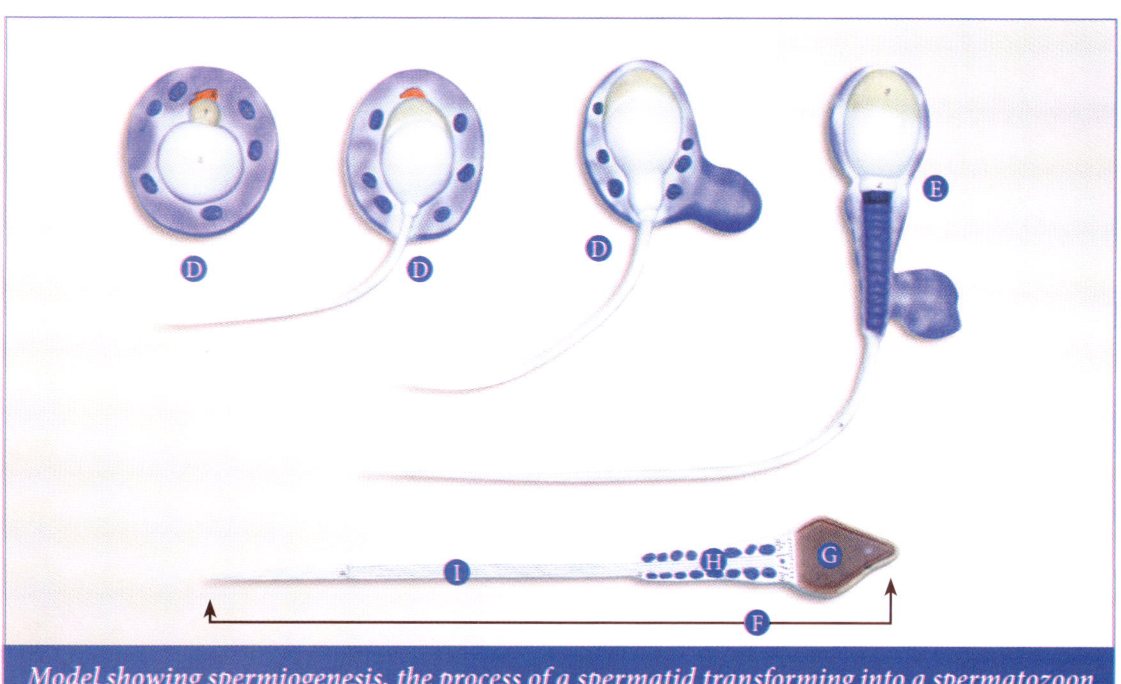

Model showing spermiogenesis, the process of a spermatid transforming into a spermatozoon

12 Chapter *Development*

Fertilization, Embryonic, and Fetal Development

Approximately every 28 days, give or take a few, one of the ovaries (they usually alternate each month) contain a mature follicle and releases an oocyte (ovulation) into the swollen uterine tube. It is here that fertilization must occur. The journey from ovulation to the uterus takes about three days. Sperm cells that enter the uterine tube anytime during the three days before ovulation to ten hours after ovulation have a good chance of fertilizing the egg. Sperm require capacitation before being able to fertilize an egg and that takes about ten hours to complete. The egg is only viable during the first 20 hours after ovulation, so the window for pregnancy is no more than five days. As the sperm and egg chromosome pair up (homologous pairs) a diploid zygote is formed. This cell will divide, giving rise to millions of trillions of cells in one's lifetime. However, it produces 100 cells and is called a blastocyst upon implantation in the uterine lining. Embryonic development gives rise to three germ layers and organ systems. The fetus grows from week eight until parturition (childbirth), around week 35.

From ovary to embryo

- Primordial follicles
- Primary follicle
- Secondary follicles
- Rupturing follicle
- Corpus luteum
- Oocyte
- Fertilization (union of sperm and egg)
- Pronuclei
- Zygote
- 2-cell stage
- 4-cell stage
- Morula
- Blastocyst
- Implanted embryo (embryoblast)
- Ectoderm
- Mesoderm
- Endoderm

Structures of one month embryo

- Head
- Primitive lens
- Pharyngeal arches
- Heart bulge
- Arm bud
- Leg bud
- Tail

Uterine development

- Embryo
- Fetus
- Umbilical cord
- Child's portion of the placenta (chorion)
- Mother's portion of the placenta (endometrium)
- Amnion
- Chorion
- Decidua parietalis of endometrium

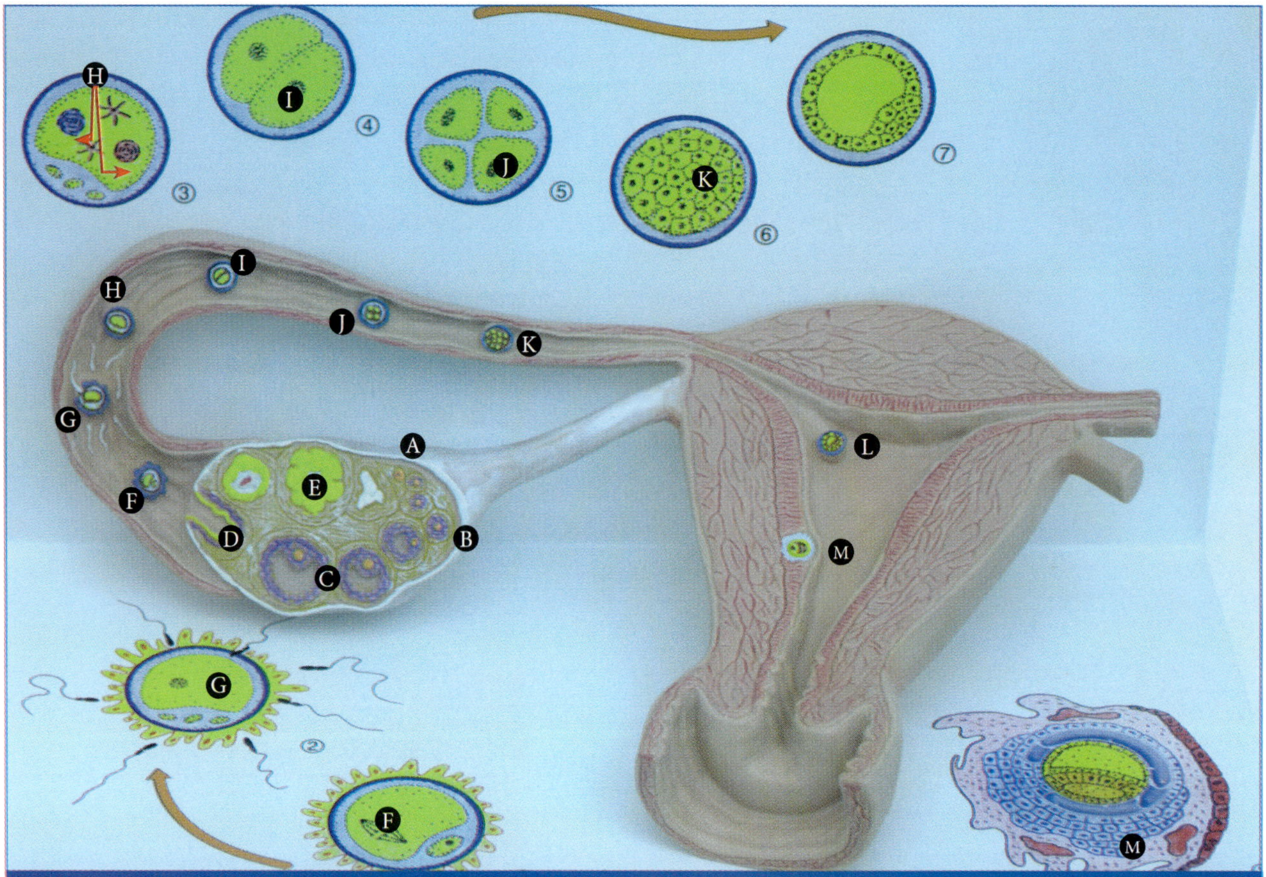

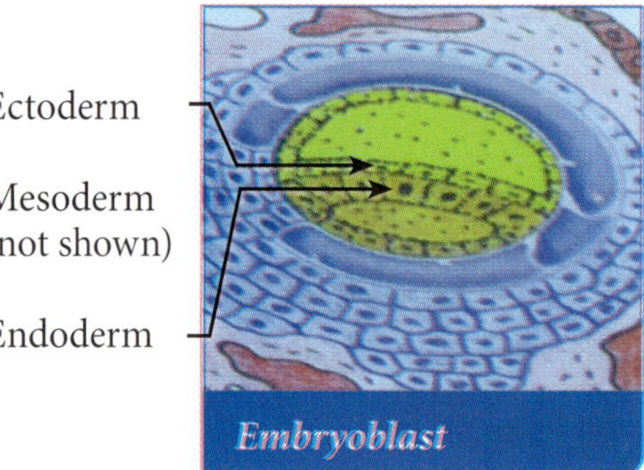

Model showing ovulation, fertilization, and implantation

- **A** Primordial follicles
- **B** Primary follicle
- **C** Secondary follicles
- **D** Rupturing follicle
- **E** Corpus luteum
- **F** Oocyte
- **G** Fertilization (union of sperm and egg)
- **H** Sperm and egg nuclei swell and become pronuclei (arrows) then join, forming the zygote
- **I** 2-cell stage
- **J** 4-cell stage
- **K** Morula
- **L** Blastocyst
- **M** Implanted embryo (embryoblast)

Fertilization must occur within the first day after ovulation in the uterine tube. A few thousand sperm cells make it to the uterine tube with an egg (about ten minutes after ejaculation) but are unable to fertilize an egg until capacitation (a ten hour process that changes membrane permeability) occurs. Once capacitation occurs, the sperm continue to live for about four days. The acrosome of sperm cells contain enzymes that digest a hole through the granulosa cells and zona pellucida. As the zona pellucida is penetrated, it initiates the egg to depolarize and undergo the cortical reaction, both of which help prevent polyspermy. The fertilized egg (zygote) continues down the uterine tube for another day or two, developing into a morula (16–32 cell cluster) by the time it arrives in the uterus. It continues dividing for a few more days in the uterus, now called a blastocyst, before implanting into the engorged uterine lining (endometrium) and transforming into an embryoblast. Embryogenesis begins and produces three primary germ layers (endoderm, mesoderm, and ectoderm) that will become the child.

Ectoderm

Mesoderm (not shown)

Endoderm

Embryoblast

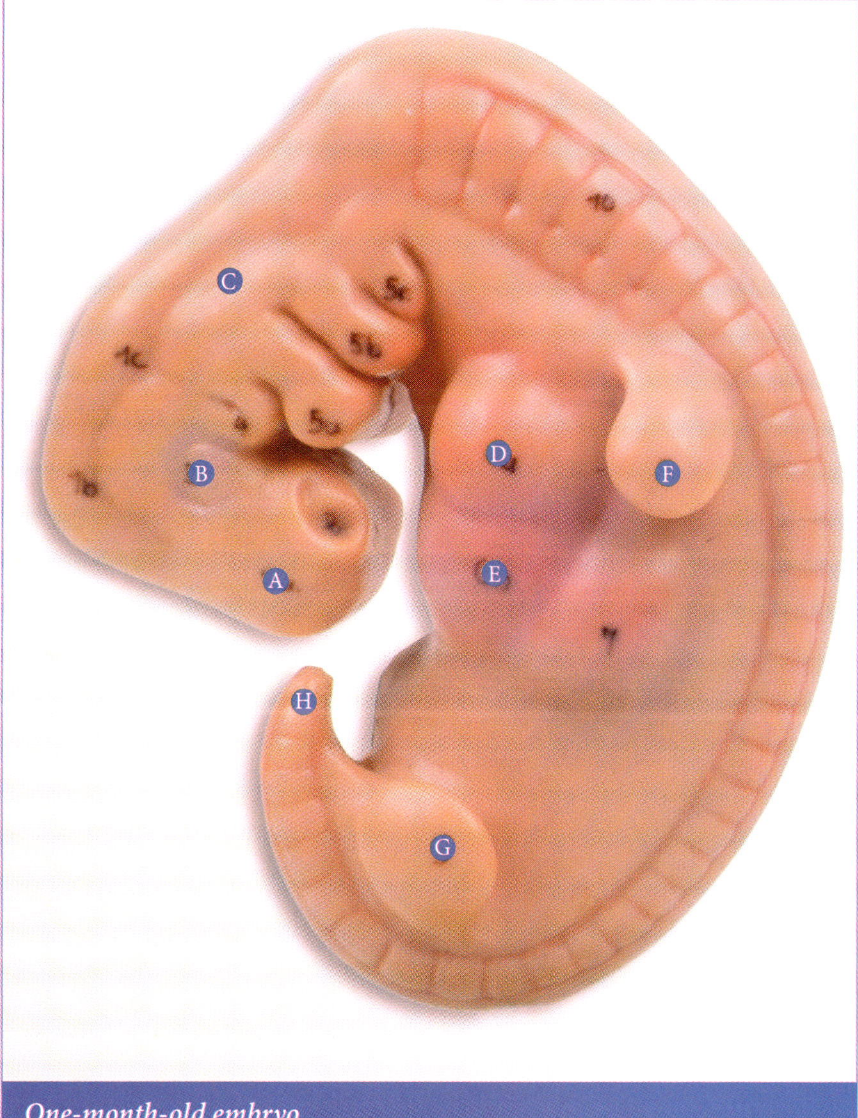

One-month-old embryo

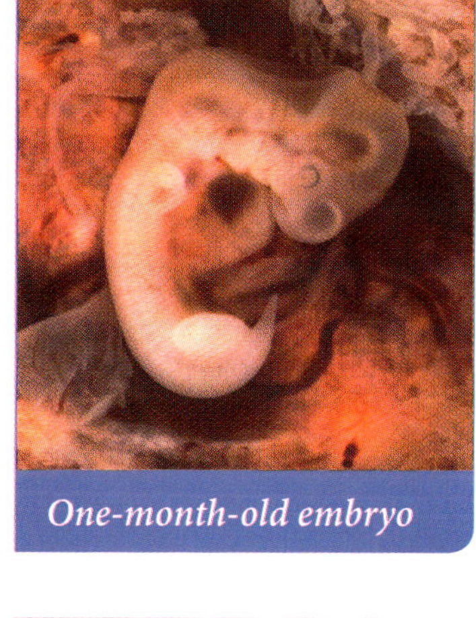

One-month-old embryo

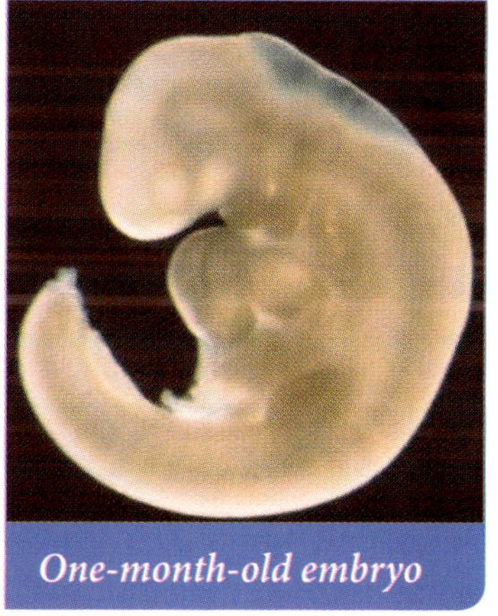

One-month-old embryo

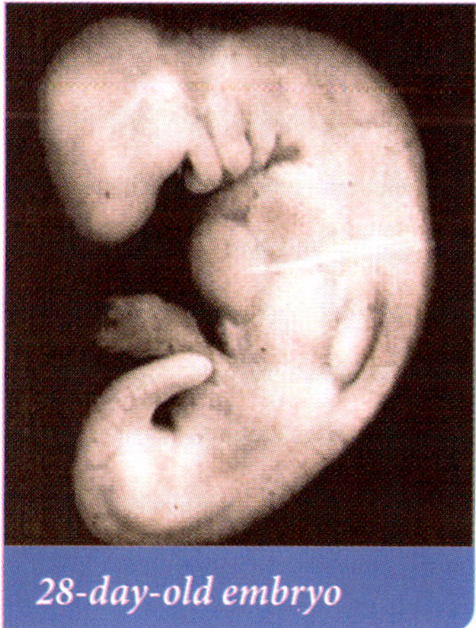

28-day-old embryo

A   Head

B   Primitive lens

C   Pharyngeal arches

D   Heart bulge

E   Liver bulge

F   Arm bud

G   Leg bud

H   Tail

About two weeks after implantation, the embryonic stage begins and continues over the next six weeks. The three germ layers begin differentiating into specific tissue and organ systems (organogenesis). The amnion and chorion are fluid filled membranes that surround and protect the embryo. Four weeks into development, the embryo's heart begins to beat, establishing separate circulatory systems (exchanging gases, nutrients and wastes in the placenta) between mother and child. A neural tube is present that will become the brain and spinal cord. At this time, pharyngeal arches, yolk sac, (not shown) and a tail are present. These structures are shared with most vertebrates and are examples of homologous structures of organisms that share common ancestry (evolutionary embryology).

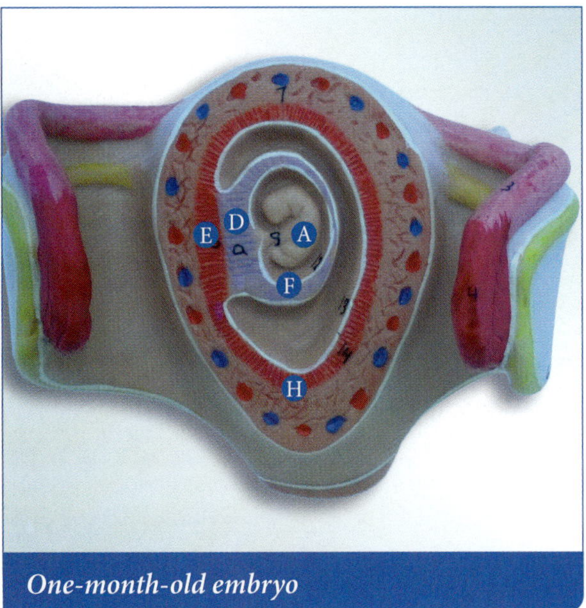

One-month-old embryo

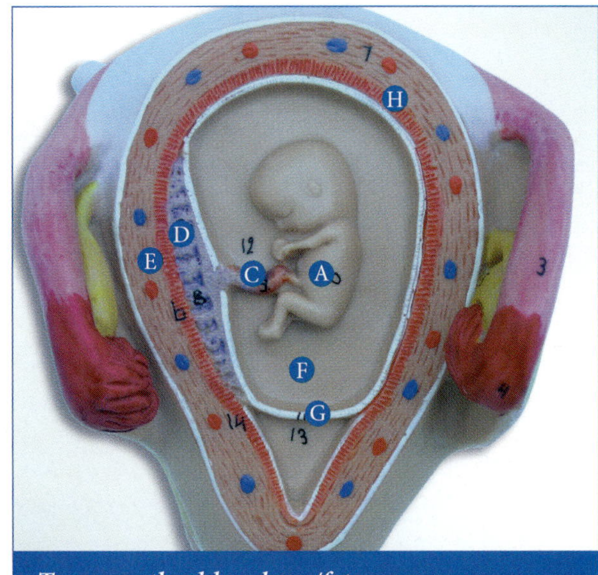

Two-month-old embryo/fetus

Ⓐ Embryo

Ⓑ Fetus

Ⓒ Umbilical cord

Ⓓ Child's portion of the placenta (chorion)

Ⓔ Mother's portion of the placenta (endometrium)

Ⓕ Amnion

Ⓖ Chorion

Ⓗ Decidua parietalis of endometrium

The developing embryo continues growth, and at about two months, is now called a fetus. The embryo and fetus float in amniotic fluid that protects and allows movement. The amnion is only penetrated by the umbilical cord, which provides the fetus with nutrients and gas exchange that occur between chorionic villi and endometrial crypts in the growing placenta. Amniotic fluid increases in volume from two months of development on, largely from fetal urine output.

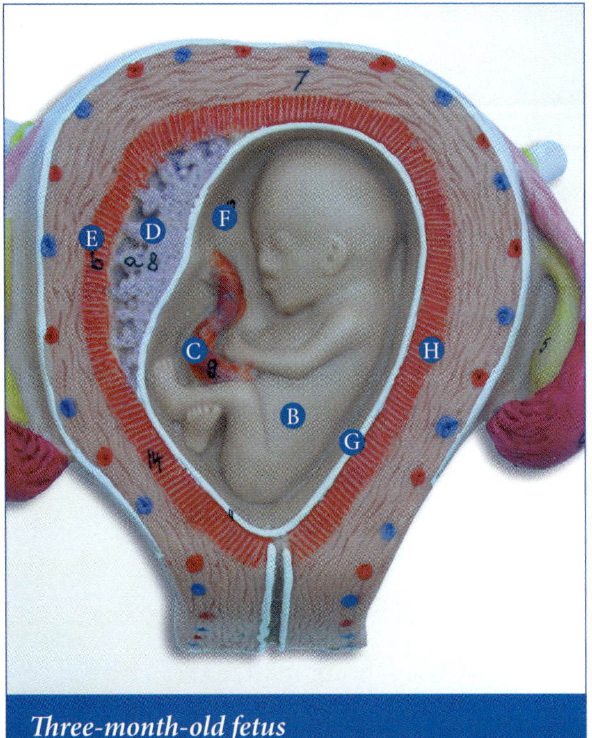

Three-month-old fetus

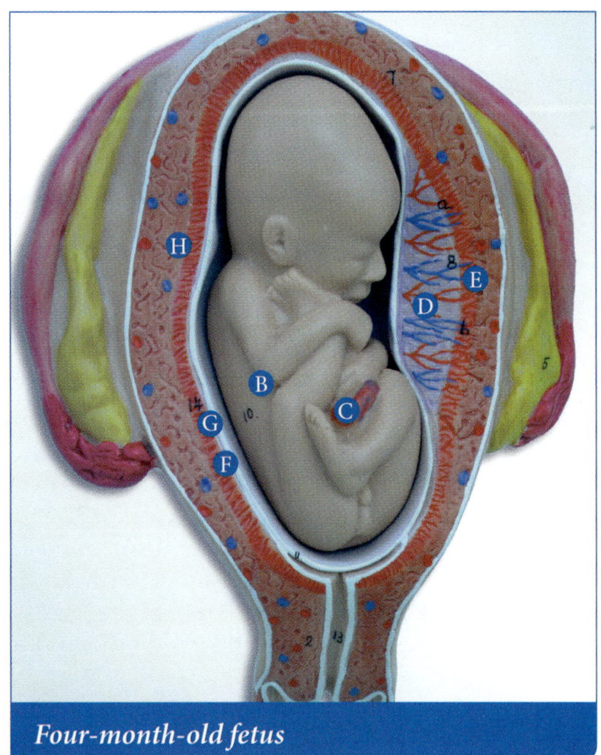

Four-month-old fetus

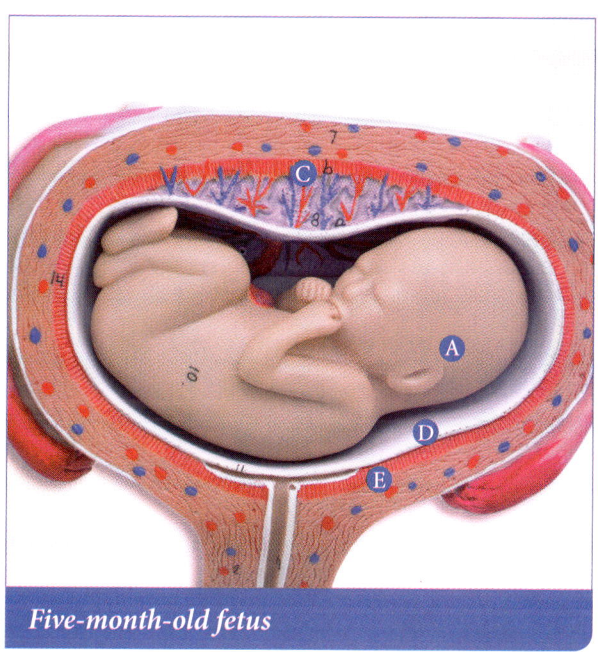

Five-month-old fetus

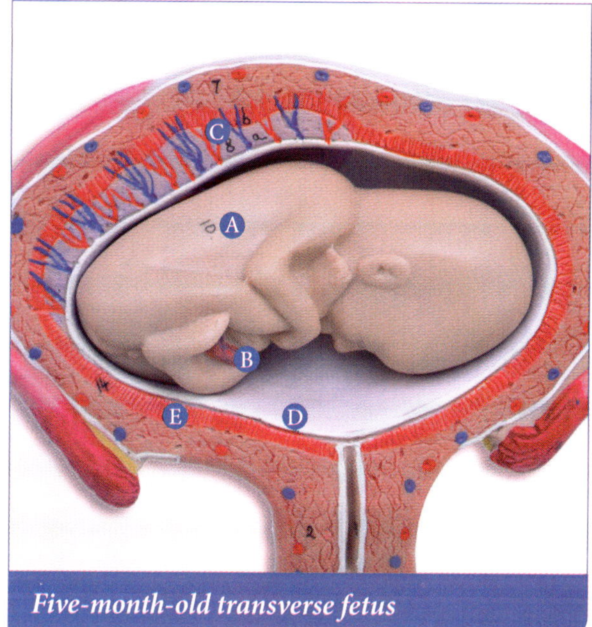

Five-month-old transverse fetus

A Fetus

B Umbilical cord

C Placenta (chorion and endometrium)

D Amnion and chorion

E Decidua parietalis of endometrium

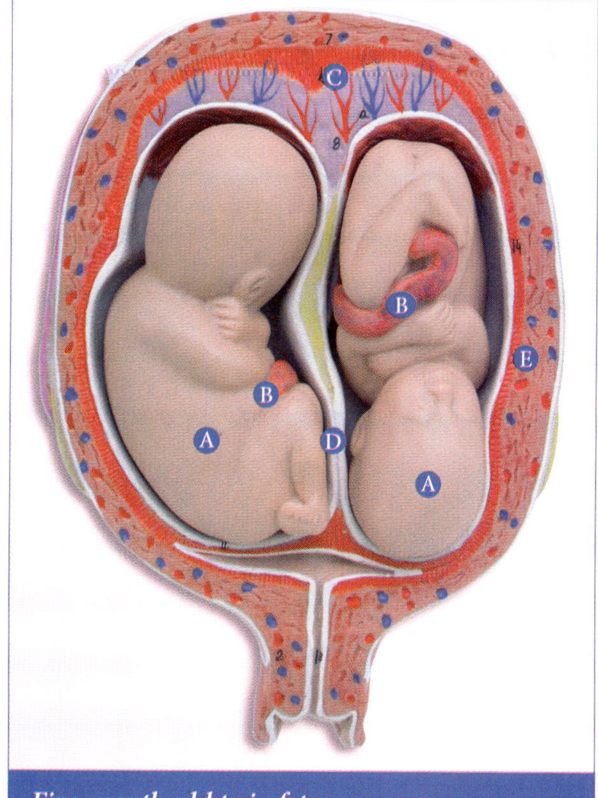

Five-month-old twin fetuses

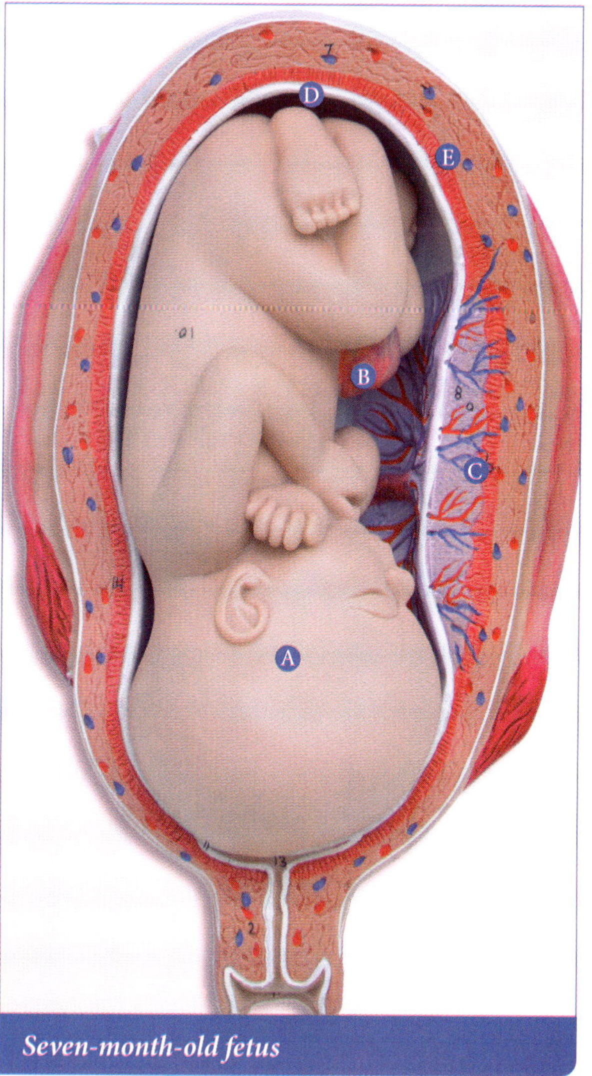

Seven-month-old fetus

Acknowledgments

Chapter 2

Page 2-2	http://www.sciencebuddies.org/science-fair-projects/project_ideas/HumBio_ p017.shtml
Page 2-3	Photoshelter
Page 2-3	http://nobelprize.org/nobel_prizes/medicine/laureates/2004/illpres/images/olfactory.jpg
Page 2-6	http://www.masseyeandear.org/research/oph- thalmology/programs/retina-research-institute/
Page 2-8	http://www.usdb.org/deafblind/db/CIT%20Web%20Lessons/The%20Auditory%20System/ossicles.gif
Page 2-10	http://www.uu.se/en/node1039
Page 2-11	Photoshelter

Chapter 4

Page 4-2	Photoshelter
Page 4-2	http://www.beltina.org/health-dictionary/plasma-blood.html
Page 4-3	Photoshelter
Page 4-3	http://www.sbceo.org/~kmguad/public_html/departments/7thScience/Pictures.htm
Page 4-6	http://www.mybloodyourblood.org/hs_biology_white.htm
Page 4-6	http://www.ncbi.nlm.nih.gov/books/NBK26919/
Page 4-7	http://scopeblog.stanford.edu/2009/08/a_beautiful_blo_1/
Page 4-7	http://www.nih.gov/researchmatters/ march2009/images/blood_l.gif

Chapter 5

| Page 5-2 | Photoshelter |

Chapter 6

| Page 6-3 | Photoshelter |
| Page 6-4 | Photoshelter |

Chapter 7

| Page 7-2 | By SEER [Public domain], via Wikimedia Commons, *http://commons.wikimedia.org/wiki/File%3Allu_lymph_capillary.png* |
| Page 7-3 | Photoshelter |

Chapter 8

Page 8-5	http://radiographics.rsna.org/content/25/2/525/F4.expansion.html
Page 8-6	Photoshelter
Page 8-8	Photoshelter
Page 8-9	Photoshelter

Chapter 9

Page 9-5	Photoshelter
Page 9-6	http://www.masseyeandear.org/research/oph-thalmology/programs/retina-research-institute/
Page 9-6	http://www.topnews.in/healthcare/content/22143us-organ-recipient-contracts-hiv-after-donor-had-post-screening-sex
Page 9-6	Photoshelter
Page 9-7	Photoshelter
Page 9-8	Photoshelter
Page 9-9	Photoshelter

Chapter 10

| Page 10-10 | Photoshelter |

Chapter 11

| Page 11-3 | http://www.asdk12.org/staff/vanarsdale_mark/pages/mrva/is9/semester2/endocrine.htmlsystem |
| Page 11-5 | Photoshelter |

Notes

Notes

Notes

Notes

Notes

Notes

Name: ____

Lab Day/Time: _________ / _________   Section#: _____________

1) T or F : Homeostasis is one of the many bodily processes that is controlled by a positive feedback loop.

2) Oxytocin is secreted by
   A. Prolactin
   B. Follicle Stimulating Hormone
   C. Neurohypophysis
   D. Somatostatin

3) The pituitary gland is made up of the ____________________ and the ____________________.

4) The endocrine system
   A. Produces effects in the body that can last up to days
   B. Releases hormones that alter physiological activities of different tissues and organs
   C. Releases chemical messengers into the bloodstream that travel throughout the body
   D. All of the above

5) Basal metabolic rate is raised by both ____________________ and ____________________ hormones secreted by the thyroid gland.

6) Diabetis mellitus, type I, occurs when ____.
   Diabetis mellitus, type II, occurs when ____.

7) The pancreatic cells that secrete insulin are the
   A. Alpha cells
   B. Beta cells
   C. Glucagon cells
   D. All of the above

8) A patient with known adrenal gland problems that has increased blood sodium levels most likely has an issue with the
   A. Zona faciculata
   B. Zona glomerulosa
   C. Zona reticularis
   D. All of the above

9) The effects of the hormone estradiol are ____.

10) T or F : Women produce testosterone.

11) The effect of the ____________________ hormones of the thyroid gland's ____________________ cells is to raise metabolic rate.

12) The anterior pituitary is ____________________ tissue and produces and secretes ____________________ major types of hormones directly into the blood stream.

13) The pancreas is a mixed organ of both ____________________ (ducts) and ____________________ (ductless) portions.

14) Hormone

1A. _______________________________

2A. _______________________________

3A. _______________________________

1B. _______________________________

2B. _______________________________

3B. _______________________________

1C. _______________________________

2D. _______________________________

3G. _______________________________

Effects

1. _______________________________

2. _______________________________

3. _______________________________

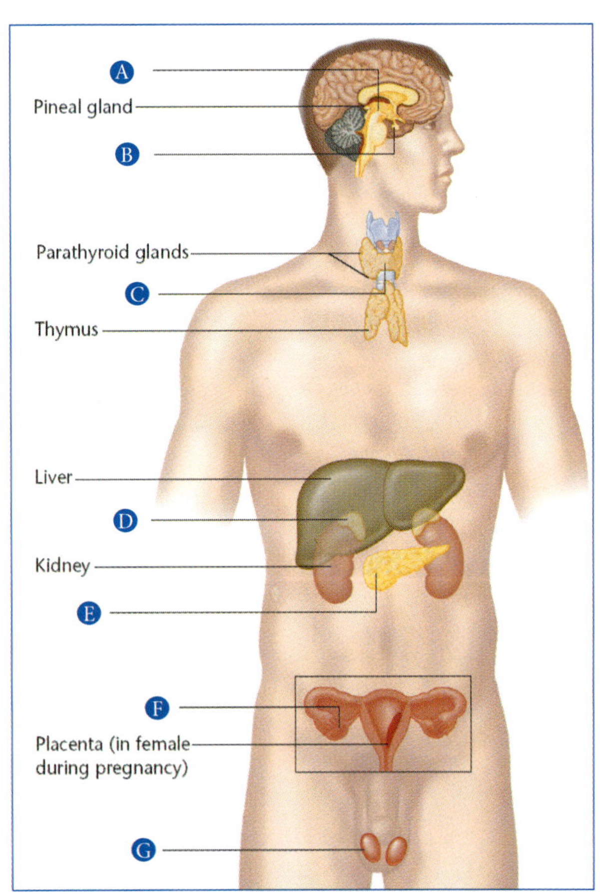

15) Name the structures labeled below.

A. _______________________________

B. _______________________________

C. _______________________________

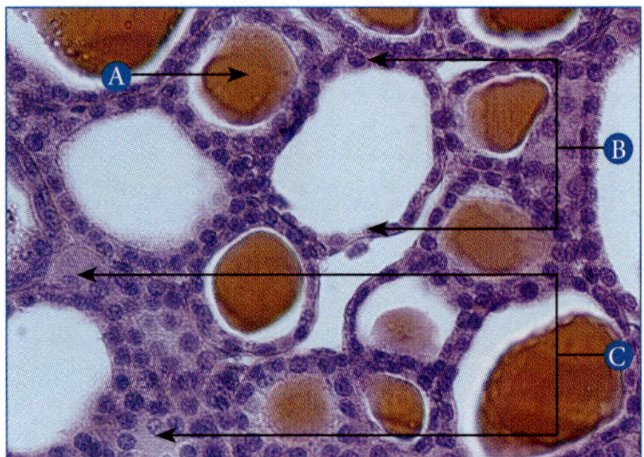

16) Name each zone of the adrenal gland and one hormone it produces.

A. _______________________________

B. _______________________________

C. _______________________________

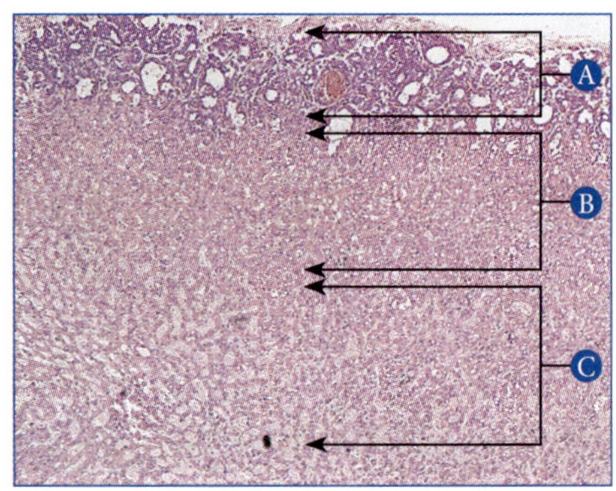

17) Name the region of the pancreas labeled.

A. _______________________________

Name two hormones it produces and their effects.

____

____

____

____

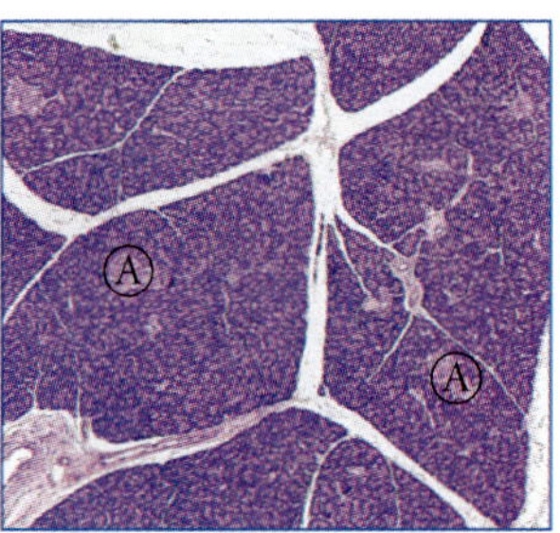

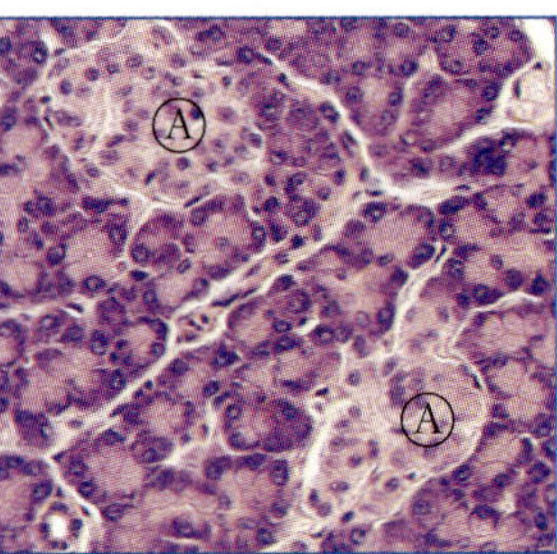

Prelab

Name: ____

Lab Day/Time: _______ / _______   Section#: _____________

1) What types of receptors are found in the eye?
   A. Cones and rods   C. Basilar membrane
   B. Semicircular ducts   D. All of the above

2) How many different types of chemoreceptors are found in taste? _______________________
   Name them: ____

3) Sound enters the ear through the auditory canal, the sound waves come in contact with the
   _____________________ which causes vibration in auditory ossicles.
   A. Stapes   C. Cochlea
   B. Tympanic membrane   D. All of the above

4) Equilibrium is maintained by the _____________________ and the _____________________.

5) What part of the eye regulates the amount of light that enters?
   A. Iris   C. Retina
   B. Pupil   D. Helix of the Auricle

6) T or F : One taste bud can only distinguish one type of compound.

7) What type of structures of olfactory neurons bind to specific odor molecules to transmit an
   action potential?
   A. Cilia   C. Neurotransmitters
   B. Olfactory bulb   D. None of the above

8) The red eye effect that occurs sometimes when a camera flashes is the blood in the
   A. Sclera   C. Macula lutea
   B. Choroid   D. None of the above

9) In our retina, _____________________ cells perceive color and _____________________ cells
   are responsible for our night vision.

10) When a stereocilia is bent by contact with the tectorial membrane what ion gates are opened
    causing the cell to depolarize and release a neurotransmitter?
    A. K^+ C. Na^+
    B. Ca^+ D. All of the above

11) Light passes through which portions of the eye? _____________________,
    _____________________, _____________________, _____________________ &
    _____________________.

12) Each cochlea has around _____________________ outer hair cells with _____________________
    stereocilia on each cell, which are embedded in the tectorial membrane.

13) Name the structures on the anterior view of the eye.

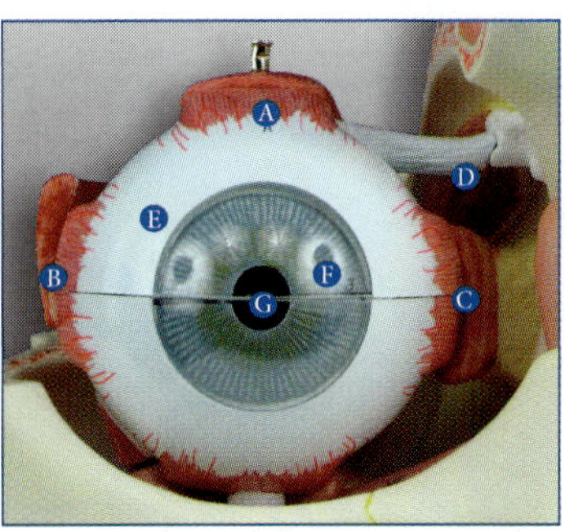

A. ___________________________
B. ___________________________
C. ___________________________
D. ___________________________
E. ___________________________
F. ___________________________
G. ___________________________

14) Name the structures labeled on a section of the cochlea.

A. ___________________
____
B. ___________________
____
C. ___________________
____
D. ___________________________
E. ___________________________
F. ___________________________
G. ___________________________
H. ___________________________
I. ___________________________
J. ___________________________

15) Name the labeled structures.

A. ___________________________
B. ___________________________
C. ___________________________

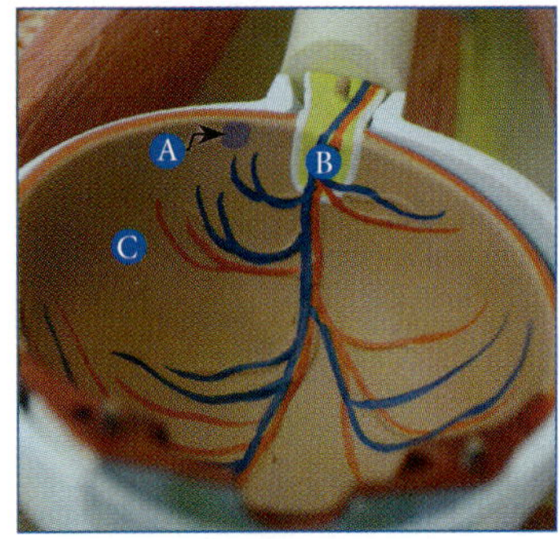

16) Name the structures on the anterior view of the ear.

A. ___________________________
B. ___________________________
C. ___________________________
D. ___________________________
E. ___________________________
F. ___________________________
G. ___________________________
H. ___________________________
I. ___________________________
J. ___________________________
K. ___________________________
L. ___________________________
M. ___________________________
N. ___________________________

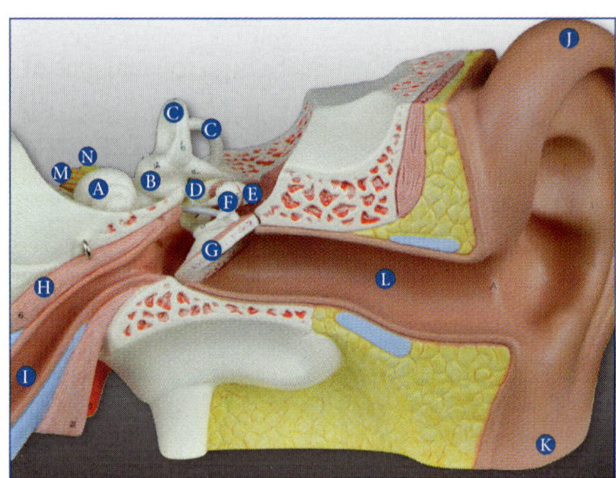

17) Write what each letter's arrow points to.

A. ___________________________
B. ___________________________
C. ___________________________
D. ___________________________
E. ___________________________
F. ___________________________
G. ___________________________
H. ___________________________
I. ___________________________
J. ___________________________

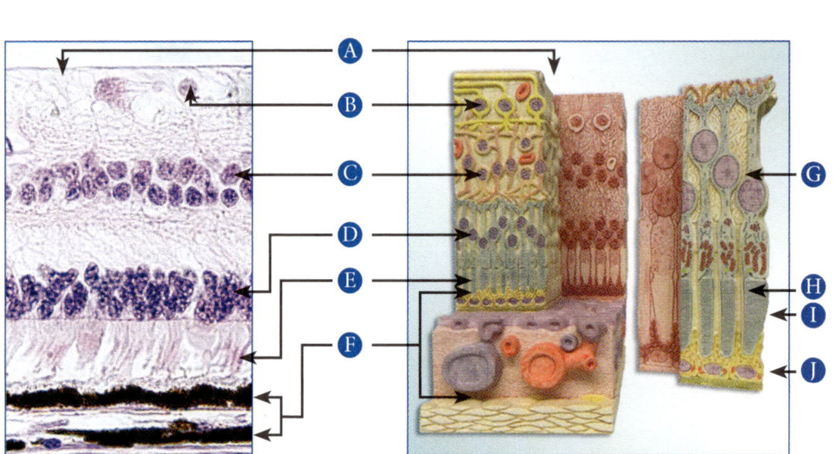

Prelab

Name: ____

Lab Day/Time: ________ / ________   Section#: ____________

1) _____________________ is an individual's unique interpretation of the environment.

2) Touch receptor _____________________ varies from region to region.

3) The process of decreasing the sensitivity of olfaction after a couple seconds of exposure to an odor molecule is known as _____________________.

4) As objects approach, the more round the lens becomes due to contraction of ciliary muscles and subsequent reduction in tension of suspensory ligaments. This is called _____________________.

5) As an object approaches, the visual axis of each eye is trained on the object, causing the eyes to cross when one looks at the object inches away. This is called _____________________.

6) The _____________________ chart measures visual accuity.

7) Asymmetries of the cornea result in an _____________________.

8) Restricted movement of the typanic membrane or ossicles results in _____________________ deafness.

9) The Weber and Rinne tests detect differences in _____________________ between each ear.

10) T or F : Men are more likely to be color blind than women.

11) The structure of the eye found in nocturnal animals that humans lack is the tapetum _____________________.

12) Side to side movement of the eyes after rotating in a chair is known as _____________________.

13) Name the labeled structures of the sheep eye dissection.

    A. _______________________________

    B. _______________________________

    C. _______________________________

    D. _______________________________

    E. _______________________________

    F. _______________________________

    G. _______________________________

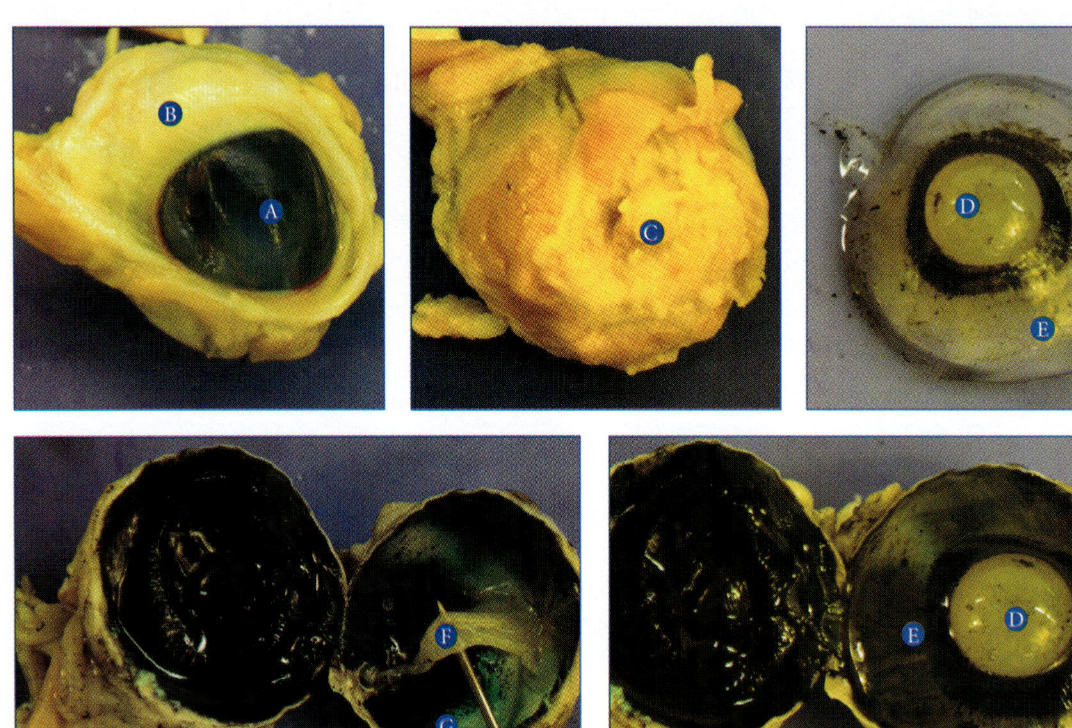

Name: ____

Lab Day/Time: ________ / ________   Section#: _____________

1) The main functions of the blood are
   A. To carry food and oxygen to our cells.
   B. To take wastes to the kidneys.
   C. To collect residual carbon dioxide and wastes from our cells.
   D. All of the above

2) Most of the plasma in blood is comprised of _________________.

3) The most abundant blood cell is the
   A. Erythrocytes
   B. Leukocytes
   C. Electrolytes
   D. Platelets

4) What is the difference between donating plasma and "whole" blood?
   A. Donating plasma can be done about twice a week.
   B. Donating blood (erythrocytes) can be done once every few weeks.
   C. Donating plasma can be done with heparin, so cells float to the bottom and plasma floats to the top once centrifuged.
   D. All of the above

5) What is the name of the molecule that transports oxygen in the blood?
   A. Erythrocyte
   B. Hemoglobin
   C. Leukocytes
   D. None of the above

6) Which type of granular leukocytes kills bacteria?
   A. Eosinophils
   B. Monocytes
   C. Neutrophils
   D. None of the above

7) T or F : One job of a basophil is to secrete heparin to increase and maintain blood flow.

8) Platelets arise from a huge cell called a _____________________________________.

9) The most abundant blood cells in the body are called _________________.

10) Plasma proteins consist of _________________, _________________ & _________________, of which _________________ is involved in coagulation or forming a clot.

11) There are approximately _________________ red blood cells in one drop of blood.

12) _____________________________________ are responsible for our immunity.

13) Write the names and approximate percentages of each of the labeled components of blood.

    A. _______________________________________

    B. _______________________________________

    C. _______________________________________

14) Name the structures labeled in the blood smear.

    A. _______________________________________

    B. _______________________________________

    C. _______________________________________

    D. _______________________________________

    E. _______________________________________

15) Describe the final five steps of coagulation.

    A. _______________________________________

    B. _______________________________________

    C. _______________________________________

    D. _______________________________________

    E. _______________________________________

16) Name the formed elements.

    A. _______________________________________

    B. _______________________________________

    C. _______________________________________

    D. _______________________________________

    E. _______________________________________

    F. _______________________________________

17) Name the three formed elements.

    A. _______________________________________

    B. _______________________________________

    C. _______________________________________

 Human Anatomy & Physiology

CHAPTER

5 Prelab

Name: _______________________________________

Lab Day/Time: _______ / _______   Section#: _____________

1) The heart receives blood from the
   A. Pulmonary veins
   B. Vena cava
   C. Bicuspid valve
   D. Both A. and B.

2) T or F : The right ventricle delivers blood at a relatively high blood pressure (120 mmHg) to the lungs, and the left ventricle supplies the rest of the body at low pressure (40 mmHg).

3) In terms of heart pumping action, systole occurs when the heart ____________________ and diastole occurs when the heart ____________________.

4) The SA node fires, generating the ____________________. The ____________________ depolarizes and delivers the signal to ventricular myocardium to depolarize, causing the QRS complex.

5) A heart muscle cell, also called a ____________________, is striated/unstriated (circle one), voluntary/involuntary (circle one), contains one/two nuclei (circle one), and contains relatively large ____________________.

6) ____________________ connect two cells with tube-like protein complexes that allow ion flow between the cells. These can be seen as ____________________.

7) Blood flows from the left atrium to the left ventricle through the ____________________ valve.

8) Shortly after ventricular systole, the ____________________ appears as the ventricles ____________________.

9) The second heart sound, the "dupp," is
   A. Aortic pressure is greater left ventricle pressure
   B. Bicuspid and tricuspid closing
   C. Increase in blood pressure by ventricular contraction
   D. Both A. and B.

10) The fetal heart has two unique structures that would be fatal in the adult heart. The structures are the ____________________ and the ____________________.

11) When a child is born, it takes its first breath causing
   A. A drop in pressure in the right atrium
   B. Momentary flow of blood from left atrium to right atrium
   C. Separation between the two atria and elimination of left to right shunt
   D. All of the above

12) In terms of electrical activity, what is happening at the following phases:
   P-wave ____________________
   QRS complex ____________________
   T-wave ____________________

13) Name the structures labeled on the anterior view of the heart models below.

A. _______________________________________

B. _______________________________________

C. _______________________________________

D. _______________________________________

E. _______________________________________

F. _______________________________________

G. _______________________________________

H. _______________________________________

I. _______________________________________

J. _______________________________________

K. _______________________________________

L. _______________________________________

M. _______________________________________

N. _______________________________________

O. _______________________________________

P. _______________________________________

Q. _______________________________________

R. _______________________________________

S. _______________________________________

T. _______________________________________

U. _______________________________________

V. _______________________________________

W. _______________________________________

X. _______________________________________

Y. _______________________________________

Z. _______________________________________

14) Name the structures labeled on the coronal section of a plastinated sheep's heart.

A. _______________________________________

B. _______________________________________

C. _______________________________________

D. _______________________________________

E. _______________________________________

F. _______________________________________

G. _______________________________________

H. _______________________________________

I. _______________________________________

J. _______________________________________

K. _______________________________________

L. _______________________________________

M. _______________________________________

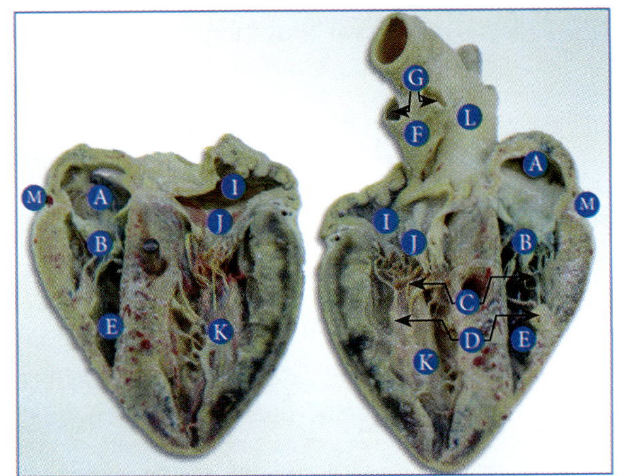

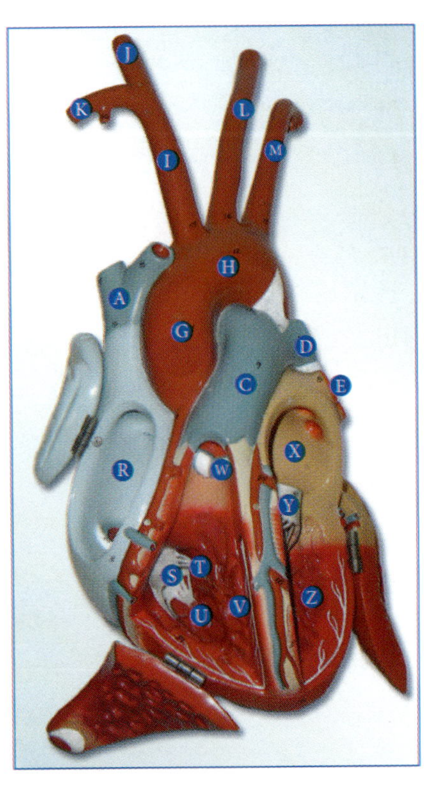

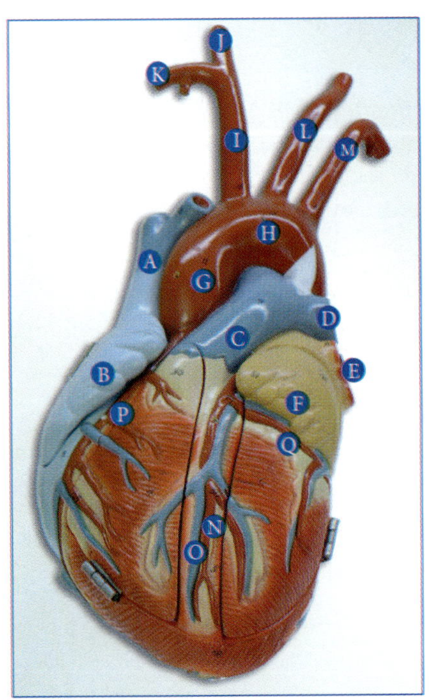

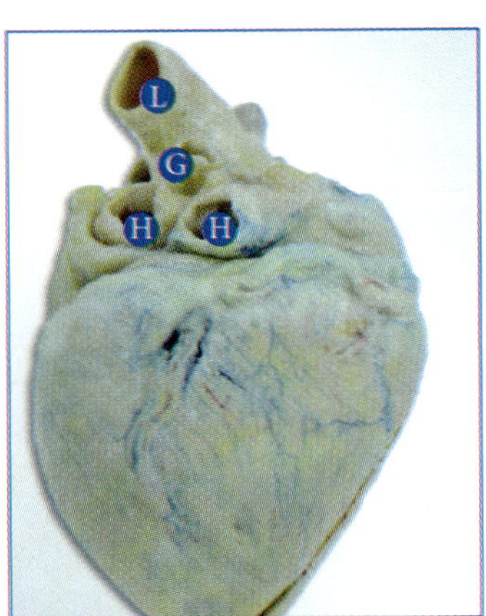

Name: ____

Lab Day/Time: _________ / _________   Section#: _______________

1) Relative to the heart, arteries carry blood ____________________ whereas veins take blood ____________________.

2) Vessels that allow exchange of gases, nutrients, and wastes are called
   A. Venules
   B. Vertebral arteries
   C. Maxillary arteries
   D. Capillaries

3) T or F : A temporary occlusion to the brain of one artery will deprive half the brain of its supply of blood.

4) T or F : The narrow diameter of the capillary increases blood flow and allows diffusion and fluid exchange to take place.

5) ____________________ arteries control overall systemic blood pressure by changes in their diameter.

6) An artery with a diameter of 0.3 mm has a flow of 160 ml/minute. Vasoconstriction occurs and the artery's diameter is reduced to 0.1 mm. What is the flow through this vessel? ____________________ ml/minute.

7) ____________________ contain valves.

8) The largest artery is the ____________________.

9) Blood transports the blood gases ____________________ and ____________________.

10) The ____________________ helps prevent loss of blood flow to the cerebrum.

11) ____________________ are the vessels that provide filtration and reabsorption of blood plasma.

12) There are about ____________________ capillaries with a total surface area of ____________________ in the human body.

13) Capillaries can be ____________________ (tightly bound cells with very narrow clefts), ____________________ (small pores), or ____________________ (blood filled spaces).

14) Name the blood vessels of the inferior half of the body.

Arteries

A. _______________________________

B. _______________________________

C. _______________________________

D. _______________________________

E. _______________________________

Veins

V. _______________________________

W. _______________________________

X. _______________________________

Y. _______________________________

Z. _______________________________

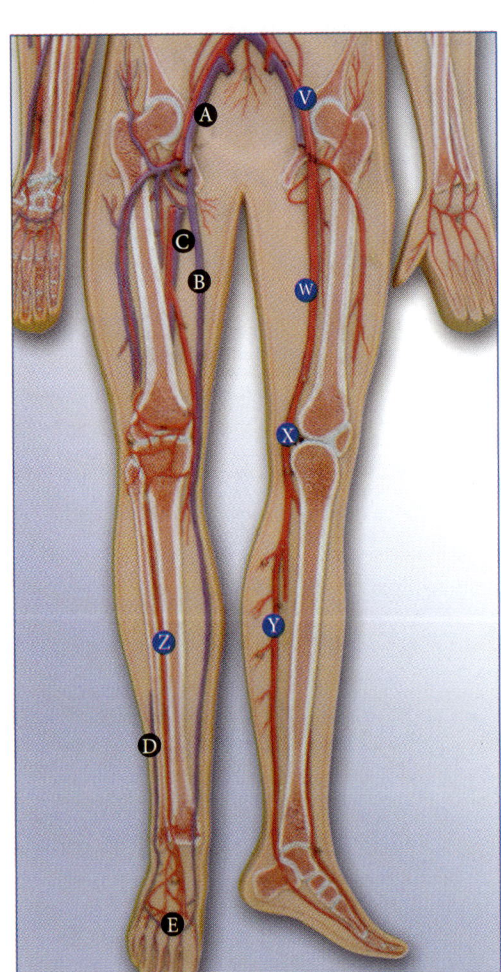

15) Name the blood vessels of the superior half of the body.

Arteries

A. _______________________________

B. _______________________________

C. _______________________________

D. _______________________________

E. _______________________________

F. _______________________________

G. _______________________________

H. _______________________________

I. _______________________________

J. _______________________________

Veins

Q. _______________________________

R. _______________________________

S. _______________________________

T. _______________________________

U. _______________________________

V. _______________________________

W. _______________________________

X. _______________________________

Y. _______________________________

Z. _______________________________

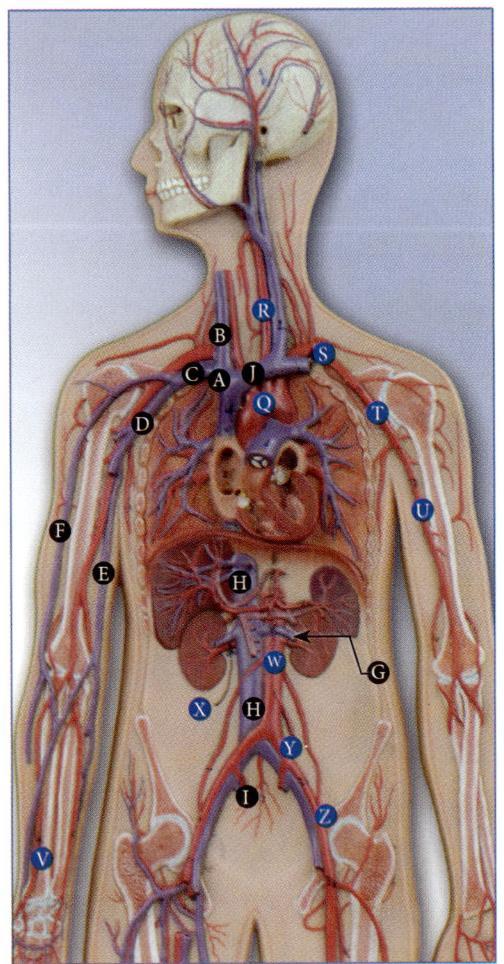

7 CHAPTER *Prelab*

Name: ____

Lab Day/Time: _________ / _________   Section#: _____________

1) Helper T cells (TH) secrete ___________________________ in response to non-self antigens and are necessary for most immune responses.

2) A lymphatic capillary with large opening involved in lipid absorption is called a(n)
   A. Lacteal
   B. Lymphocyte
   C. Occipital
   D. None of the above

3) T or F : The three primary functions of the lymphatic systems are fluid recovery, lipid absorption, and immunity.

4) Lipid absorption occurs in the ___________________________ within villi of the small intestine.

5) In order for cytotoxic T cells to function they must be activated by
   A. APC
   B. Monocyte
   C. Eosinophil
   D. None of the above

6) Elephantiasis is an extreme example of ___________________________ as a result of blockage of lymph flow.

7) The ___________________________ is the largest lymphatic organ.

8) ___________________________ are phagocytic machines that also play a role in the immune system by acting as an antigen presenting cell (APC).

9) About how many lymph nodes does a person have? ___________________________

10) ___________________________ filter lymphatic fluid returned from tissue and contain lymphocytes, plasma cells, macrophages and reticular cells.

11) Name the six major types of lymphocytes. ___________________________
____
____

12) Where are most of the formed elements produced? ___________________________

 Human Anatomy & Physiology

13) Circulatory capillary bed and lymphatic capillary network.

A. _______________________________

B. _______________________________

C. _______________________________

D. _______________________________

E. _______________________________

F. _______________________________

G. _______________________________

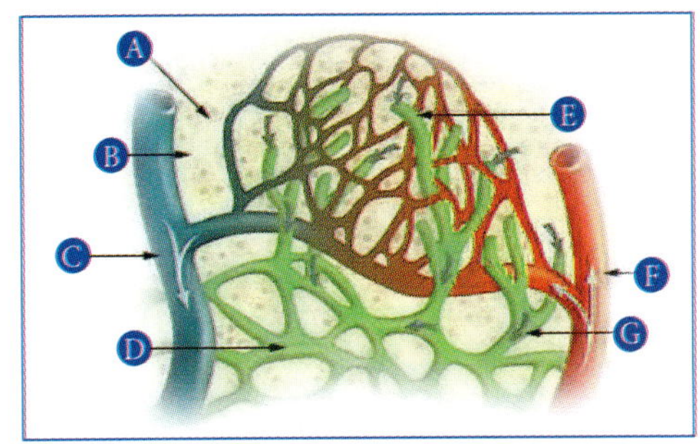

14) Name the labeled structures.

A. _______________________________

B. _______________________________

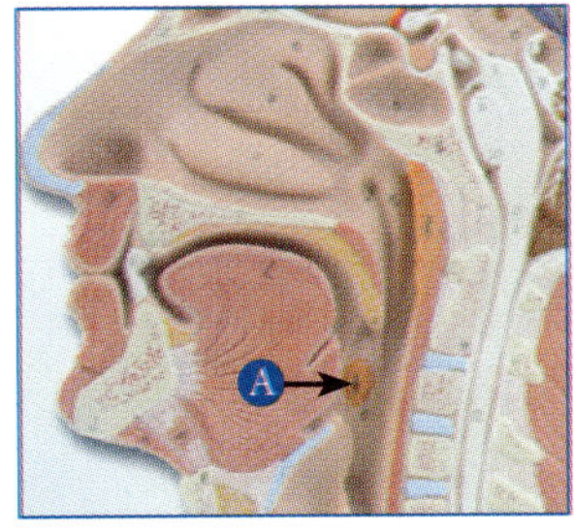
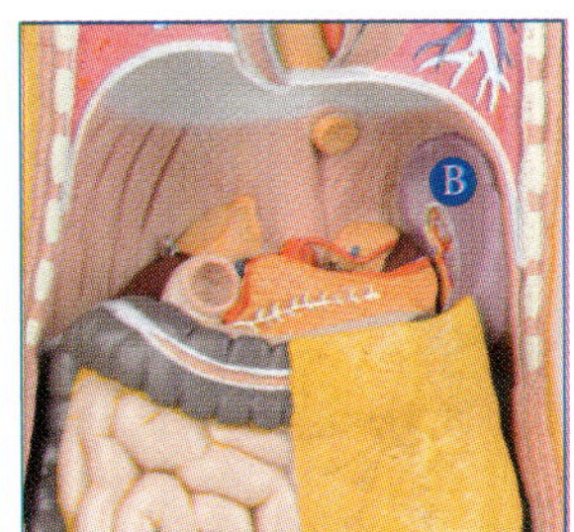

15) Name the labeled components of the lymphatic system.

A. _______________________________

B. _______________________________

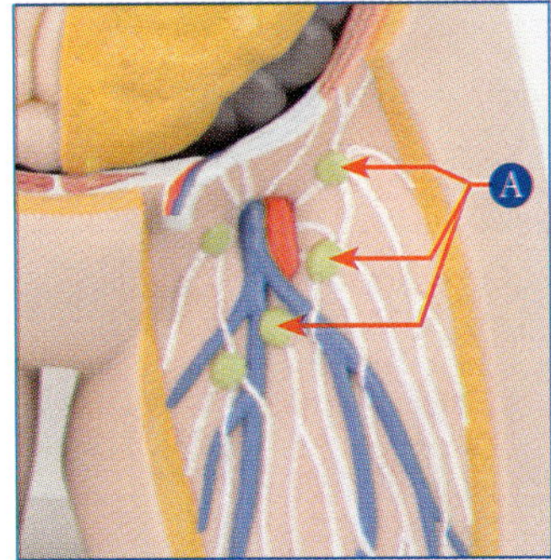
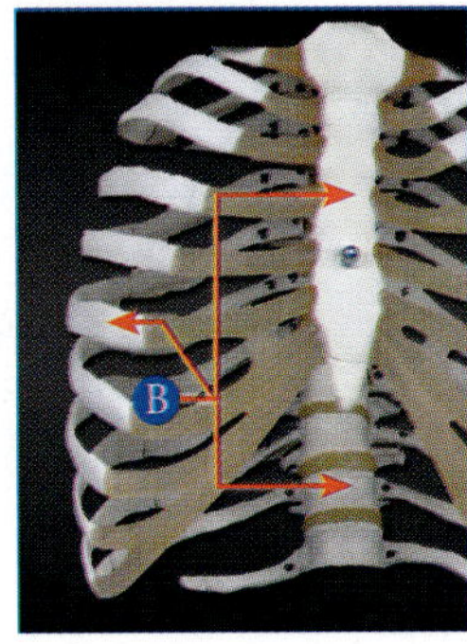

16) Lymph nodes

A. _______________________________

B. _______________________________

C. _______________________________

D. _______________________________

E. _______________________________

F. _______________________________

G. _______________________________

H. _______________________________

I. _______________________________

J. _______________________________

K. _______________________________

L. _______________________________

Lymphatic vessels

M. _______________________________

N. _______________________________

O. _______________________________

P. _______________________________

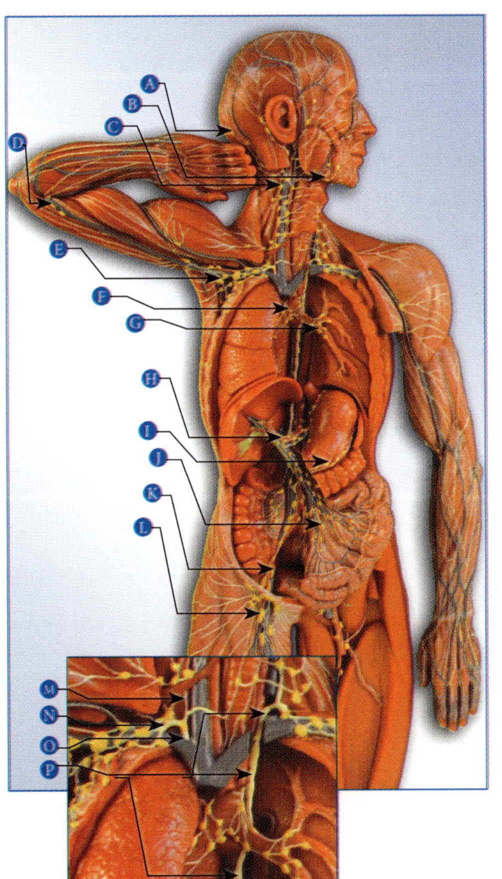

Name: ____

Lab Day/Time: _________ / _________   Section#: _____________

1) Name 3 muscles of inspiration _____________________, _____________________ &
_____________________.

2) Name 3 muscles of expiration _____________________, _____________________ &
_____________________.

3) The need to breath is regulated by changes in _____________________, which is altered by the
amount of _____________________ released by _____________________ in our cells.

4) The false vocal cord is called the _____________________ and is found
_____________________ the vocal cord.

5) T or F : The right and left lungs are symmetrical.

6) The thin layer where gas exchange occurs within the alveoli is called _____________________.

7) There are around _____________________ alveoli in our lungs with about
_____________________ of surface area.

8) The primary gas exchange units of the lungs are called the
A. Alveoli
B. Carina
C. Hyoid bone
D. All of the above

9) The respiratory membrane is the barrier that _____________________ and
_____________________ cross to get into and out of the bloodstream.

10) T or F : The respiratory membrane is extremely thin.

11) Each tertiary bronchus supplies an independent unit of the lung called a bronchopulmonary
segment. The right lung contains _____________________ of these and the left lung has
_____________________.

12) A device used to measure lung volumes is a _____________________.

13) What does COPD stand for? ____
Name three types: ____

14) _____________________ is the amount of air moved into and out of the lungs at rest.
About 500 ml.

15) Vital Capacity (VC) is the _____________________ that can be moved into and out of the lungs
in one breath. About 4,800 ml.

16) Total lung capacity (TLC) is the total volume of air the lungs can hold. About
_____________________ ml.

17) Name the labeled structures of the midsagittal section of the head showing the upper respiratory tract.

A. _______________________________________
B. _______________________________________
C. _______________________________________
D. _______________________________________
E. _______________________________________
F. _______________________________________
G. _______________________________________
H. _______________________________________
I. _______________________________________
J. _______________________________________
K. _______________________________________
L. _______________________________________
M. _______________________________________

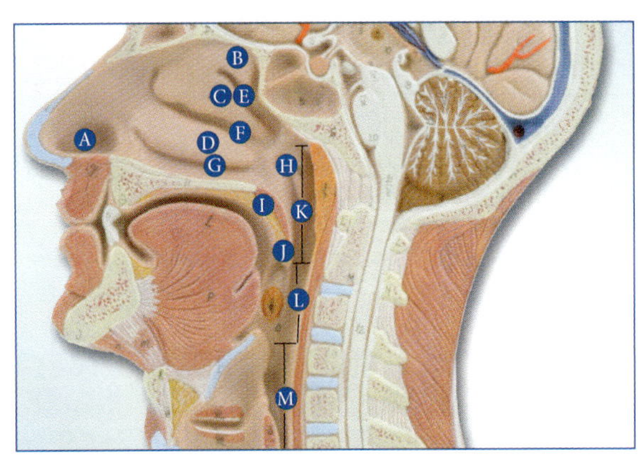

18) Name the labeled structures.

A. _______________________________________
B. _______________________________________
C. _______________________________________
D. _______________________________________
E. _______________________________________
F. _______________________________________

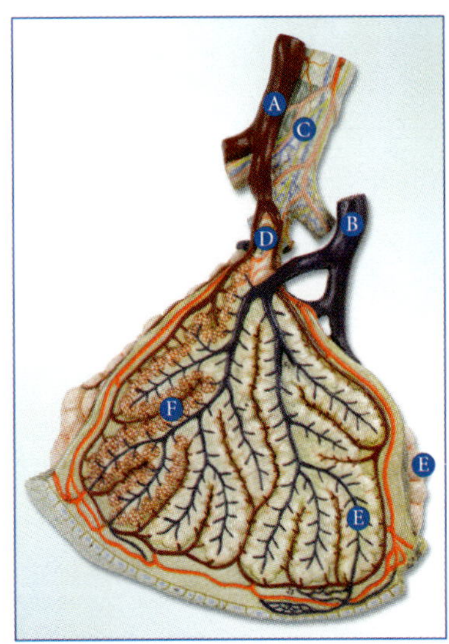

19) Name the labeled structures of the larynx.

A. _______________________________________
B. _______________________________________
C. _______________________________________
D. _______________________________________
E. _______________________________________
F. _______________________________________
G. _______________________________________
H. _______________________________________
I. _______________________________________
J. _______________________________________
K. _______________________________________
L. _______________________________________
M. _______________________________________

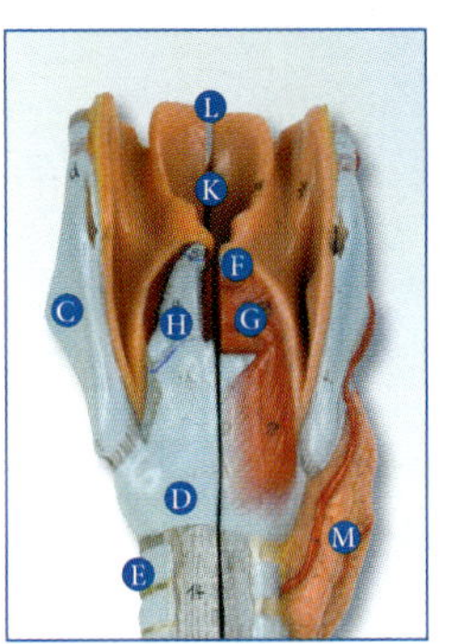
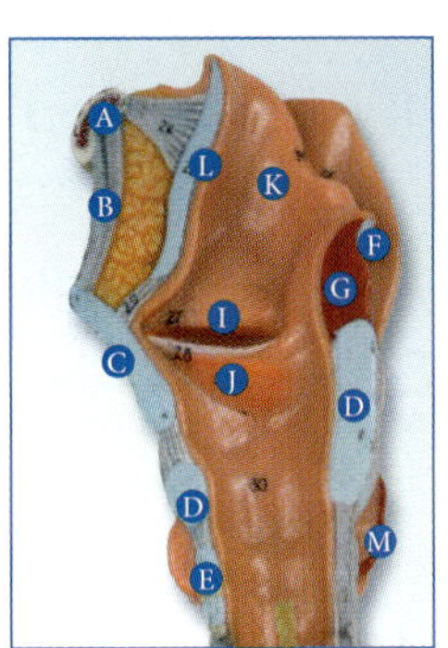

20) Name the labeled structures of the lower respiratory tract.

A. _______________________________________
B. _______________________________________
C. _______________________________________
D. _______________________________________
E. _______________________________________
F. _______________________________________
G. _______________________________________
H. _______________________________________

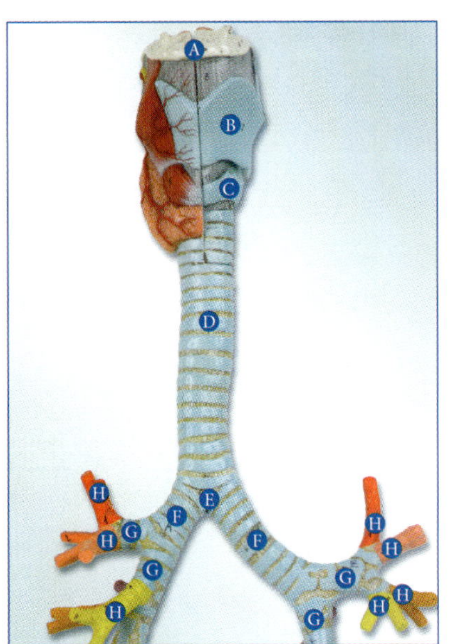

 Human Anatomy & Physiology

Name: ____

Lab Day/Time: ________ / ________   Section#: ____________

1) The _____________________ has the highest blood pressure of any capillary bed, about 60 mmHg.

2) Filtration slits that wrap around glomerular capillaries are created by _____________________.

3) During dehydration, _____________________ is secreted from the _____________________ gland.

4) T or F : The DCT is responsible for modifying and monitoring the filtrate.

5) What happens to the concentration of filtrate as it travels down the collecting duct?

6) The primary site of blood filtration occurs in the _____________________.

7) The glomerulus and the Bowman's capsule collectively are known as the
    A. Capillary bed
    B. Renal Corpuscle
    C. Waste and mineral deposits
    D. None of the above

8) Sequence the following terms of urine from the formation to secretion out of the body.
    ___, ___, ___, ___, ___, ___.
    A. Major calyx
    B. Minor calyx
    C. Nephron
    D. Urethra
    E. Ureter
    F. Collecting duct

9) Aldosterone secretion
    A. Promotes sodium reabsorption
    B. Water retention
    C. Increase in JG cells activity
    D. A. and B. are correct
    E. B. and C. are correct

10) _____________________ from the posterior pituitary causes protein "holes" called aquaporins to be incorporated into the collecting duct.

11) The _____________________ ensures that arterial and venous vasa recta vessels do not reduce the saltiness of the medulla.

12) About 60% of filtrate formed by glomeruli is reabsorbed in the _____________________.

13) The Juxtaglomerular cells are _____________________ that are in contact with, and are influenced by, the macula densa.

14) JG cells secrete _____________________.

15) _____________________ causes an increase in systemic blood pressure by vasoconstriction and aldosterone secretion which promotes sodium and water retention.

16) Name the labeled structures of the kidney model.

A. ___________________________________

B. ___________________________________

C. ___________________________________

D. ___________________________________

E. ___________________________________

F. ___________________________________

G. ___________________________________

H. ___________________________________

I. ___________________________________

J. ___________________________________

K. ___________________________________

L. ___________________________________

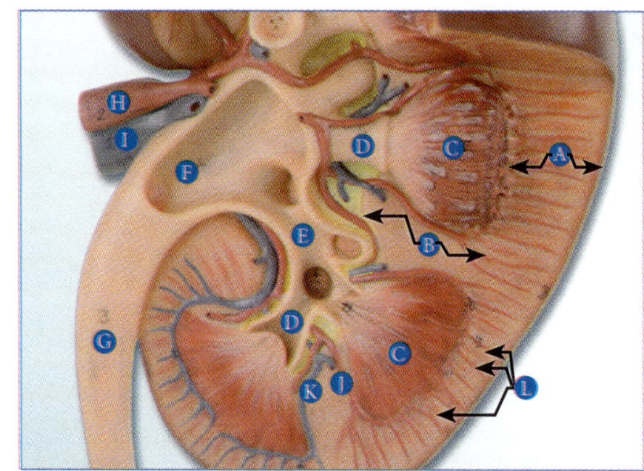

17) Name the labeled structures of the urinary system.

A. ___________________________________

B. ___________________________________

C. ___________________________________

D. ___________________________________

E. ___________________________________

F. ___________________________________

G. ___________________________________

H. ___________________________________

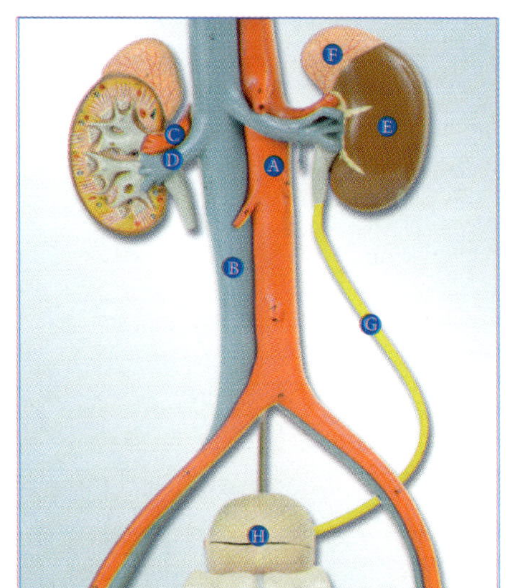

18) Name the labeled structures of the renal corpuscle and associated structures.

A. ___________________________________

B. ___________________________________

C. ___________________________________

D. ___________________________________

E. ___________________________________

F. ___________________________________

G. ___________________________________

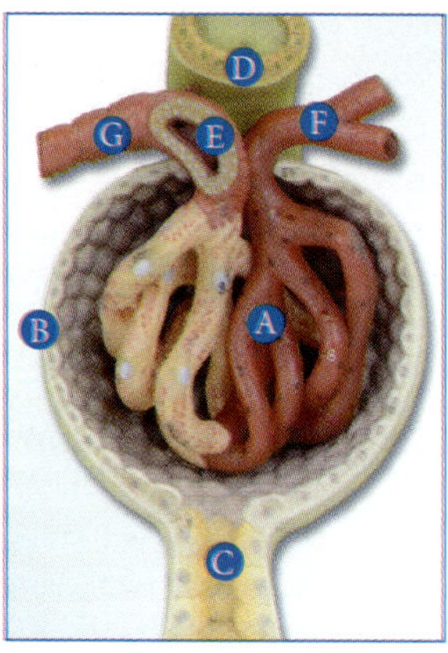

19) Name the labeled structures of the urinary system.

A. ___________________________________

B. ___________________________________

C. ___________________________________

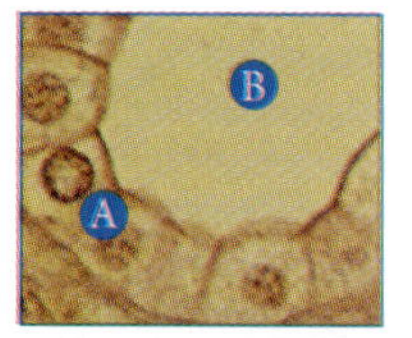
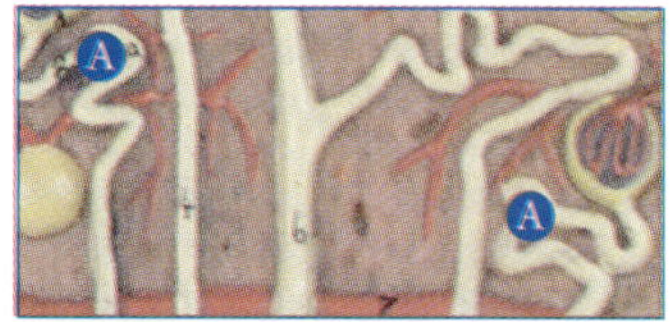

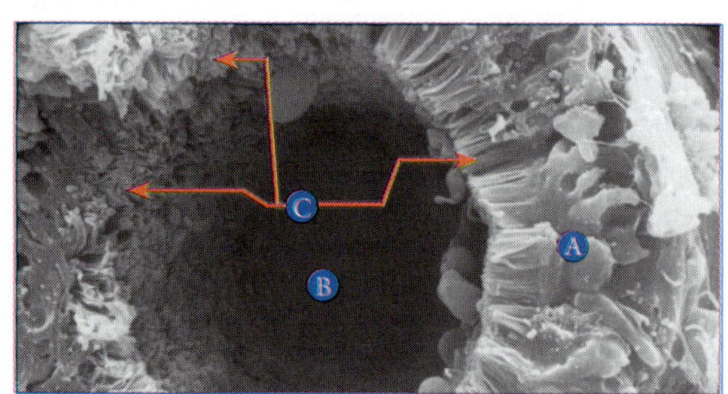

 Human Anatomy & Physiology

Prelab

Name: ____

Lab Day/Time: _________ / _________   Section#: _____________

1) Food is digested into macromolecules prior to absorption. Some of these are
   A. Glucose   C. Nucleic acids
   B. Fatty acids   D. All of the above

2) ___________________ digestion occurs in the mouth by ___________________.

3) T or F : Peristalsis is the muscular contractions of the muscularis externa.

4) In your mouth the incisors
   A. Are specialized for cutting
   B. Are the 4 top and 4 bottom center of the smile
   C. Are specialized for ripping and tearing
   D. Both A. and B.

5) "Bad breath" is caused by
   A. Lack of saliva in the mouth
   B. Excessive bacteria buildup on the tongue
   C. Poor dental hygiene
   D. All of the above

6) The four layers of the digestive tract are the ___________________, ___________________, ___________________, and ___________________.

7) T or F : The esophagus secretes mucous to aid in reducing friction as food is swallowed.

8) The wrinkles in the stomach's interior are called _______________.

9) The secretions from the stomach are from
   A. Mucous cells   C. Chief cells
   B. Parietal cells   D. All of the above

10) The order of the small intestine regions are
   A. Duodenum → Jejunum → Ileum
   B. Jejunum → Ileum → Duodenum
   C. Ileum → Duodenum → Jejunum
   D. None of the above

11) T or F : The pancreas produces all the enzymes of digestion.

12) T or F : Villi and microvilli greatly decrease the surface area of the small intestines.

13) The ___________________ extends from the ileocecal valve to the anus.
   A. Jejunum   C. Anus
   B. Large intestine   D. None of the above

14) Water reabsorption occurs primarily in which part of the intestine? ___________________

15) Digestion enzymes are produced by which organ? ___________________

16) Food that enters the stomach is called ___________________.

17) Name two functions of hydrochloric acid: ___________________ & ___________________

18) Name the labeled structures of the small intestine.

A. _______________________________________

B. _______________________________________

D. _______________________________________

E. _______________________________________

F. _______________________________________

G. _______________________________________

H. _______________________________________

I. _______________________________________

J. _______________________________________

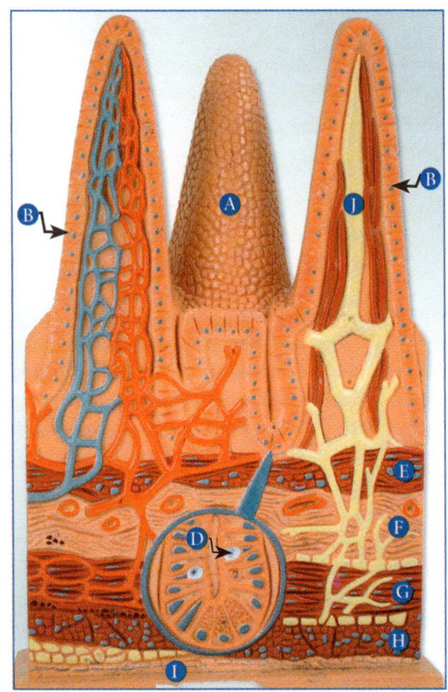

19) Name the labeled structures of the stomach.

A. _______________________________________

B. _______________________________________

C. _______________________________________

D. _______________________________________

E. _______________________________________

F. _______________________________________

G. _______________________________________

H. _______________________________________

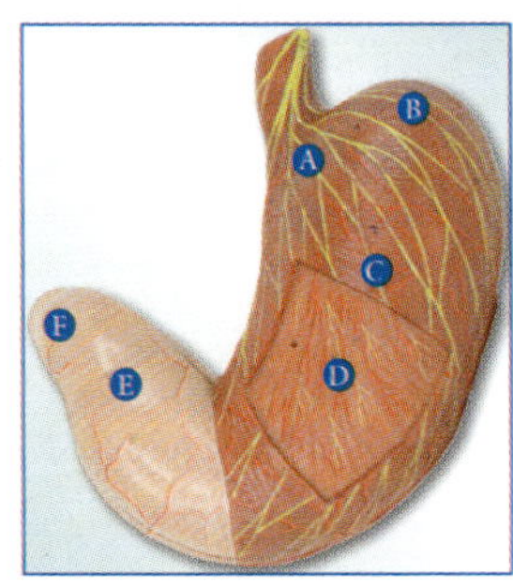 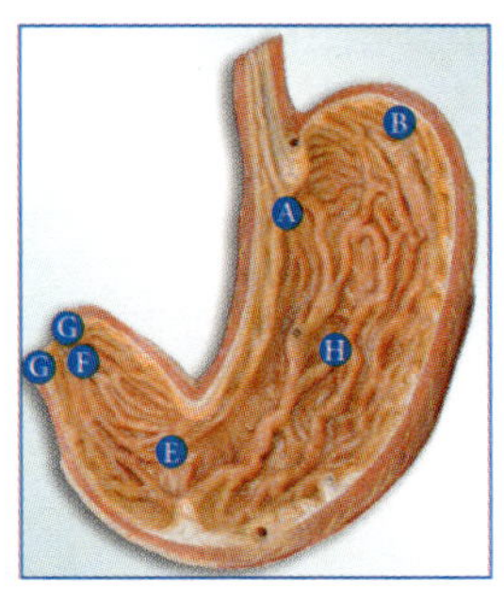

20) Name the labeled structures of the digestive tract.

A. _______________________________________

B. _______________________________________

C. _______________________________________

D. _______________________________________

E. _______________________________________

F. _______________________________________

G. _______________________________________

H. _______________________________________

I. _______________________________________

J. _______________________________________

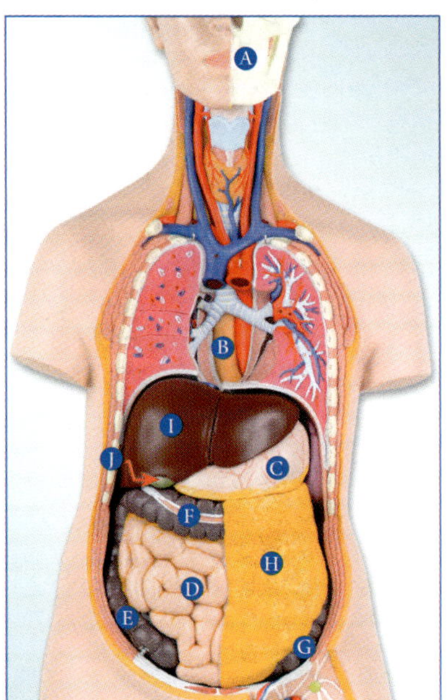

21) Name the labeled gland and its function for the digestive system.

F. _______________________________________

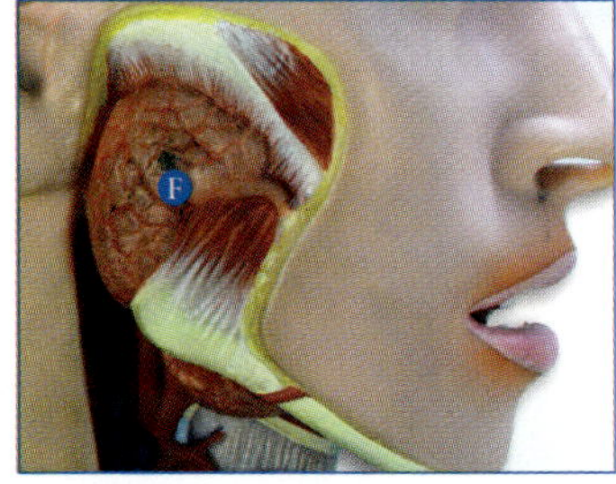

22) Name the labeled structure of the digestive track.

I. _______________________________________

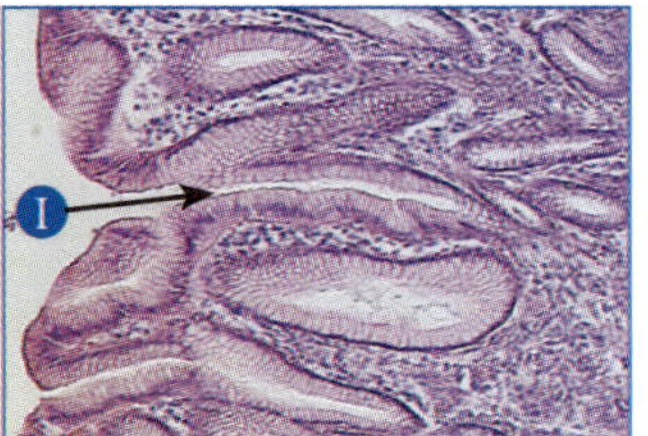

Prelab

Name: ____

Lab Day/Time: ________ / ________   Section#: ______________

1) T or F : Humans (*Homo sapiens*) perform external fertilization.

2) Which part of the female reproductive system produces the eggs?
   - A. Ovary
   - B. Vagina
   - C. Uterus
   - D. None of the above

3) A single male gamete (spermatozoa) contains
   - A. 13 chromosomes
   - B. 23 chromosomes
   - C. 46 chromosomes
   - D. Both A. and B.

4) The path that sperm travels to fertilize an egg is
   - A. Vagina → Cervix → Uterine Tube → Uterus
   - B. Vagina → Uterus → Uterine Tube → Cervix
   - C. Vagina → Cervix → Uterus → Uterine Tube
   - D. None of the above

5) T or F : The female reproductive cycle involves pituitary hormones called Luteinizing Hormone and Follicle Stimulating Hormone.

6) The event that triggers ovulation is a(n)
   - A. Increase in LH
   - B. Decrease in LH
   - C. Decrease in FSH
   - D. Both A. and C.

7) Spermatozoa are produced in the
   - A. Seminiferous tubules
   - B. Ureter
   - C. Corpus cavernosum
   - D. All of the above

8) T or F : The only unique cells in a male's body to undergo meiosis are called spermatocytes.

9) The menstrual cycle from start to finish is
   - A. 7 days
   - B. 28 days
   - C. 20 days
   - D. 40 days

10) Using the chart on page 11-3, which days of the month seem most likely to increase the chances of viable offspring?
   - A. During menstruation
   - B. Between 12 and 16 days
   - C. Between 25 and 28 days
   - D. After 20 days

11) Recent research has shown that female eggs may not have a finite (set) amount that you are born with, unlike with male gametes that are produced from
   - A. Puberty to death
   - B. Birth to death
   - C. Puberty to around 50
   - D. None of the above

12) Name the labeled structures of the female reproductive system.

A. _______________________________

B. _______________________________

C. _______________________________

D. _______________________________

E. _______________________________

F. _______________________________

G. _______________________________

H. _______________________________

I. _______________________________

J. _______________________________

K. _______________________________

L. _______________________________

M. _______________________________

14) Name the labeled structures of the male reproductive system.

A. _______________________________

B. _______________________________

C. _______________________________

D. _______________________________

E. _______________________________

F. _______________________________

G. _______________________________

H. _______________________________

I. _______________________________

J. _______________________________

K. _______________________________

L. _______________________________

M. _______________________________

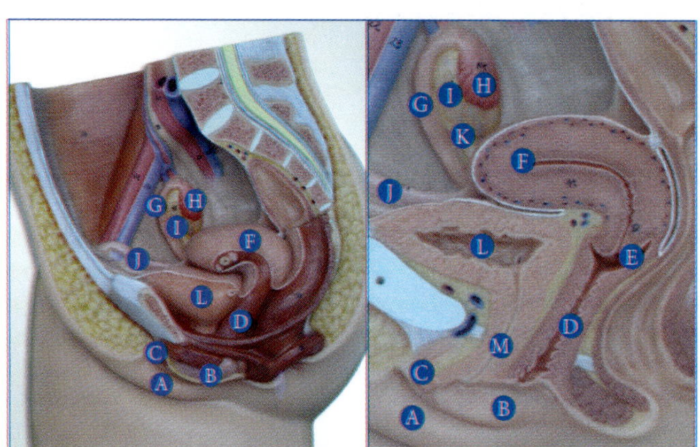

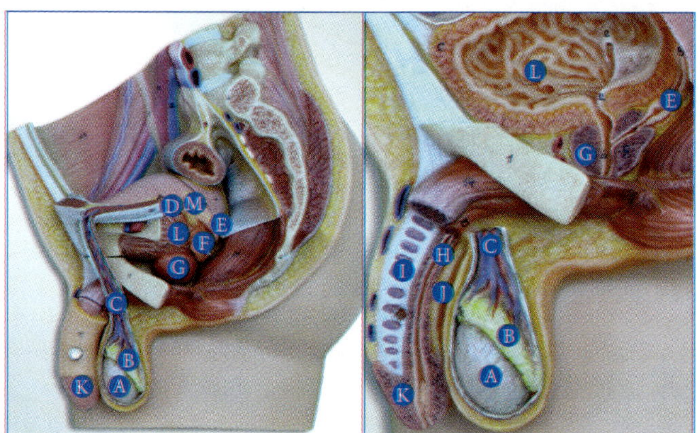

13) Name the labeled structures of the ovary.

A. _______________________________

B. _______________________________

D. _______________________________

E. _______________________________

F. _______________________________

G. _______________________________

H. _______________________________

15) Name the labeled structures of the sperm.

D. _______________________________

E. _______________________________

F. _______________________________

G. _______________________________

H. _______________________________

I. _______________________________

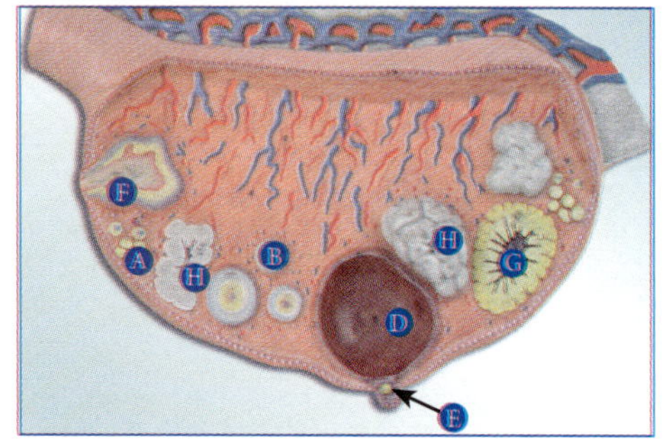

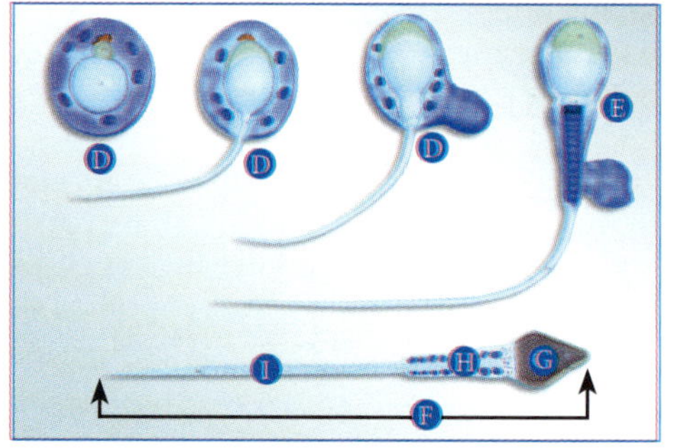

Name: ____

Lab Day/Time: _________ / _________   Section#: _______________

1) T or F : Sperm require capacitation (roughly ten hours) before being able to fertilize an egg.

2) As sperm and egg chromosomes pair up, a __________________ zygote is formed.
   A. Triploid
   B. Diploid
   C. Haploid
   D. Both B. and C.

3) What term is used to describe the fusion of sperm and egg?
   A. Gametion
   B. Fertilizaion
   C. Combination
   D. None of the above

4) T or F : Once sperm reach the uterine tube it is only minutes until the egg is fertilized.

5) The part of a sperm cell that is used to digest a hole through the granulose cells and zona pellucid is called the
   A. Oocyte
   B. Blastocyst
   C. Acrosome
   D. Both B. and C.

6) T or F : When the zona pellucida is penetrated, many sperm can then enter the egg.

7) The cells produced in the first stage of embryonic development are called
   A. Morula
   B. Mesoderm
   C. Embryoblast
   D. Both A. and B.

8) The three germ layers that form tissues and organs are the __________________, __________________, and the __________________.

9) The fluid filled membrane(s) that surround and protect the embryo is (are) called the
   A. Amnion
   B. Chorion
   C. Blastocyst
   D. Both A. and B.

10) T or F : The neural tube will become the brain and the spinal cord in the embryo.

11) The umbilical cord provides the fetus with __________________ and __________________.

12) T or F : Twins can be monozygotic (identical), originated from one single ovum, or dizygotic (fraternal), those originated from two separate ova fecundated by two different sperm cells.

13) Name the labeled events of ovulation, fertilization, and implantation.

    A. _______________________________________
    B. _______________________________________
    C. _______________________________________
    D. _______________________________________
    E. _______________________________________
    F. _______________________________________
    G. _______________________________________
    H. _______________________________________
    I. _______________________________________
    J. _______________________________________
    K. _______________________________________
    L. _______________________________________
    M._______________________________________

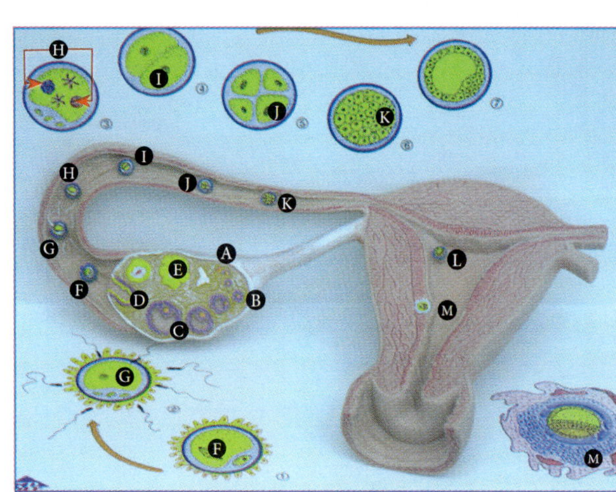

14) Name the labeled structures of the development process.

    A. _______________________________________
    B. _______________________________________
    C. _______________________________________

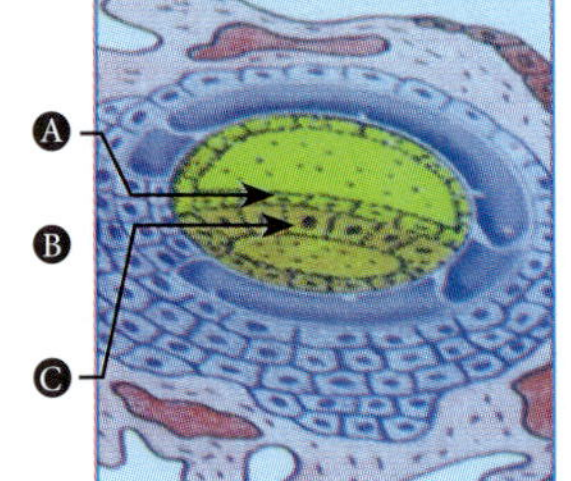

15) Name the labeled structures of the one-month embryo.

    A. _______________________________________
    B. _______________________________________
    C. _______________________________________
    D. _______________________________________
    E. _______________________________________
    F. _______________________________________
    G. _______________________________________
    H. _______________________________________

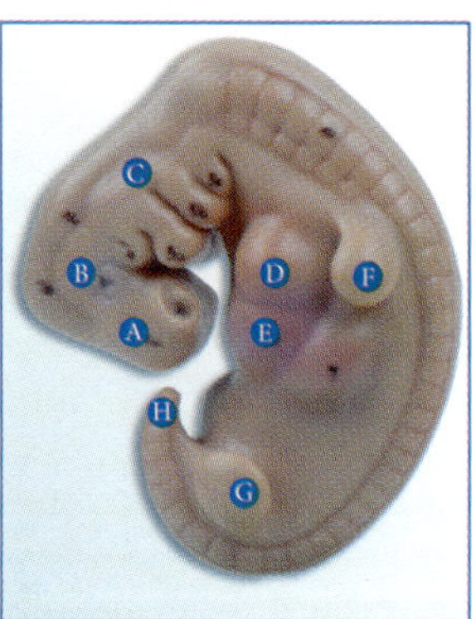

16) Name the labeled structures of the three-month-old fetus and surrounding structures.

    B. _______________________________________
    C. _______________________________________
    D. _______________________________________
    E. _______________________________________
    F. _______________________________________
    G. _______________________________________
    H. _______________________________________

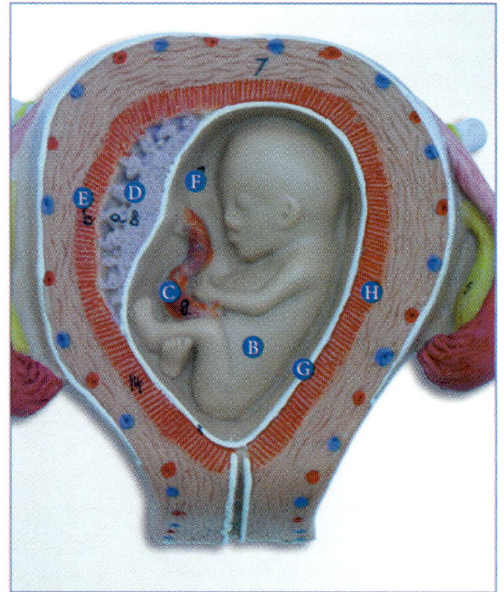

THE ENDOCRINE SYSTEM

You are a first year medical technician at a endocrinology clinic. Thus, it is your job to do the grunt work that the senior technicians don't want to do. This includes reviewing the test results of patients and "flagging" any results that seem abnormal. On each patient's results, the first column lists what hormone was tested for. The column labeled "result" is the concentration of that hormone for that patient. It is your responsibility to compare this hormone level with the hormone level that is physiologically normal for that patient's age and sex. The normal hormone level is listed in the column labeled "reference interval." If the value in the "result" column is out of the range specified in the "reference interval" column, then you need to flag that hormone. Also if the value in the "reference interval" column is an approximate value and not a range, then only flag results that are more than 5% different from the approximate value.

Once you have flagged all of the abnormal hormone levels, another part of your job is to list the organ or tissue in which that hormone is produced. For this you may reference your Human Anatomy & Physiology Laboratory Manual that you saved from your days as an undergrad. Then you are required to include a brief description of the possible physiological effects resulting from that specific hormone abnormality, in the column labeled "Possible physiological effect." Additionally, you are expected to list any anatomical structures that may be directly affected by this abnormal hormone level; for this you may reference your Human AAnatomy & Physiology Laboratory Manual and models in class.

Finally, you are required to look at all of the abnormal hormone levels as a whole and provide a conclusion of possible health implications for that patient. In some cases the patient has come in because of a particular ailment and in other cases they may be getting a routine checkup. Each patient has information relating to their age, sex, height, and weight so be sure to take this into account when forming your final conclusion.

Today's patients include:

Adult male – Complaining of fatigue and feeling dizzy frequently.

Adult female – Routine check up

Female child – Her parents have brought her in as she has only recently started puberty but is having very irregular cycles.

Once the activity is complete, please list any additional applications of the concepts learned in this exercise.

Hormone	Results	Flag	Units	Reference interval		Complete if flagged	
						Tissue or organs of production	Possible physiological effect
Thyroxine (T^4)	3		ug/ml	4.5–12.0			
T^3	20		ug/ml	24–39			
Calcitonin	11		ng/L	~12			
Glucagon	70		pg/ml	40–130			
Insulin	1.4		ug/ml	3.3–22.1			
CRH	34		pg/ml	~ 34			
ACTH	7		pg/ml	6.0–48			
GnRH	6		mcg/dl	5.0–12.0			
FSH	5		pg/ml	2–9.2			
LH	4		mIU/ml	1.5–9			
GHRH	6		mcg/dl	3.0–10.0			
GH	4		ng/L	0.2–14.8			
PRL	3		ng/ml	2.3–11			
TSH	0.3		mIU/L	0.58–4.07			
TRH	1		mIU/L	2.5–20			
Somatosatin	1		mIU/L	0.25–2.12			
ADH	0.2		pg/ml	0.7–3.8			
Aldosterone	8		ng/dl	7.0–30			
Cortisol	45		ng/L	11.0–84			
Estrogen	5		ng/dl	2.0–8.0			
Norepinepherine	176		pg/ml	217–1109			
Epinephrine	43		pg/ml	~50			
Estradiol	1.7		ng/dl	1.5			
Progesterone	11		ng/dl	10.0–11.0			
Testosterone	60		ng/dl	128–430 / 20–39 yrs 95–350 / 40–49 yrs 95–285 / 50–69 yrs 60–240 / 70–79 yrs			

PATIENT 2: ADULT FEMALE   Age: 25   Height: 5′4″   Weight: 120 lbs

Hormone	Results	Flag	Units	Reference interval		Tissue or organs of production	Possible physiological effect
Thyroxine (T^4)	7		ug/ml	4.5–12.0			
T^3	30		ug/ml	24–39			
Calcitonin	4.6		ng/L	~ 5			
Glucagon	66		pg/ml	40–130			
Insulin	12		ug/ml	3.3–22.1			
CRH	67		pg/ml	~ 34	non pregnant		
				~40	1st trimester		
				~153	2nd trimester		
				~847	3rd trimester		
ACTH	53		pg/ml	6.0–48			
GnRH	7		mcg/dl	5.0–12.0			
FSH	3		pg/ml	1.8–11.2	pre and post ovulation		
				6.0–35	mid-cycle		
				30–120	post menopausal		
LH	5		mIU/ml	2.0–9.0	pre ovulation		
				18–49	mid-cycle		
				2.0–11.0	post ovulation		
				20–70	post menopausal		
GHRH	14		mcg/dl	3.0–10.0			
GH	20		ng/L	0.2–14.8			
PRL	25		ng/ml	2.3–19	non pregnant		
				> 19	1st trimester		
				>19	2nd trimester		
				>19	3rd trimester		
TSH	3.4		mIU/L	0.58–4.07			
TRH	10.5		mIU/L	2.5–20			
Somatosatin	5.55		mIU/L	0.25–2.12			
ADH	7.89		pg/ml	0.7–3.8			
Aldosterone	35.1		ng/dl	7.0–30			
Cortisol	40.1		ng/L	10.0–34.0	non pregnant		
				16–60	pregnancy		
Estrogen	16.5		ng/dl	6.0–20.0	pre ovulation		
				16–40	post ovulation		
				~5	post menopausal		
Norepinepherine	2000		pg/ml	217–1109			
Epinephrine	53		pg/ml	~50			
Estradiol	9.6		ng/dl	3.0–10.0	pre ovulation		
				7.0–30.0	post ovulation		
				~1.5	post menopausal		
Progesterone	15.2		ng/dl	10.0–17.0	cycle days 1–6		
				10.0–135.0	cycle days 7–12		
				10.0–1563.0	cycle days 13–15		
				10.0–2555.0	cycle days 16–28		
				~10	post menopausal		
Testosterone	11.7		ng/dl	1.1–14.3			

PATIENT 3: FEMALE CHILD   Age: 12   Height: 62″   Weight: 99 lbs

Hormone	Results	Flag	Units	Reference interval	Complete if flagged	
					Tissue or organs of production	**Possible physiological effect**
Thyroxine (T^4)	8.8		ug/ml	4.5–12.0		
T^3	27		ug/ml	24–39		
Calcitonin	22		ng/L	15–40		
Glucagon	50		pg/ml	40–130		
Insulin	10.2		ug/ml	3.3–22.1		
CRH	338		pg/ml	~ 338		
ACTH	36		pg/ml	6.0–48		
GnRH	0.2		mcg/dl	5.0–12.0		
FSH	0.3		pg	1–4.2		
LH	0		pg/ml	0.02–0.3		
GHRH	6		mcg/dl	3.0–10.0		
GH	9		ng/L	7.5–42		
PRL	1.6		ng/ml	2.3–11		
TSH	3.5		mIU/L	0.58–4.07		
TRH	0.369		pg/ml	~0.3620		
Somatosatin	1.8		mIU/L	0.25–2.12		
ADH	0.9		pg/ml	0.7–3.8		
Aldosterone	77		ng/dl	5.0–80		
Cortisol	8		ng/L	3.0–9.0		
Estrogen	1.4		ng/dl	~2.5		
Norepinepherine	1221		pg/ml	~1251		
Epinephrine	444		pg/ml	~464		
Estradiol	0.1		ng/dl	1.5		
Progesterone	4		ng/dl	10.0–26.0		
Testosterone	1.3		ng/dl	0.2–1.3		

 Human Anatomy & Physiology

Class Activity

Name: _______________________________________

Lab Day/Time: _______ / _______   Section#: _____________

THE SENSES

You are a nurse in a medical refugee camp for a natural disaster in a foreign country. You have a long list of patients with ear, nose, and throat ailments, but because you are short on staff, they have also assigned you patients with eye ailments.

On your list there is a basic ailment, the patient number, and their complaints and/or symptoms. Before the doctor will even see these patients, he wants you to write all the areas that may be affected so he can operate quickly. These should be listed in the column labeled "Anatomical structures possibly affected."

Most importantly, the doctor will not have to meet with patients that need a specialist. The medical treatment the camp has is minimal, and if the ailment will require invasive surgery beyond basic sticking or additional testing with specific equipment, then the patient will need a specialist. If the patient needs a specialist, write "yes" in the appropriate column, and if not, write "no" and a brief explanation of why.

Note: Because the resources are minimal, you will need to err on the side of caution and indicate a wide range of possibly affected structures.

EXAMPLE:

Ailment	Patient #	Complaint/symptoms	Anatomical structures possibly affected	Specialist needed?
Focusing	239	Difficulty focusing on objects	Pupil Lens Suspensory Ligaments Ciliary muscles Iris Cornea	Yes. The exact cause is ambiguous but will require additional testing and possibly invasive treatment.

EAR AILMENTS

Ailment	Patient #	Complaint/symptoms	Anatomical structures possibly affected	Specialist needed?
Cannot hear out of left ear	1131	Pain in skull behind ear. Perfect balance.		
Cannot hear out of left ear	1214	Pulsing feeling directly below back of ear where head meets neck.		
Cannot hear out of right ear	973	Ears feel plugged up but cannot reach with finger/Q-tip.		
Cosmetic damage	1112	Severe burn left side of head.		
Tinnitus	1354	Persistent ringing in both ears but perfect hearing and balance.		
Tinnitus	1175	Persistent ringing in and loss of hearing in right ear, no pain.		

EYE AILMENTS

Ailment	Patient #	Complaint/symptoms	Anatomical structures possibly affected	Specialist needed?
Directional control	135	Pain when looking up and to the right.		
Directional control	446	Pain when looking down and to the left. Difficulty looking in this direction.		
Allergies	698	Eyes are very bloodshot.		
Focusing	239	Difficulty focusing on objects.		
Focusing	1234	Difficulty focusing on objects. Mostly when driving into oncoming traffic.		
Color blind	1111	Difficulty distinguishing color, specifically when looking directly at something, but not in peripheral area.		

Class Activity

Name: _______________________________________

Lab Day/Time: ________ / ________   Section#: _____________

The two-point discrimination test

Follow the procedure on page 3-2. Be sure that both points touch the skin at the same time. (Use calipers very carefully.)

Minimum distance on palm   _______________________

Minimum distance on back of forearm _______________

Minimum distance on fingertip _________________________

Minimum distance on back of neck _________________

Olfactory discrimination

Vial No.	Detection	Recognition	Identification	I.D. with list
1	_______	_______	_______	_______
2	_______	_______	_______	_______
3	_______	_______	_______	_______
4	_______	_______	_______	_______
5	_______	_______	_______	_______
6	_______	_______	_______	_______
7	_______	_______	_______	_______
8	_______	_______	_______	_______
9	_______	_______	_______	_______
10	_______	_______	_______	_______

The Weber test

Sound perception (right or left ear perceived as louder or equal in both ears) _______________________

The Rinne tests

Right ear   _______________________

Left ear _______________________

The Barany test

Accurately detects the direction of rotation? _______________________

After rotating for 10 seconds, can the subject tell which way they are spinning? _______________

Does nystagmus occur after rotation ceases? _____________ In which direction? _______________

Near point determination (use the white accommodation rule)

    Near point of right eye _______________________________________

    Near point of left eye _______________________________________

Color blindness (use Ishihara's test for color deficiency)

Color plates identified incorrectly ______________________________

Snellen test

    Right eye ______________

    Left eye ______________

Astigmatism test

    Astigmatism numbers ___________________________________

Measurement of visual field (use the vision disk)

    Angle where object is perceived ___________________________

    Angle where color of object is perceived ___________________

Determination of the blind spot – The blind spot occurs on an area of the retina that lacks photoreceptors due to the optic nerve and blood vessels that penetrate this area. Cover your left eye and stare at the X below, paying attention to the O on the right in your peripheral vision. Move your head towards or away from the page until the O disappears in your peripheral vision.

Cover your right eye and stare at the O below, paying attention to the X on the right in your peripheral vision. Move your head towards or away from the page until the X disappears in your peripheral vision.

Distance between your right eye and the manual _______________

Distance between your left eye and the manual _______________

       X   O

METHODS TO VIEW THE HEART

There are several non-invasive procedures to produce images of the heart and blood vessels. These have the obvious advantage of viewing the heart without surgery. Different techniques and equipment allow unique perspectives of the heart.

Echocardiography uses ultrasonic waves for real time imaging of the heart and blood flow. It is the most common diagnosis tool for heart problems, needs no preparation and is harmless. Echo is often combined with color Doppler to evaluate blood flow across the heart's valves. It is the most inexpensive and easiest procedure but produces low quality images and is subject to interpretation.

Computed tomography (CT) uses X-rays and a computer to produce still images and short 3-D computer generated clips, showing the beating heart, valve function, and show calcium and blockages in the heart arteries.

Coronary computed tomography angiography (CCTA) uses an injection of contrast material and CT scanning to view the coronary arteries.

Coronary calcium computed tomography (CCCT) shows the location and amount of calcified plaques that narrow or close the arteries that supply blood to the heart.

Magnetic resonance imaging (MRI) uses a magnetic field and radio waves processed by a computer to generate high resolution imagery of the heart and blood vessels. High power magnets are used instead of X-ray radiation and produce the highest detailed imagery.

In this activity we will look at various images of the heart capture by echocardiography, CT, CCTA, CCCT and MRI.

Working in groups, describe each image in regard to all visible anatomical structures of the heart and the type of technique used to produce the image. Draw and label each image in regard to chambers, valves, and vessels of the heart.

9

10

11

12

5

6

7

8

1

2

3

4

Class Activity

Name: ____

Lab Day/Time: _________ / _________   Section#: _____________

BLOOD VESSEL MRA AND DISORDERS

A magnetic resonance angiogram (MRA) is a type of MRI that allows imagery of blood vessels. The procedure may or may not require an injection of contrast material such as gadolinium. It is a non-invasive procedure that can produce images to detect:

Aneurysms—bulging or ballooning of an artery. Tunica intima, media and externa are affected. It may cause the vessels to rupture, causing internal bleeding.

Atherosclerosis—fatty deposits form initially in the tunica intima and eventually affect all layers. Increasing size of the deposit decreases the vessel's lumen diameter and dramatically decreases blood flow.

Dissection of the aorta—blood can become trapped between layers of the aorta and cause an aneurysm and eventually rupture or flow between layers and cause hypoxia or internal bleeding.

Embolism—a blockage of a vessel caused by a thrombus, fat globule or gas bubble. This results in a lack of blood flow to the affected tissue. If it happens in the brain it may cause a stroke.

Thrombosis—the formation of a blood clot (thrombus) within the tunica intima that restricts blood flow. If it breaks free it may result in an embolism.

Stenosis—general term for narrowing of a vessel

In this activity we will look at various images of MRAs.

Draw the arteries that you are responsible to know from chapter 6 from each image .

1

2

3

4

5

6

7

8

9

10

11

12

Name: ____

Lab Day/Time: _________ / _________   Section#: _____________

The Simpsons all have cancer. Specifically, Grampa, Homer, and Bart have all been experiencing symptoms of lymphatic disease, and they have a known history of lymphoma (cancer of the lymphatic system) among the men in their family. Generally, the Simpsons carry two types of cancer.

1. A highly metastatic cancer (I)– Characterized by its ability to spread throughout the major regions of the body where it originates **independent of the directional flow of lymphatic fluid** (head/neck, upper torso, lower torso/pelvis).

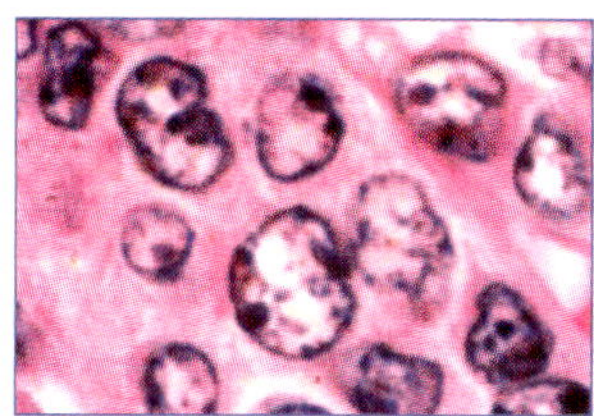

2. A mildly metastatic cancer (II) – Characterized by its ability to **spread ONLY through the components of capillary bed that are neighboring the component in which the cells are discovered (the arteriole or venule).**

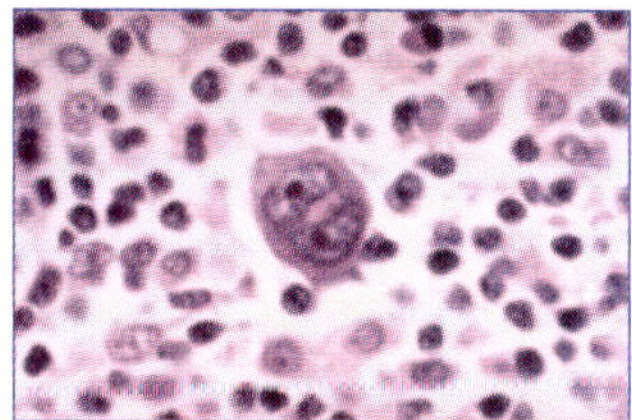

The doctors and technicians at Springfield Hospital have taken cell samples from either the arteriole or the venule of the capillary bed surrounding five of the lymph nodes throughout the Simpsons' bodies. They now need you to interpret the results and make recommendations about which anatomical structures within the lymphatic system should be sampled and tested for the presence of cancer.

Using your lab manuals and these instructions as a reference, identify all the cell types in the histology results for Homer, Grampa, and Bart based on each cell's morphology. Then ID the structures using the models throughout the classroom.

Example of correct data interpretation and activity completion:

Homer (10)	#
Submandibular	4
Deep cervical	5
Reasoning	
The cancer is type I (highly metastatic) so it will spread to all other lymph nodes in that region of the body it was discovered. It was discovered in the occipital lymph node which is in the head/neck so the submandibular and deep cervical may be affected	

Arteriole	Venule	Lymph node
		Occipital

Arteriole	Venule	Lymph node	Homer (10)
		Occipital	
		Axillary	
		Thoracic duct	
		Jugular trunk	
		Inguinal	

Reasoning

GRAMPA SIMPSON'S HISTOLOGY RESULTS

Arteriole	Venule	Lymph node	Grampa (16)
		Occipital	
		Axillary	
		Thoracic duct	
		Jugular trunk	
		Inguinal	

Reasoning

BART SIMPSON'S HISTOLOGY RESULTS

Arteriole	Venule	Lymph node
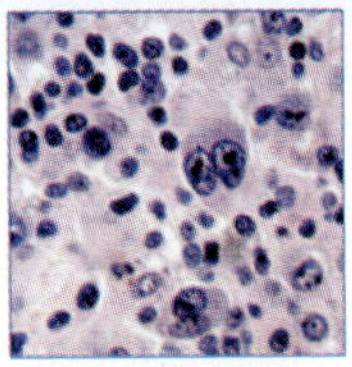	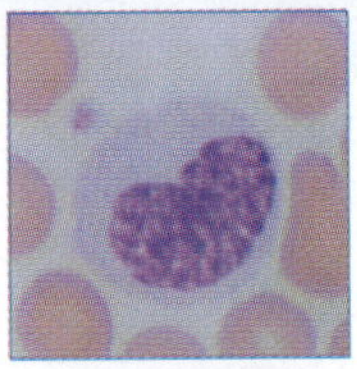	Occipital
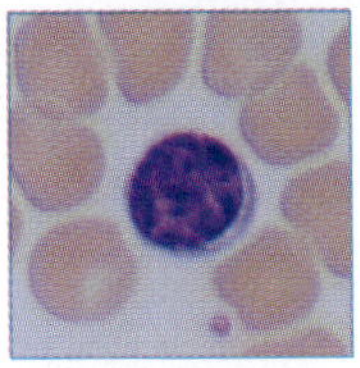	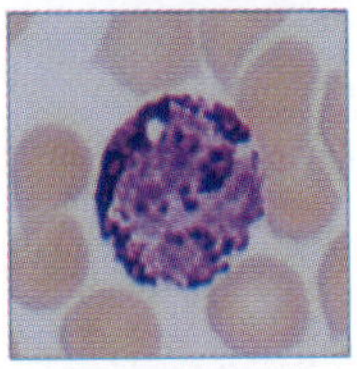	Axillary
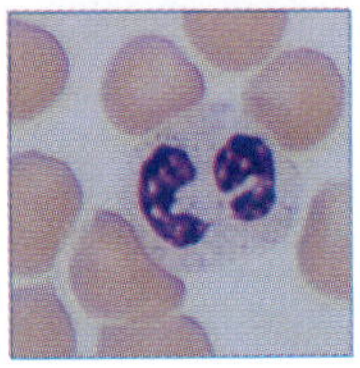	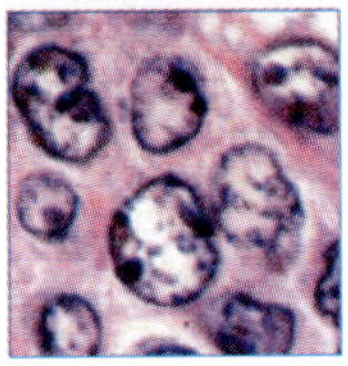	Thoracic duct
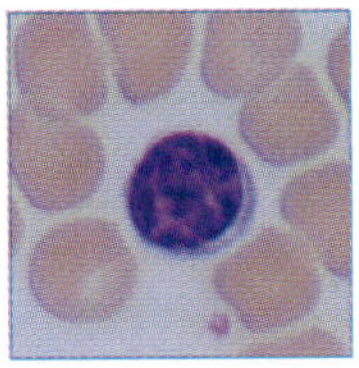	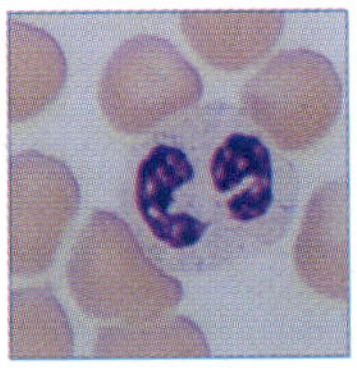	Jugular trunk
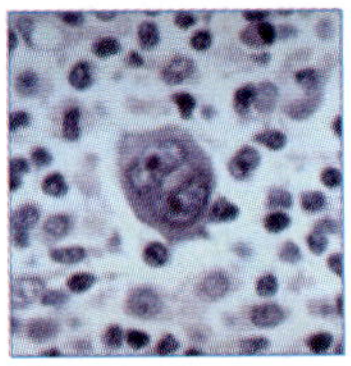	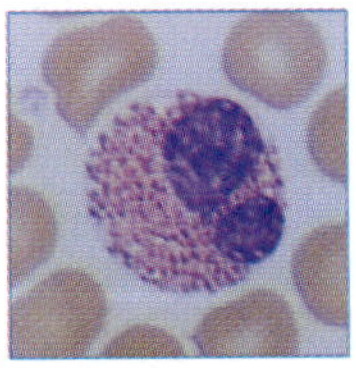	Inguinal

Bart (14)

Reasoning

You are the chief medical examiner for the city of Akron in a high profile case involving a series of murders carried out by a network of organized crime. The Akron law enforcement has long been eluded by the killers because they have not been able to make a connection between the methodology of the killings. Recently, there have been five murders, all of which targeted the respiratory system. Your assistant performed a basic autopsy and filed a standard medical examination form. It is your job to look at the report, consider how exactly the victim died, and create your own "report."

Because this is such a high profile case, the district attorney's office wants a clearly detailed report that includes a list of all anatomical structures whose damage, dysfunction, or failure lead to the victim's death.

VICTIM #1

Anatomical structure

VICTIM #2

Anatomical structure

Anatomical structure

Anatomical structure

Anatomical Structure

The following bold terms are a list of possible conditions and tests that are used for evaluating human kidneys.

Kidney stones

A kidney stone is a solid piece of material that forms in a kidney when substances that are normally found in the urine (such as calcium) become highly concentrated. A persistent sharp pain in the abdomen frequently when urinating is a common symptom of a kidney stone. Kidney stones vary in size and can either remain in the kidney affecting or damaging the major tissues of the kidney as well as the major pathways of urinary excretion. Other kidney stones pass through the kidney and cause pain, discomfort, and possibly damage to the structures of the urinary tract.

Urine salinity

The sodium urine test measures the amount of salt (sodium) in a urine sample. This test is used to determine the kidney's ability to conserve or remove sodium from urine. Normal values are generally 40–220 mEq/L/day. Results greater or less than the reference interval of 40–220 mEq/L/Day indicate possible damage or dysfunction of some or all nephron structures related to the filtration of ions between the salty medulla and the concentrating urine.

High blood pressure

Over time, high blood pressure can damage blood vessels and wear down a harder working heart. When the blood vessels of the kidneys experience damage due to high blood pressure, they can no longer remove wastes and extra fluid from the body effectively. This becomes a dangerous cycle as the buildup of extra fluid also increases blood pressure. Determining if someone has high blood pressure requires the use of a blood pressure cuff. This measurement is described by two numbers. The bottom number, called the diastolic pressure, represents pressure when the heart is resting between beats. The top number, called systolic pressure, shows the pressure when the heart is beating. Normal blood pressure is near 120/80. If the diastolic pressure is greater than 90 or if the systolic blood pressure is greater than 140 then it is considered to be high blood pressure and may indicate damage in blood vessels or vascular structures related to the kidneys.

Proteinuria

This is a condition in which there is an abnormal amount of protein in the urine. Most proteins are too large to move through the kidney's filters into the urine. However, one or more sub components of the kidney's filter/nephrons can be damaged when proteins from the blood leak into urine. Healthy kidneys have a concentration of ~45 mg/mmol, and levels equal to or greater than 100 mg/mmol indicate damage in one or more sub components of the filters/nephrons of the kidney.

Creatinine

Muscle activity generates a number of waste products, one of which is creatinine. The kidneys normally remove this from the blood, but as kidneys process fluid more slowly, the concentration of creatinine rises. The concentration of creatinine is used to calculate the glomerular filtration rate (GFR). The National Kidney Foundation has divided Chronic Kidney Disease into 3 stages based on GFR. The chart below describes how GFR can indicate damage in one or more sub components of the filters/nephrons of the kidney.

GFR	Stage	Description
90 mL/min or more	1	No Kidney damage
30–89 mL/min	2	Kidney damage only at the subcellular structures
Less than 30 mL/min	3	Kidney damage to nephron, capillary, and subcellular

NOTE: If two tests indicate the same affected structure, list that structure under the "anatomical structures affected or damaged" column for each test (meaning you will write it twice).

Clue: (6) terms need to be identified.

Test	Results	Reference interval	Anatomical structures affected or damaged
Pain in abdomen	no	Yes – could be kidney stone No – probably no kidney stone	
Sodium	437	44–220 mEg/L/day	
Blood pressure	120/80	Systolic/Diastolic 120/80 is normal systolic = 140 (high) or diastole = 90 (high)	
Protein concentration	~45	~ 45 mg/mmol	
Creatinine concentration	1	See GFR table.	

PATIENT NAME: Clue: (8) terms need to be identified.

Test	Results	Reference interval	Anatomical structures affected or damaged
Pain in abdomen	yes	Yes – could be kidney stone No – probably no kidney stone	
Sodium	56	44–220 mEg/L/day	
Blood pressure	120/80	Systolic/Diastolic 120/80 is normal systolic = 140 (high) or diastole = 90 (high)	
Protein concentration	45	~ 45 mg/mmol	
Creatinine concentration	1	See GFR table.	

Test	Results	Reference interval	Anatomical structures affected or damaged
Pain in abdomen	no	Yes – could be kidney stone No – probably no kidney stone	
Sodium	60	44–220 mEg/L/day	
Blood pressure	160/100	Systolic/Diastolic 120/80 is normal systolic = 140 (high) or diastole = 90 (high)	
Protein concentration	45	~ 45 mg/mmol	
Creatinine concentration	1	See GFR table.	

Name: ____

Lab Day/Time: _________ / _________   Section#: _____________

                                    Clue: (2) terms need to be identified.

Test	Results	Reference interval	Anatomical structures affected or damaged
Pain in abdomen	no	Yes – could be kidney stone No – probably no kidney stone	
Sodium	177	44–220 mEg/L/day	
Blood pressure	120/80	Systolic/Diastolic 120/80 is normal systolic = 140 (high) or diastole = 90 (high)	
Protein concentration	45	~ 45 mg/mmol	
Creatinine concentration	2	See GFR table.	

Clue: (9) terms need to be identified.

Test	Results	Reference interval	Anatomical structures affected or damaged
Pain in abdomen	no	Yes – could be kidney stone No – probably no kidney stone	
Sodium	177	44–220 mEg/L/day	
Blood pressure	120/80	Systolic/Diastolic 120/80 is normal systolic = 140 (high) or diastole = 90 (high)	
Protein concentration	100	~ 45 mg/mmol	
Creatinine concentration	3	See GFR table.	

CHAPTER 10

Name: _______________________________________

Lab Day/Time: _________ / _________   Section#: _____________

A gastroenterologist is a physician who specializes in diseases of the gastrointestinal tract. Often times these physicians can form a diagnosis using the patient's comments alone, but in order to be certain they use a variety of techniques to diagnose a wide range of problems that patients may have with their digestive tract. In some cases they will use a technique known as manometry to identify exactly what structures in the GI tract need to be treated.

During a manometry procedure, a catheter is run through the nose or mouth into the various regions of the GI tract. The patient then consumes either food or water, and the catheter is then used to record changes in pressure that correlate to the contractions of that anatomical structure. When contractions are abnormally high, they are represented by sharp spikes as shown in the image below.

In some cases of extreme ambiguity, they will use a version of manometry to monitor the GI tract as a whole and then record the duration of time after the food is consumed in which pressure changes occur. Knowing roughly how long it takes food to pass through different features of the GI tract, physicians can then estimate which structures need treated. By knowing what structures are affected, they can form a more conclusive diagnosis of the problem based on the patient's comments and diseases known to affect that structure.

The list below contains common diseases affecting the GI tract:

<u>Acid reflux</u> is a condition in which stomach acids rise up into the esophagus because the valve that separates the stomach contents from the esophagus is faulty. Some symptoms include chest pain, dental/oral erosion, difficulty swallowing, heartburn, hoarseness, and regurgitation.

<u>Irritable bowel syndrome (IBS)</u> is a common condition that occurs in the large intestine. Symptoms of irritable bowel syndrome include abdominal pain, cramping, diarrhea, bloating, and constipation.

<u>Celiac</u> (SEE-lee-ak) disease is a digestive condition triggered by consumption of the protein gluten. Celiac disease can cause abdominal pain and diarrhea. This pain results from a lack of digestive <u>enzymes, zymogens, sodium bicarbonate or bile salts</u>. Eventually, the decreased absorption of nutrients leads to weight loss.

<u>Crohn's disease</u> is an inflammatory bowel disease. This condition results in an inflammation in the lining of the gastrointestinal tract that can create severe diarrhea, abdominal pain, and even malnutrition. Areas affected by this type of inflammation can vary greatly between individuals.

<u>Ulcerative colitis</u> is a chronic inflammatory disease of the rectum and/or large intestine characterized by recurrent episodes of fever, chills, abdominal pain, diarrhea, and in some instances, blood in the stool.

In this week's lab you will be analyzing the results of several patients who have undergone a manometry procedure. In order to assess what anatomical structures need treatment, you will first have to correctly label a timeline of digestion using the table and the timeline provided on the next page. To do this you will need to identify the correct interval of time for EVERY term in the table using arrows or brackets. Once you are certain that your timeline is correct, you will use it as a reference in analyzing the results of the patients. Once you think you know what general part of the GI tract is affected, use your books to identify the specific anatomical features that are affected. Then list all structures in the blank table to the right of the patient's results and proceed to identify those structures on the models in the classroom.

Use the patient's comments, the manometry results, and the list of diseases in the background information to identify which disease or diseases you think may be affecting that patient. Then write this conclusion below that patient's manometry data where it says "possible diseases:"

Clue: With the exception of the terms on page 10-4, as well as "greater omentum" and "appendix," EVERY term in Chapter 10 should be identified once and only once in this activity. The terms on page 10-4, greater omentum, and appendix should not be identified at all for this activity.

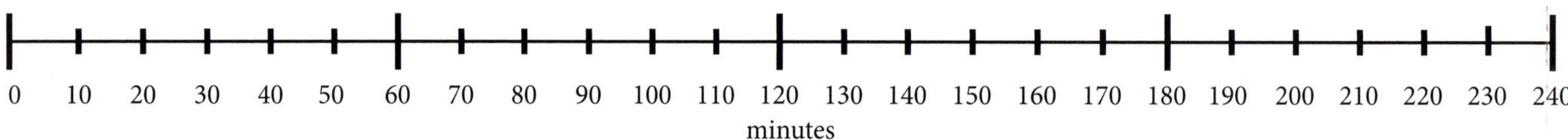

Name	Time (minutes)
Tongue, Teeth, Esophagus	1–5
Stomach	5–65
Duodenum	65–95
Jejunum*	95–160
Ileum*	160–180
Ileocecal Valve, Cecum, Appendix	180–190
Ascending Colon	190–205
Transverse Colon	205–220
Descending/Sigmoid Colon	220–230
Rectum, Anus	230 →

Name	Time (minutes)
Cardiac Region Fundic Region Circular Muscle Layer Oblique Muscle Layer Gastric Rugae Gastric Pit	15–35
Antrum Pylorus Pyloric Sphincter	35–36

*Small intestine and all surface area increasing sub-structures

1. Patient's comments: Pain/heartburn 5–10 minutes after eating.
   Frequent throat clearing throughout the day, especially at night after dinner.

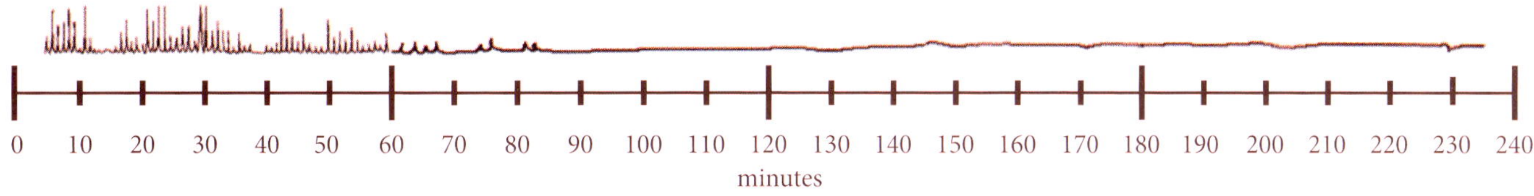

Damaged or dysfunctional anatomical structures	Possible diseases

2. Patient's comments: Pain ~1 hour after eating. Diarrhea approximately every 24 hours, sometimes more often.
   Few if any solid bowel movements. Difficulty gaining/maintaining weight.

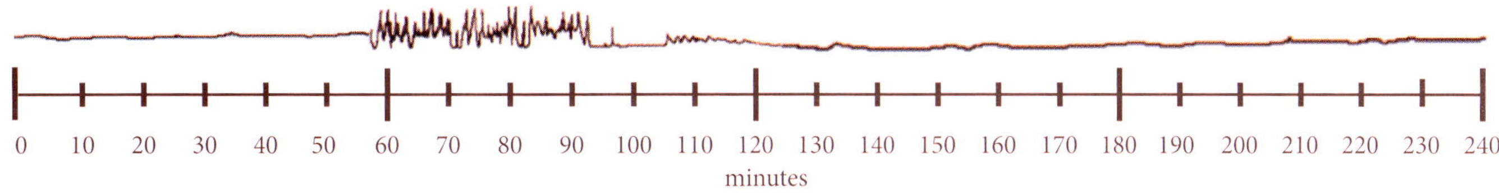

Damaged or dysfunctional anatomical structures	Possible diseases

3. Patient's comments: Abdominal pain, bouts of diarrhea.

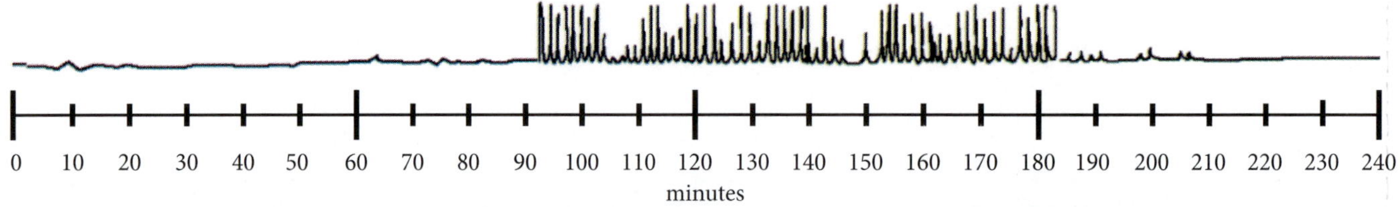

Damaged or dysfunctional anatomical structures	Possible diseases

4. Patient's comments: Daily diarrhea, blood in stool.

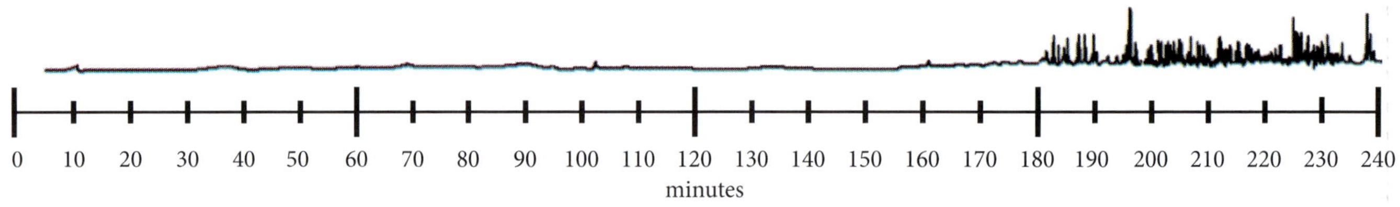

Damaged or dysfunctional anatomical structures	Possible diseases

Class Activity

MALE INFERTILITY

Causes: Low sperm counts and poor sperm quality are the reason for more than 90% of male infertility. Various other anatomical factors account for the remaining 10% of infertility.

1. **Sperm Abnormalities**
   Sperm abnormalities can be caused by a range of factors including chemical exposure, congenital birth defects, disease, and life style habits. However, the cause of sperm abnormalities is often unknown.

   Sperm abnormalities are categorized by whether they affect sperm shape, sperm count, or sperm movement. These abnormalities include:

   Low Sperm Count (Oligospermia) – A sperm count is considered low if it is less than 20 million cells/mL. The complete absence of sperm from the ejaculate is called azoospermia and is the case for 10–15% of male infertility. Obstruction or deformity in any part of the long passages through which sperm pass can decrease sperm counts.

   Poor Sperm Motility (Asthenospermia) – Sperm motility is the sperm's ability to propel itself forward. If the movement is not in a straight line or is slowed, then the sperm have difficulty passing through the cervical mucous or penetrating the egg. Sperm motility is considered average if there is healthy motility in at least 60% of the cells in a sample. Sperm motility is considered abnormal if less than 40% of the cells fail to move in a straight line. This is indicative of problems with the structures of the sperm cells that contribute to its motion.

   Abnormal Sperm Morphology (Teratospermia) – The morphology of the sperm refers to the structure and shape of the cell. Sperm that are abnormally shaped cannot fertilize an egg. If 40% or more of the sperm are not normal in size or shape, it can contribute to infertility and be indicative of anatomical defects in the substructures of the testis.

2. **Retrograde Ejaculation**
   When the anatomical structures near the bladder wall do not function correctly during orgasm and sperm is pushed into the bladder rather than into the urethra, it is called retrograde ejaculation. This is most easily diagnosed by the presence of sperm in the urine.

3. **Structural Abnormalities**
   Structural abnormalities that block or damage the tubes, testis, or other reproductive structures can have a big effect on fertility. From birth, some men have a blockage in the ejaculatory path that affects fertility later in life. A normal volume of semen is about 0.5–1 teaspoons (2.5–5 ml), and volumes less than this can be indicative of structural abnormalities in the tubes transporting the sperm.

1. **Causes of Failure to Ovulate:** Roughly 30% of female infertility can be attributed to ovulatory disorders that are categorized as follows:

   Hormonal Problems – Ovulation depends on a balance of hormones that if disrupted, will impede ovulation. The reasons for anovulation are as follows:

   Failure to Produce Mature Eggs – Half of all anovulation is caused by the production of abnormal structures needed for the implantation and fertilization of eggs. The ovaries' failure to produce such structures is reflected in the increased concentrations of estrogen, testosterone, and LH as well as a decrease in the concentration of FSH.

   Scarred Ovaries – Another cause of failed ovulation is physical damage to the ovaries. Exposure to invasive, extensive, or multiple surgeries for reoccurring ovarian cysts may cause <u>substructures of the ovary to become damaged or scarred</u>. The result is that ovulation will not occur or that fertilized cells will not mature properly. Additionally, infections in the ovaries will have the same effect.

2. **Causes of Poorly Functioning Female/Reproductive Anatomy:** Roughly a fourth of all infertile couples suffer from a range of tubal diseases that can constitute complete tubal blockage or simply mild tubal adhesions. These conditions result in a reduced function of <u>all-major reproductive structures critical to implantation</u> and <u>fertilization</u>.

   Infection – Caused by both viruses and bacteria, infections affecting fertility are usually transmitted sexually and commonly cause inflammation that results in <u>damage and scarring to reproductive structures exposed to an infected partner's epithelium</u>.

   Abdominal Diseases – Colitis is the most common and causes inflammation in the abdominal cavity that can affect the <u>tubes of egg implantation and urination</u>. This can lead to blockages in these orifices and tubes.

   Scarring – Abdominal or pelvic surgery for nodes or cysts can result in adhesions that change the tubes such that eggs will not travel through to be correctly fertilized and implanted.

   Endometriosis – This condition results in excessive growth in the lining of the <u>uterine reproductive structures critical to implantation and fertilization</u>. Diagnosis requires diagnostic laparoscopy. The long-term pregnancy rates for patients with minimal endometriosis and normal anatomy can still be normal.

This week you will be analyzing results from randomly assorted patients at a fertility clinic who have had difficulty conceiving. Using the background information for this activity, your lab manual, and the patient's results, identify which anatomical structures may be the root of their difficulty. List these structures in the "affected or damaged structures column."

Name: ______________________________________

Lab Day/Time: _________ / _________   Section#: ______________

PATIENT 1: FEMALE

Test	Results	Reference	Affected or damaged structures
FSH	29	1.8–35 pg/ml	
LH	4	2–11 mIU/ml	
Estrogen	30	6–40 ng/dl	
Progesterone	0.09	10–2555 ng/dl	
Testosterone	4	1.1–14.3 ng/dl	
Reproductive infection STD or other?	No	Yes/No	
History of appendicitis or colitis	No	Yes/No	
History of cysts or nodules? If yes did you have them surgically removed?	Yes Yes	Yes/No Yes/No	
More than minimal case of endometriosis?	Yes	Yes/No	

PATIENT 2: MALE

Test	Results	Reference			Affected or damaged structures
Motility	5%	poor <40%	fair 40–60%	good >60%	
Count	20 × 10^6	< 20 × 10^6/ml			
Morphology	<60%	poor <60%		good >60%	
Count in Urine	0	~ 0%			
Semen Volume	2.5	2.5–5 ml			

PATIENT 3: MALE

Test	Results	Reference			Affected or damaged structures
Motility	99%	poor <40%	fair 40–60%	good >60%	
Count	1 × 10^6	< 20 × 10^6/ml			
Morphology	76	poor <60%		good >60%	
Count in Urine	0%	~ 0%			
Semen Volume	1.4	2.5–5 ml			

PATIENT 4: MALE

Test	Results	Reference			Overall affected or damaged structures
Motility	42	poor <40%	fair 40–60%	good >60%	
Count	20×10^6	$< 20 \times 10^6$/ml			
Morphology	>60%	poor <60%		good >60%	
Count in Urine	5×10^6	~ 0%			
Semen Volume	2.5	2.5–5 ml			

PATIENT 5: FEMALE

Test	Results	Reference	Affected or damaged structures
FSH	0.2	1.8–35 pg/ml	
LH	21	2–11 mIU/ml	
Estrogen	80	6–40 ng/dl	
Progesterone	1458	10–2555 ng/dl	
Testosterone	5	1.1–14.3 ng/dl	
Reproductive infection STD or other?	No	Yes/No	
History of appendicitis or colitis	No	Yes/No	
History of cysts or nodules? If yes did you have them surgically removed?	No	Yes/No Yes/No	
More than minimal case of endometriosis?	No	Yes/No	

PATIENT 6: FEMALE

Test	Results	Reference	Affected or damaged structures
FSH	22	1.8–35 pg/ml	
LH	5	2–11 mIU/ml	
Estrogen	34	6–40 ng/dl	
Progesterone	1000	10–2555 ng/dl	
Testosterone	5.6	1.1–14.3 ng/dl	
Reproductive infection STD or other?	Yes	Yes/No	
History of appendicitis or colitis	Yes	Yes/No	
History of cysts or nodules? If yes did you have them surgically removed?	No	Yes/No Yes/No	
More than minimal case of endometriosis?	No	Yes/No	

Though there are many causes of a miscarriage, some are more common than others. In an effort to reduce the incidence of future miscarriage, physicians will use the two following tests:

hCG levels– hCG is a growth hormone produced by the embryo during the first month of development. It is essential for the correct development of the embryo's essential substructures, and if hCG levels are low, it means that the embryo and all of its substructures are not developing properly. This imbalanced development may then result in miscarriage.

Days from LMP	hCG range for singleton pregnancy	hCG range for multiple pregnancy
28	9.4–120	9.5–120
33	300–600	200–1,800
36	1,200–1,800	2,400–36,000
40	2,400–4,800	8,700–108,000
45	12,000–60,000	72,000–180,000
70	96,000–144,000	348,000–480,000

Ultrasound machines use sound waves through the vagina or abdomen to get an approximate image of the developing fetus in the uterus and gestational sac.

Cystic placenta: When an overgrowth of placenta prevents the normal development of the fetus.

Chorioamniotic separation: A condition that occurs late in pregnancy at a stage in which the Amnion and Chorion should be fused. The condition can be observed in an ultrasound as the visible separation of these two layers around a fetus.

Endometrial or Uterine abnormalities: Abnormalities usually affect the shape of the uterine cavity. One of the most common abnormalities is fibroids, which are abnormal growths of smooth muscle tissue in the wall of the uterus or in the endometrial lining. These appear as pronounced black spots within what is normally a layer of white, produced by the endometrial lining.

You will be examining some medical records of pregnancies that ended in miscarriage. See if you can identify what caused the miscarriage. Then write the approximate age of the fetus, what anatomical structures may have contributed to the miscarriage, and if there was an embryo or a fetus that was affected.

PATIENT 1: APPROXIMATE AGE _______________

Anatomical structures

PATIENT 2: APPROXIMATE AGE _______________

Anatomical structures

PATIENT 3: APPROXIMATE AGE _______________

Anatomical structures

PATIENT 4: APPROXIMATE AGE _______________

Anatomical structures

Quiz 1 *Chapters 1 & 2*

Name: _________________________________ T.A.'s Name: _______________

Day/Time of Quiz: _____________ / _____________ Lab Time: _______________

1) _______________________   21) _______________________

2) _______________________   22) _______________________

3) _______________________   23) _______________________

4) _______________________   24) _______________________

5) _______________________   25) _______________________

6) _______________________   26) _______________________

7) _______________________   27) _______________________

8) _______________________   28) _______________________

9) _______________________   29) _______________________

10) _______________________   30) _______________________

11) _______________________   31) _______________________

12) _______________________   32) _______________________

13) _______________________   33) _______________________

14) _______________________   34) _______________________

15) _______________________   35) _______________________

16) _______________________   36) _______________________

17) _______________________   37) _______________________

18) _______________________   38) _______________________

19) _______________________   39) _______________________

20) _______________________   40) _______________________

Name: ____ T.A.'s Name: ____________________

Day/Time of Quiz: ______________ / ______________ Lab Time: ____________________

1) ________________________________
2) ________________________________
3) ________________________________
4) ________________________________
5) ________________________________
6) ________________________________
7) ________________________________
8) ________________________________
9) ________________________________
10) ________________________________
11) ________________________________
12) ________________________________
13) ________________________________
14) ________________________________
15) ________________________________
16) ________________________________
17) ________________________________
18) ________________________________
19) ________________________________
20) ________________________________

21) ________________________________
22) ________________________________
23) ________________________________
24) ________________________________
25) ________________________________
26) ________________________________
27) ________________________________
28) ________________________________
29) ________________________________
30) ________________________________
31) ________________________________
32) ________________________________
33) ________________________________
34) ________________________________
35) ________________________________
36) ________________________________
37) ________________________________
38) ________________________________
39) ________________________________
40) ________________________________

Quiz 3 *Chapters 5 & 6*

Name: ________________________________ T.A.'s Name: ________________

Day/Time of Quiz: ______________ / ______________ Lab Time: ________________

1) ________________	21) ________________
2) ________________	22) ________________
3) ________________	23) ________________
4) ________________	24) ________________
5) ________________	25) ________________
6) ________________	26) ________________
7) ________________	27) ________________
8) ________________	28) ________________
9) ________________	29) ________________
10) ________________	30) ________________
11) ________________	31) ________________
12) ________________	32) ________________
13) ________________	33) ________________
14) ________________	34) ________________
15) ________________	35) ________________
16) ________________	36) ________________
17) ________________	37) ________________
18) ________________	38) ________________
19) ________________	39) ________________
20) ________________	40) ________________

Quiz 4 *Chapters 7 & 8*

Name: ____ T.A.'s Name: _________________

Day/Time of Quiz: _________________ / _________________ Lab Time: _________________

1) _______________________   21) _______________________

2) _______________________   22) _______________________

3) _______________________   23) _______________________

4) _______________________   24) _______________________

5) _______________________   25) _______________________

6) _______________________   26) _______________________

7) _______________________   27) _______________________

8) _______________________   28) _______________________

9) _______________________   29) _______________________

10) _______________________   30) _______________________

11) _______________________   31) _______________________

12) _______________________   32) _______________________

13) _______________________   33) _______________________

14) _______________________   34) _______________________

15) _______________________   35) _______________________

16) _______________________   36) _______________________

17) _______________________   37) _______________________

18) _______________________   38) _______________________

19) _______________________   39) _______________________

20) _______________________   40) _______________________

Quiz 5 Chapters 9 & 10

Name: _________________________________ T.A.'s Name: _________________

Day/Time of Quiz: _____________ / _____________ Lab Time: _________________

1) _________________________ 21) _________________________

2) _________________________ 22) _________________________

3) _________________________ 23) _________________________

4) _________________________ 24) _________________________

5) _________________________ 25) _________________________

6) _________________________ 26) _________________________

7) _________________________ 27) _________________________

8) _________________________ 28) _________________________

9) _________________________ 29) _________________________

10) _________________________ 30) _________________________

11) _________________________ 31) _________________________

12) _________________________ 32) _________________________

13) _________________________ 33) _________________________

14) _________________________ 34) _________________________

15) _________________________ 35) _________________________

16) _________________________ 36) _________________________

17) _________________________ 37) _________________________

18) _________________________ 38) _________________________

19) _________________________ 39) _________________________

20) _________________________ 40) _________________________

Name: ____ T.A.'s Name: _____________________

Day/Time of Quiz: _______________ / _______________ Lab Time: _____________________

1) _________________________	21) _________________________
2) _________________________	22) _________________________
3) _________________________	23) _________________________
4) _________________________	24) _________________________
5) _________________________	25) _________________________
6) _________________________	26) _________________________
7) _________________________	27) _________________________
8) _________________________	28) _________________________
9) _________________________	29) _________________________
10) _________________________	30) _________________________
11) _________________________	31) _________________________
12) _________________________	32) _________________________
13) _________________________	33) _________________________
14) _________________________	34) _________________________
15) _________________________	35) _________________________
16) _________________________	36) _________________________
17) _________________________	37) _________________________
18) _________________________	38) _________________________
19) _________________________	39) _________________________
20) _________________________	40) _________________________

Midterm Practical Exam

Name: _________________________________ T.A.'s Name: _________________

Day/Time of Midterm Practical Exam: _________ / _________ Lab Time: _____________

1) _____________________________
2) _____________________________
3) _____________________________
4) _____________________________
5) _____________________________
6) _____________________________
7) _____________________________
8) _____________________________
9) _____________________________
10) ____________________________
11) ____________________________
12) ____________________________
13) ____________________________
14) ____________________________
15) ____________________________
16) ____________________________
17) ____________________________
18) ____________________________
19) ____________________________
20) ____________________________
21) ____________________________
22) ____________________________
23) ____________________________
24) ____________________________
25) ____________________________

26) ____________________________
27) ____________________________
28) ____________________________
29) ____________________________
30) ____________________________
31) ____________________________
32) ____________________________
33) ____________________________
34) ____________________________
35) ____________________________
36) ____________________________
37) ____________________________
38) ____________________________
39) ____________________________
40) ____________________________
41) ____________________________
42) ____________________________
43) ____________________________
44) ____________________________
45) ____________________________
46) ____________________________
47) ____________________________
48) ____________________________
49) ____________________________
50) ____________________________

Final Practical Exam

Name: _______________________________ T.A.'s Name: _______________________

Day/Time of Final Practical Exam: ___________ / ___________ Lab Time: _______________

1) _______________________________ 26) _______________________________

2) _______________________________ 27) _______________________________

3) _______________________________ 28) _______________________________

4) _______________________________ 29) _______________________________

5) _______________________________ 30) _______________________________

6) _______________________________ 31) _______________________________

7) _______________________________ 32) _______________________________

8) _______________________________ 33) _______________________________

9) _______________________________ 34) _______________________________

10) _______________________________ 35) _______________________________

11) _______________________________ 36) _______________________________

12) _______________________________ 37) _______________________________

13) _______________________________ 38) _______________________________

14) _______________________________ 39) _______________________________

15) _______________________________ 40) _______________________________

16) _______________________________ 41) _______________________________

17) _______________________________ 42) _______________________________

18) _______________________________ 43) _______________________________

19) _______________________________ 44) _______________________________

20) _______________________________ 45) _______________________________

21) _______________________________ 46) _______________________________

22) _______________________________ 47) _______________________________

23) _______________________________ 48) _______________________________

24) _______________________________ 49) _______________________________

25) _______________________________ 50) _______________________________

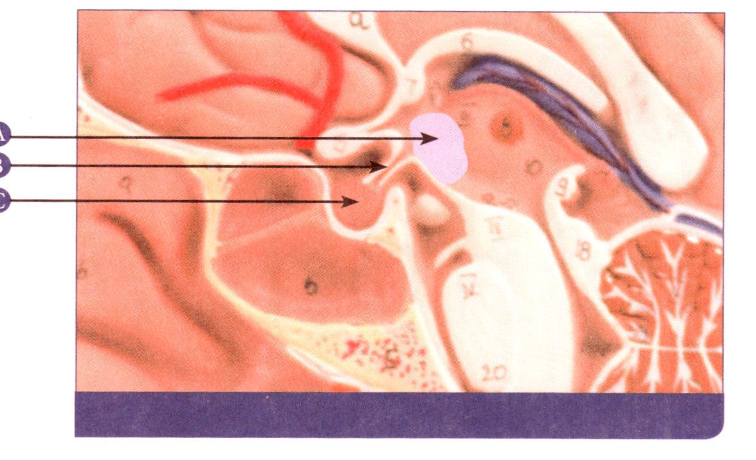

A
B
C
Chapter 1 - Endocrine

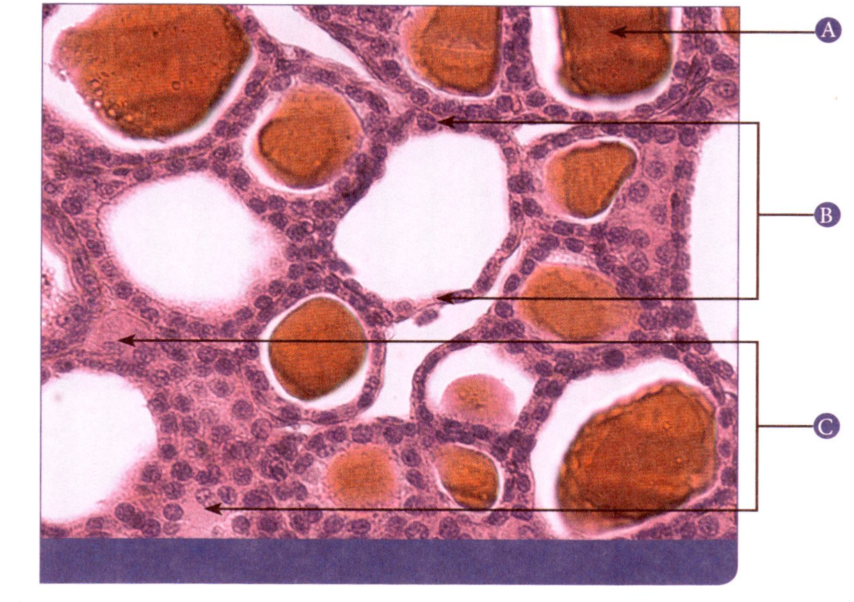

A
B
C
Chapter 1 - Endocrine

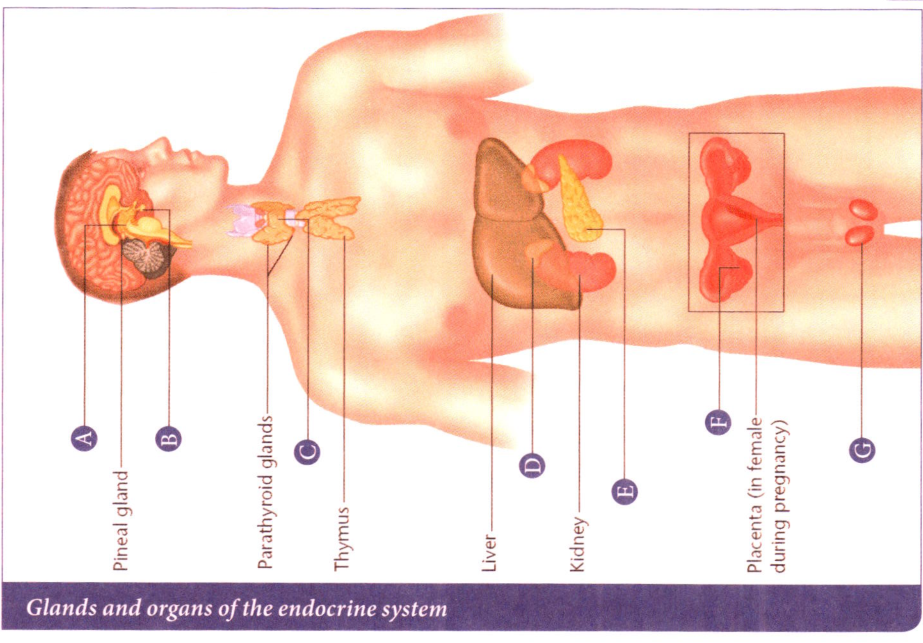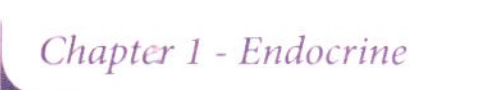

A
Pineal gland
B
Parathyroid glands
C
Thymus
D
Liver
E
Kidney
F
Placenta (in female during pregnancy)
G
Glands and organs of the endocrine system
Chapter 1 - Endocrine

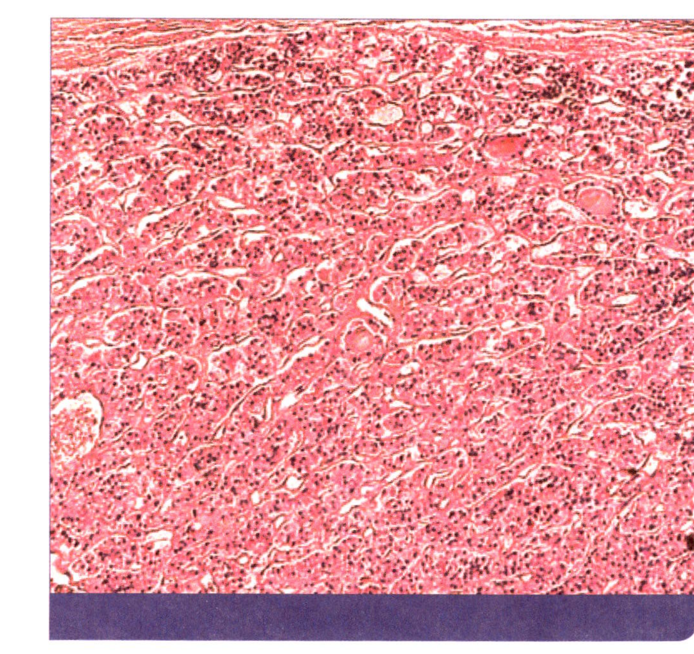

Chapter 1 - Endocrine

A **Colloid of T_3 and T_4**

Hormone	Effects
T_3 (triiodothyroxine)	Raises metabolic rate
T_4 (thyroxine)	Raises metabolic rate

B **Follicle**

C **Thyroid gland's parafollicular cells**

Hormone	Effects
Calcitonin	Bone deposition

A **Hypothalamus**

Hormone	
Corticotropin releasing hormone (CRH)	
Gonadotropin releasing hormone (GnRH)	
Growth hormone releasing hormone (GHRH)	
Prolactin inhibiting hormone (PIH)	
Thyrotropin releasing hormone (TRH)	
Somatostatin	

B **Infundibulum**

C **Adenohypophysis (anterior pituitary)**

Hormone	Target Organ
Adrenocorticotropic hormone (ACTH)	Adrenal gland
Follicle stimulating hormone (FSH)	Ovaries or testes
Luteinizing hormone (LH)	Ovaries or testes
Growth hormone (GH)	All cells
Prolactin (PRL)	Mammary glands
Thyroid stimulating hormone (TSH)	Thyroid gland
Inhibits GH and TSH	

C **Neurohypophysis (posterior pituitary)**

Hormone-stored and released here	Target Organ
Antidiuretic hormone (ADH)	Kidneys
Oxytocin (OT)	Uterus, mammary glands

Notice all the white spaces (sinusoids) that allow hormones to easily enter the bloodstream.

Photomicrograph of the pituitary

A Hypothalamus

B Pituitary gland

C Thyroid glands

D Adrenal gland

E Pancreas

F Ovary (in female)

G Testis (in male)

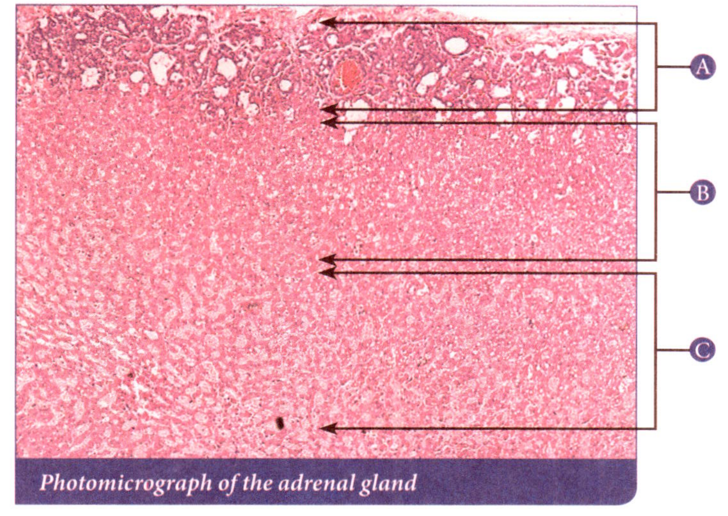

A
B
C
Photomicrograph of the adrenal gland

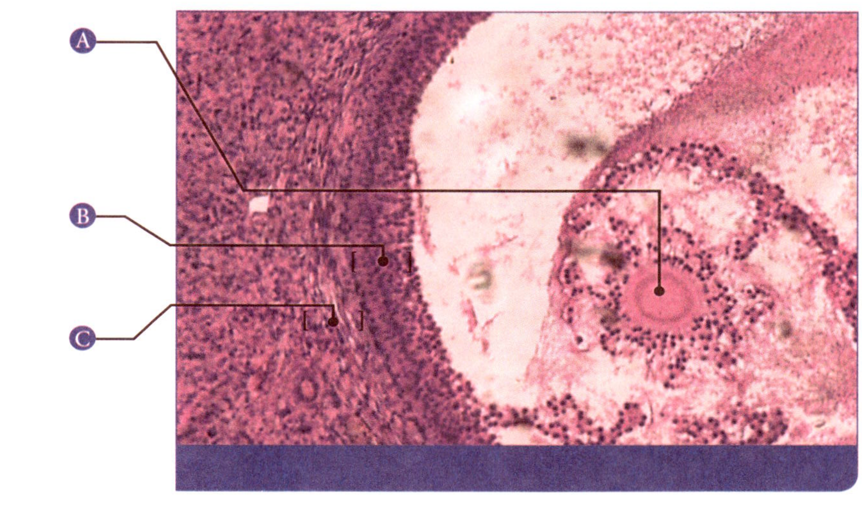

A
B
C

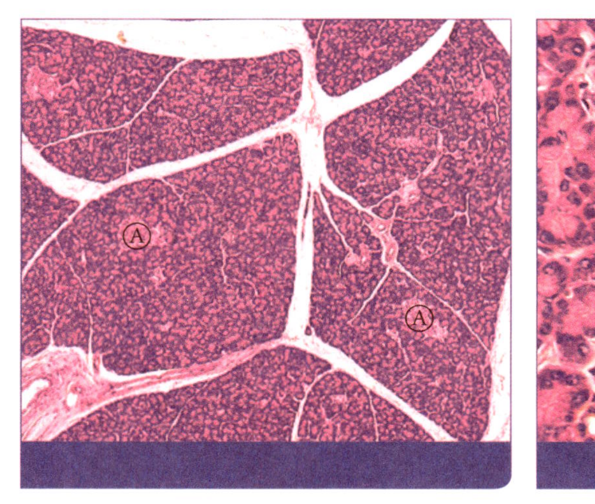

A
A

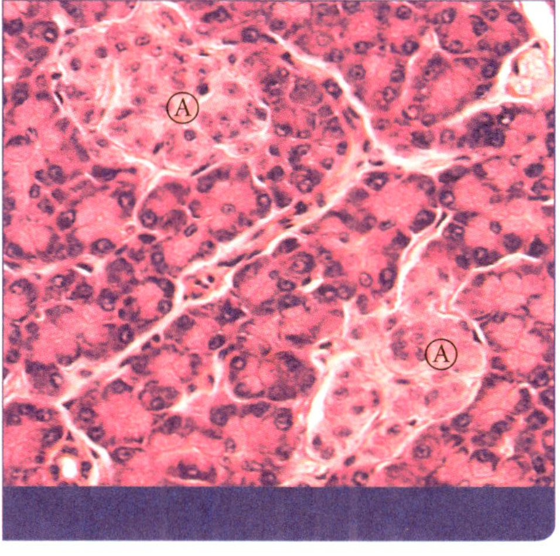

A
A

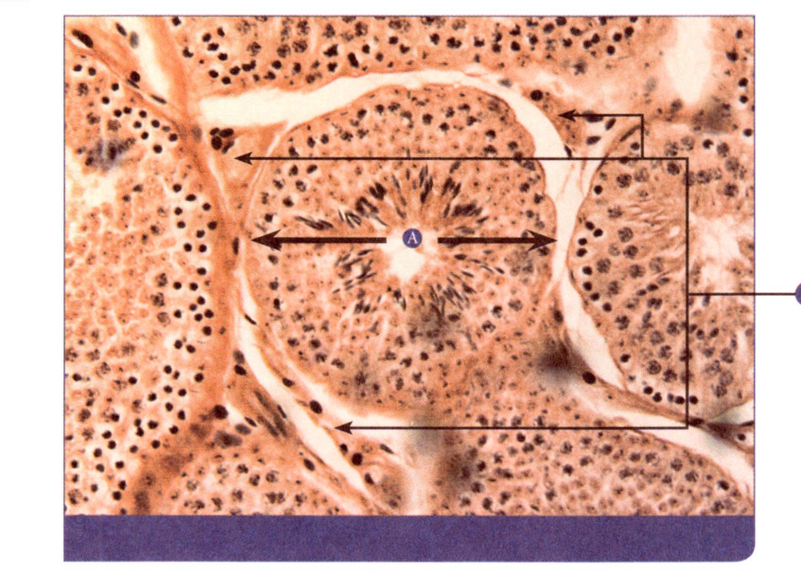

A
B

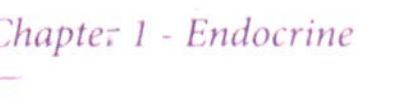

Ovaries

	Hormone	Effects
Ⓐ Oocyte		
Ⓑ Granulosa cells	Estradiol	Prepare reproductive system
Ⓒ Thecal cells	Estradiol	Prepare reproductive system

Adrenal cortex	Category	Hormone	Effects
Ⓐ Zona glomerulosa	Mineralocorticoid	Aldosterone	Water and Na^+ retention, K^+ excretion
Ⓑ Zona fasciculata	Glucocorticoid	Cortisol	Tissue repair, anti-inflammatory
Ⓒ Zona reticularis	Sex steroids	Androgens	Secondary sexual characteristics
		Estrogens	

Ⓐ Seminiferous tubule

Ⓑ **Interstitial cells of testis "cells of Leydig":**

Hormone	Effects
Testosterone	Maleness, growth, libido

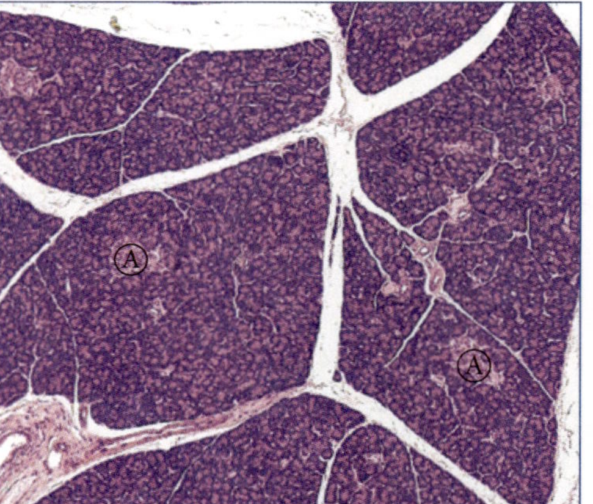

Photomicrograph of pancreatic islets (A) 100×

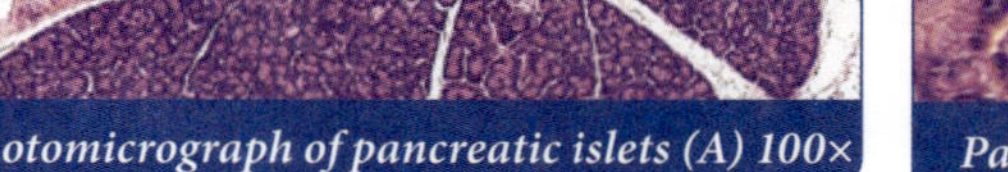

Pancreatic islets (A) 400×

Pancreatic islets

Hormone	Effects
Glucagon (alpha cells)	Raises blood sugar by breaking down glycogen inside of cells, releasing glucose
Insulin (beta cells)	Lowers blood sugar by transporting glucose from bloodstream to inside cells

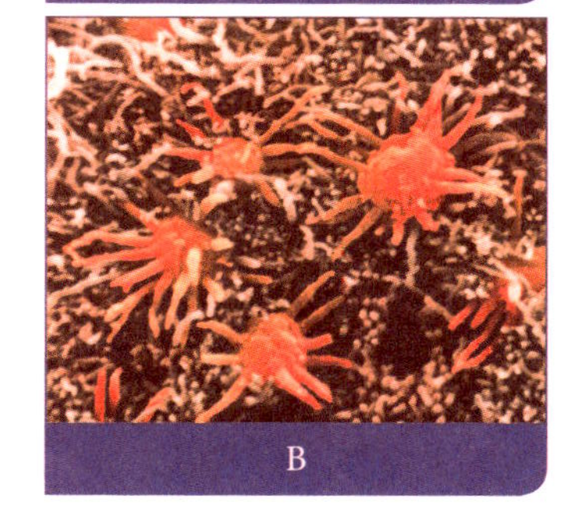

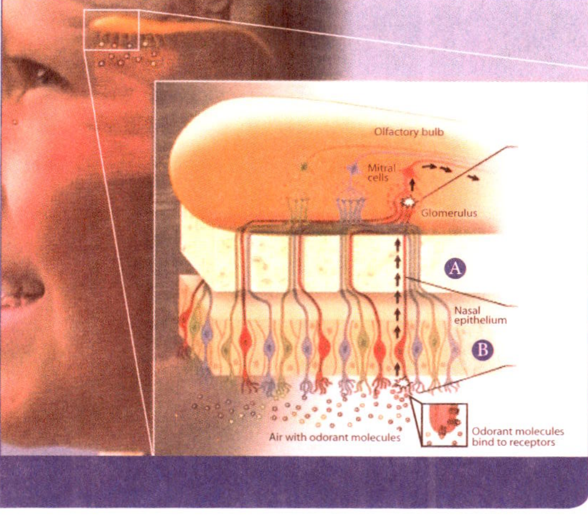

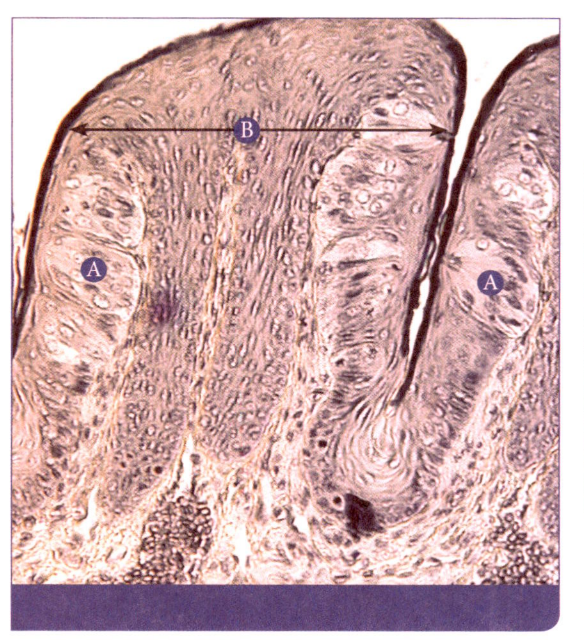

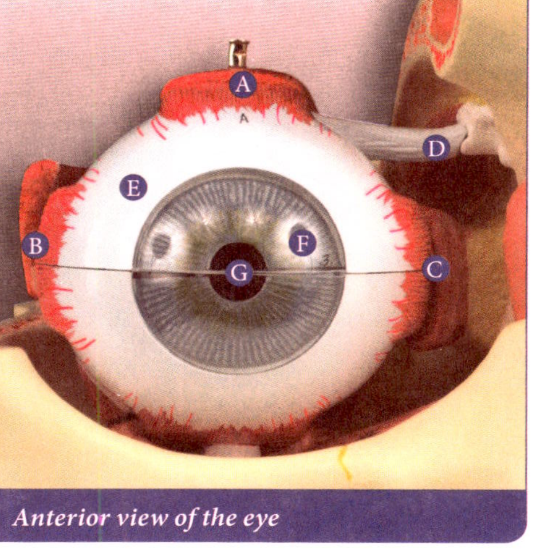

Anterior view of the eye

Anterosuperior view of the eye with the sclera removed

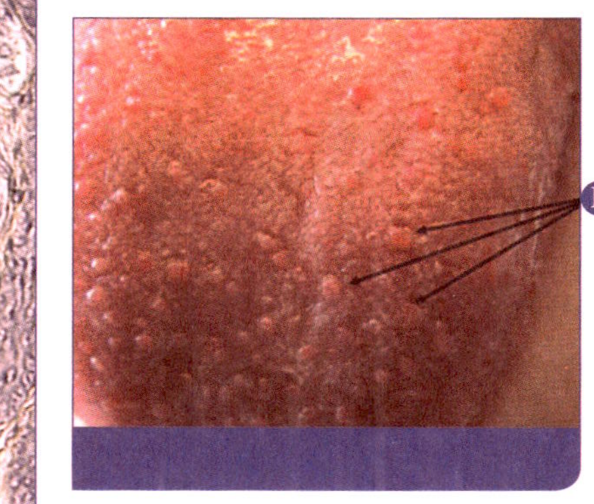

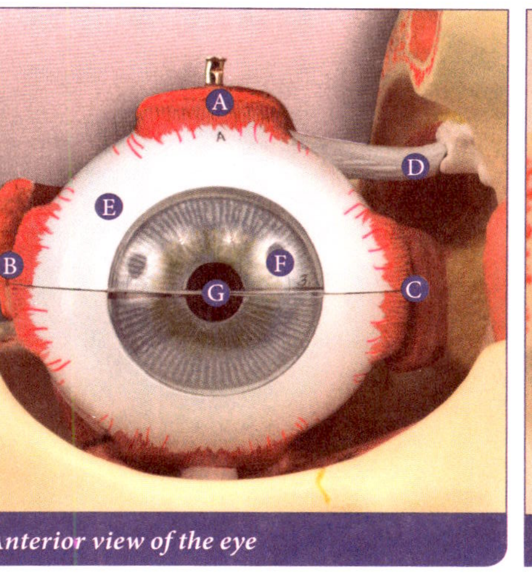

Lateral view of the eye

Ⓐ Superior rectus muscle

Ⓑ Lateral rectus muscle

Ⓒ Medial rectus muscle

Ⓓ Superior oblique rectus muscle's tendon

Ⓔ Sclera

Ⓕ Iris

Ⓖ Pupil

Ⓗ Choroid

The process of olfaction from odor molecules to nervous impulse to the brain

Ⓐ Ethmoid Bone

Ⓑ Olfactory receptor cells

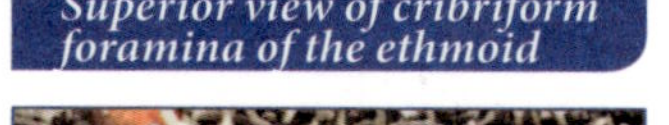

Superior view of cribriform foramina of the ethmoid

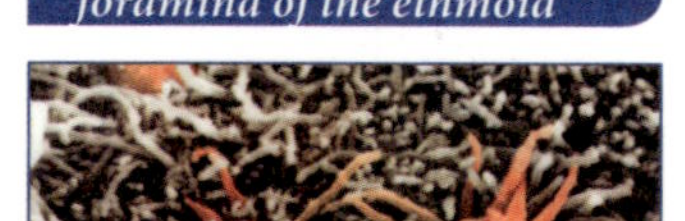

Olfactory hairs (cilia) of olfactory neurons

Ⓐ Superior rectus muscle

Ⓑ Lateral rectus muscle

Ⓔ Sclera

Ⓘ Cornea

Ⓙ Frontal sinus

Ⓐ Taste bud

Ⓑ Lingual papillae

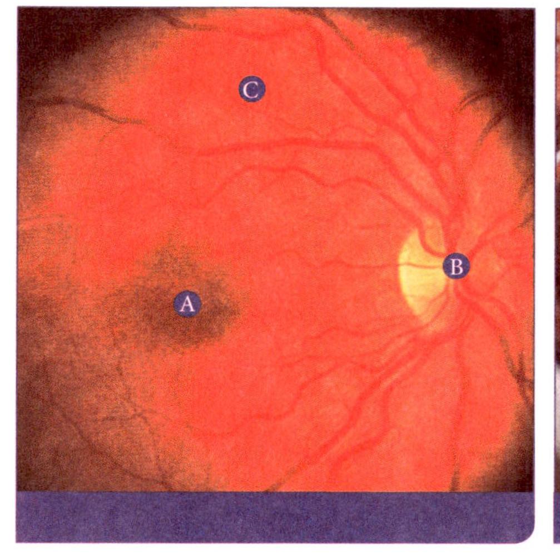

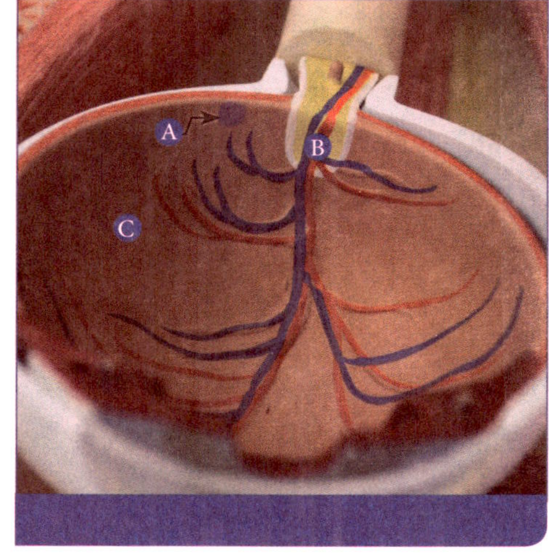

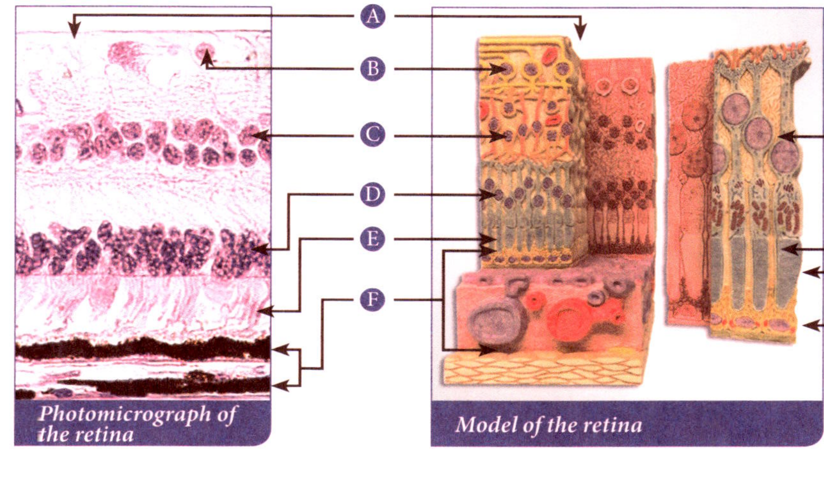
Photomicrograph of
the retina
Model of the retina

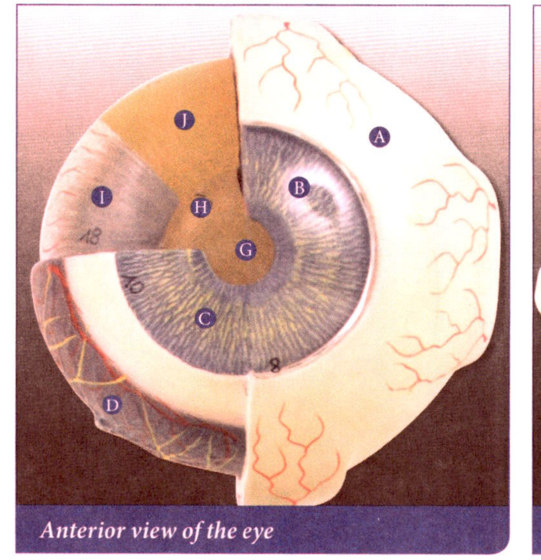
Anterior view of the eye

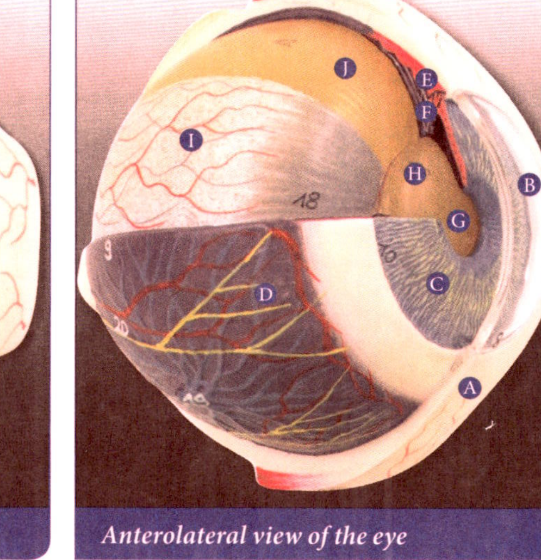
Anterolateral view of the eye

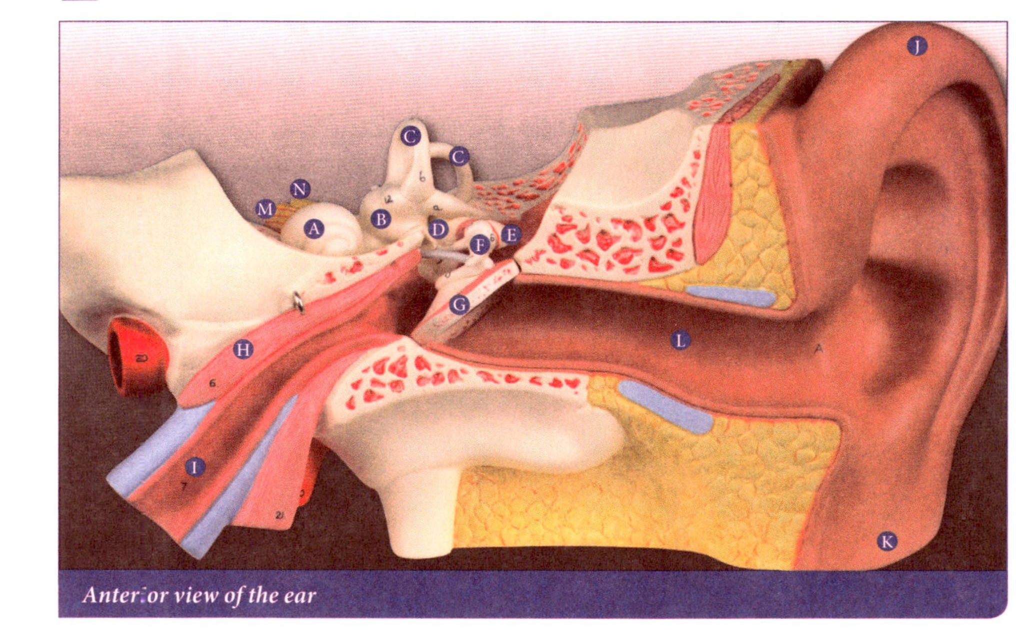
Anterior view of the ear

(A) Direction of light
(B) Ganglion cells
(C) Bipolar cells
(D) Rod and cone nuclei
(E) Rod and cone light sensitive outer segment

(F) Choroid
(G) Rod nucleus
(H) Rod
(I) Cone
(J) Pigment producing cell

(A) **Macula lutea** – Macula = spot; lutea = yellow. A small, yellow patch that an image is focused on when looking at a specific object. It contains a depression called the fovea centralis, where our highest visual acuity (resolution) occurs, as it only contains cone cells, which perceive color, and is absent of rod cells (our night vision). When looking at a faint star in your peripheral vision, it will disappear when you look directly at it.

(B) **Optic disk** – Optic nerve exits and blood vessels enter and exit

(C) **Retina** – Light sensitive, transparent membrane

(A) Cochlea
(B) Vestibule
(C) Semicircular ducts
(D) Stapes
(E) Incus
(F) Malleus
(G) Tympanic membrane

(H) Tensor tympani muscle
(I) Auditory tube
(J) Helix of the Auricle
(K) Lobule of the Auricle
(L) Auditory canal
(M) Cochlear nerve
(N) Vestibular nerve

(A) Sclera
(B) Cornea
(C) Iris
(D) Choroid
(E) Ciliary muscles
(F) Suspensory ligaments
(G) Pupil

(H) Lens
(I) Retina
(J) Vitreous body in posterior cavity

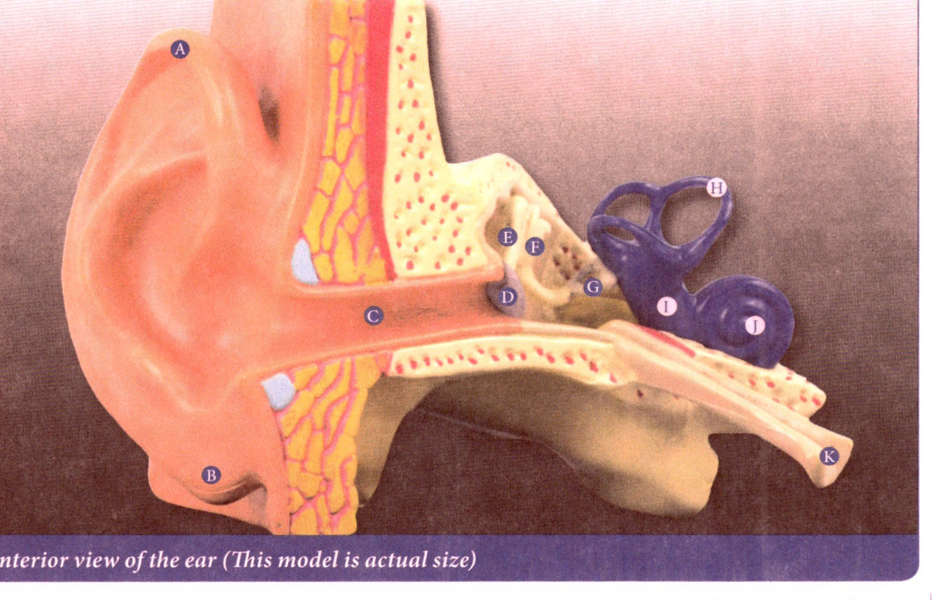

Anterior view of the ear (This model is actual size)

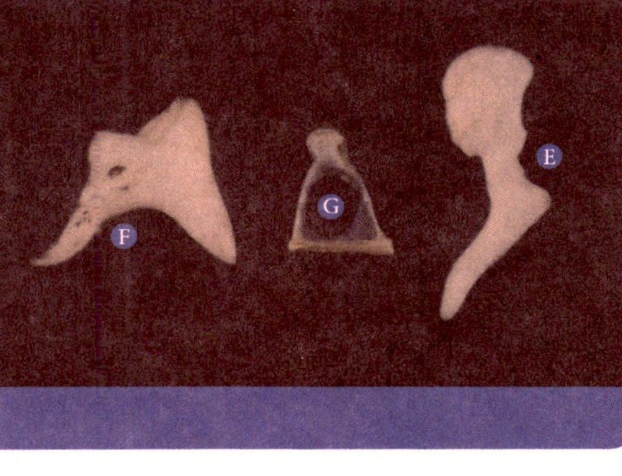

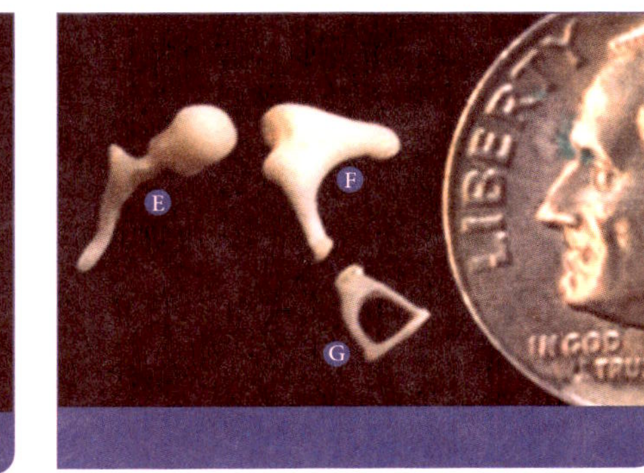

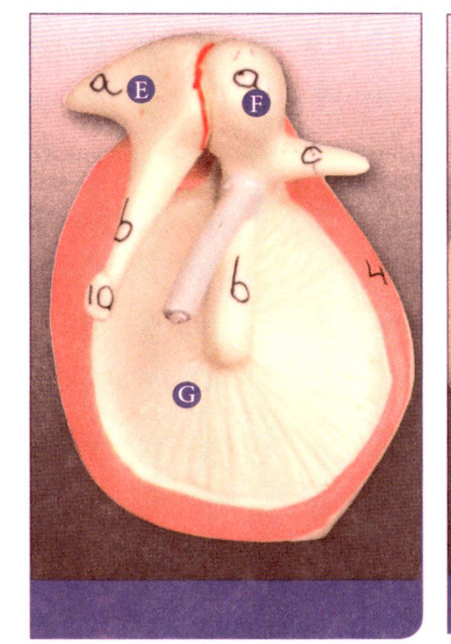

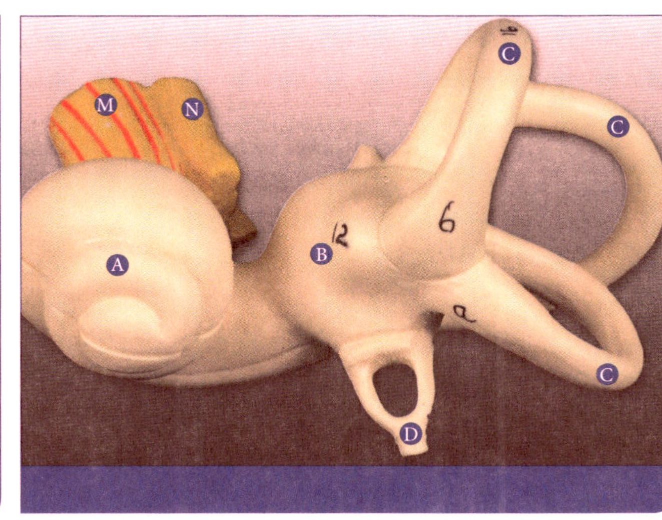

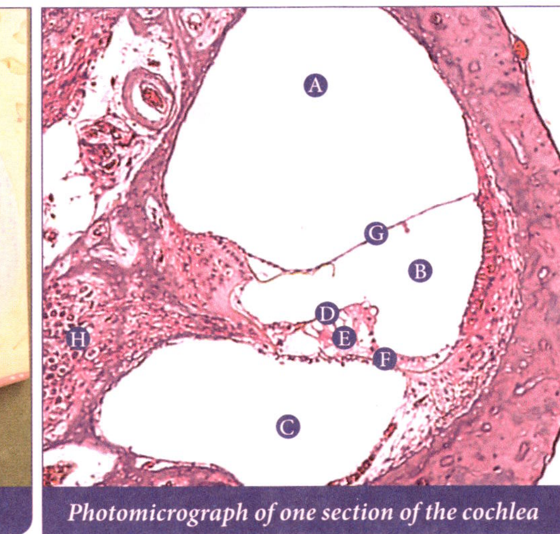

One section of the cochlea

Photomicrograph of one section of the cochlea

(E) Malleus
(F) Incus
(G) Stapes

(A) Helix of Auricle
(B) Lobule of Auricle
(C) Auditory canal
(D) Tympanic membrane
(E) Malleus
(F) Incus
(G) Stapes
(H) Semicircular ducts
(I) Vestibule
(J) Cochlea
(K) Auditory tube

(A) Scala vestibuli
(B) Cochlear duct
(C) Scala tympani
(D) Tectorial membrane
(E) Hair cells
(F) Basilar membrane
(G) Vestibular membrane
(H) Spiral ganglion

(A) Cochlea
(B) Vestibule
(C) Semicircular ducts
(D) Stapes
(E) Incus
(F) Malleus
(G) Tympanic membrane
(M) Cochlear nerve
(N) Vestibular nerve

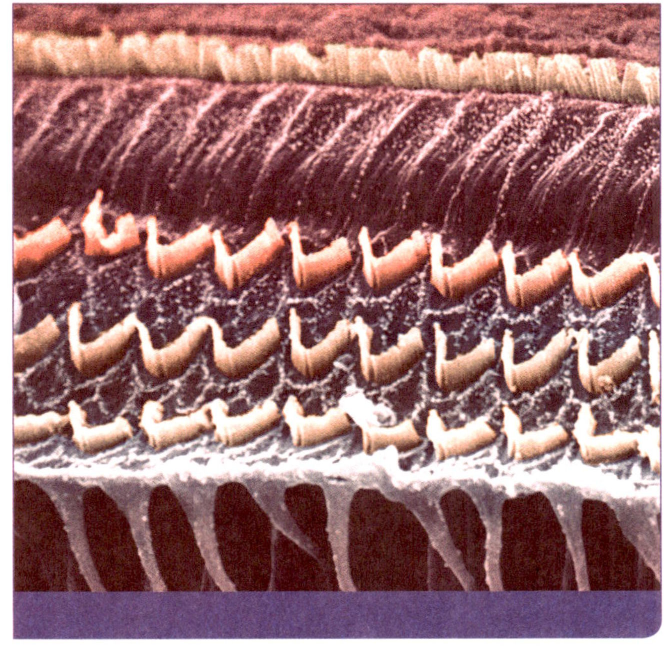

Chapter 2 - Special Senses

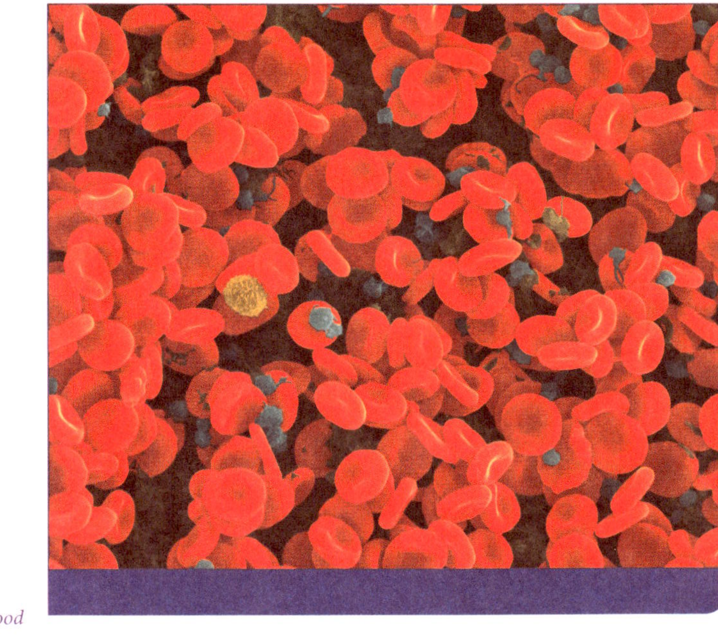

Chapter 4 - Blood

Chapter 2 - Special Senses

Chapter 4 - Blood

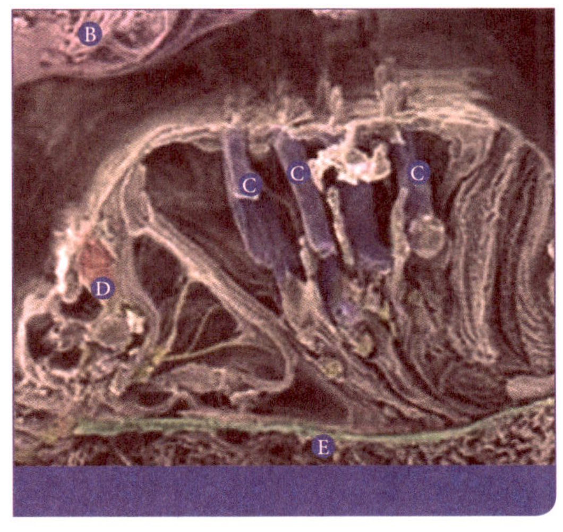

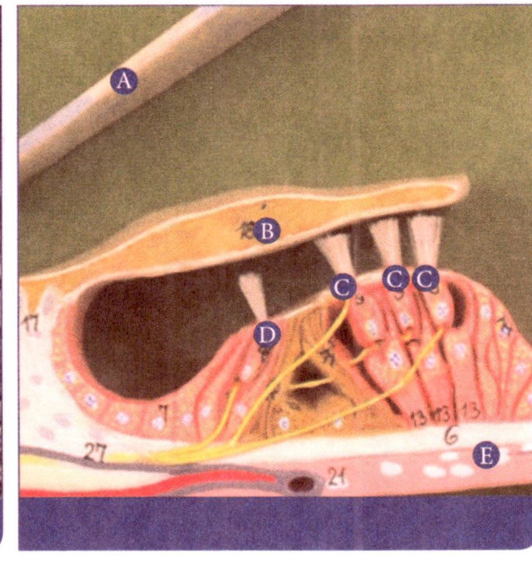

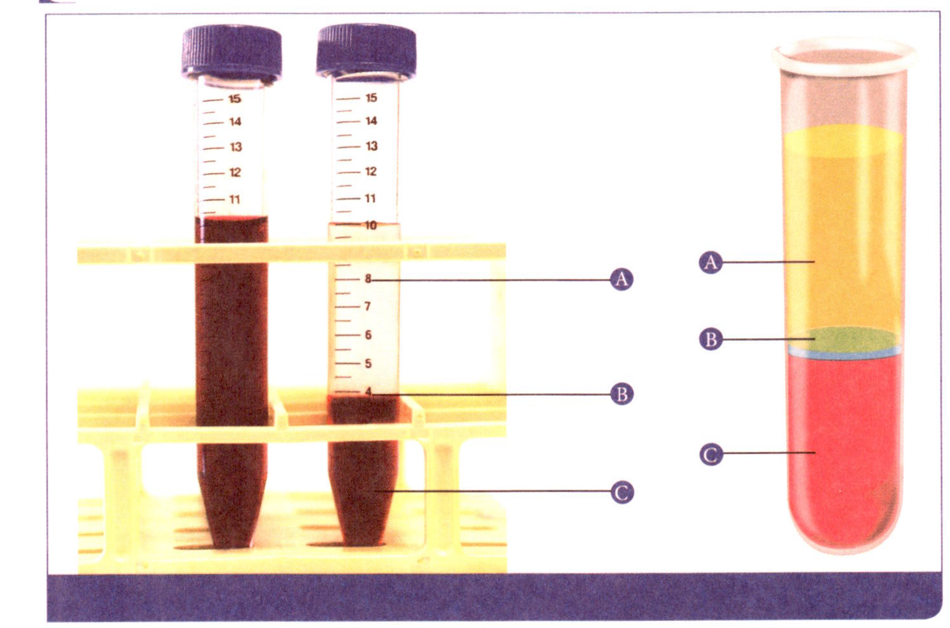

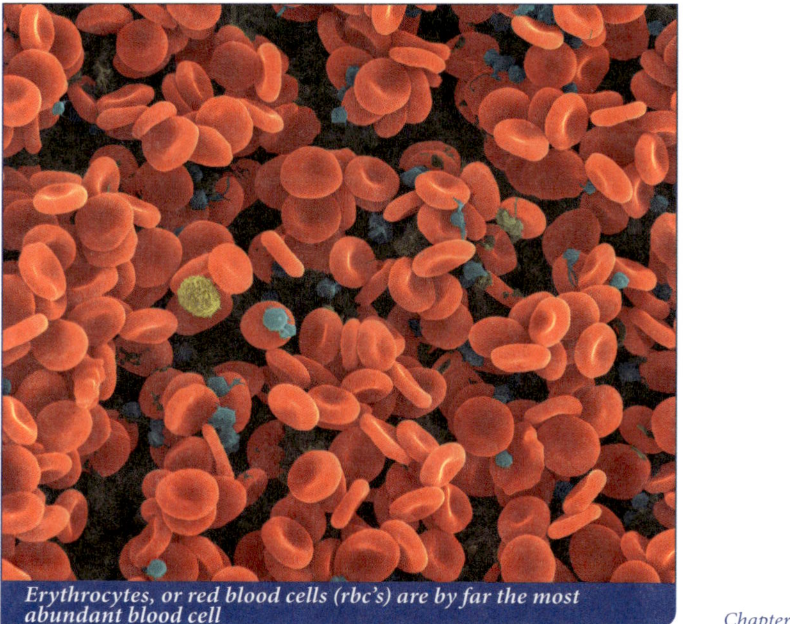

Erythrocytes, or red blood cells (rbc's) are by far the most abundant blood cell

SEM of cochlear hair cells, three rows of v-shaped outer hair cells, and one row of inner hair cells

A Plasma
≈55%

B Leukocytes
and platelets
≈1%

C Red blood cells
(hematocrit)
≈45%

A Vestibular membrane

B Tectorial membrane

C Outer hair cells

D Inner hair cells

E Basilar membrane

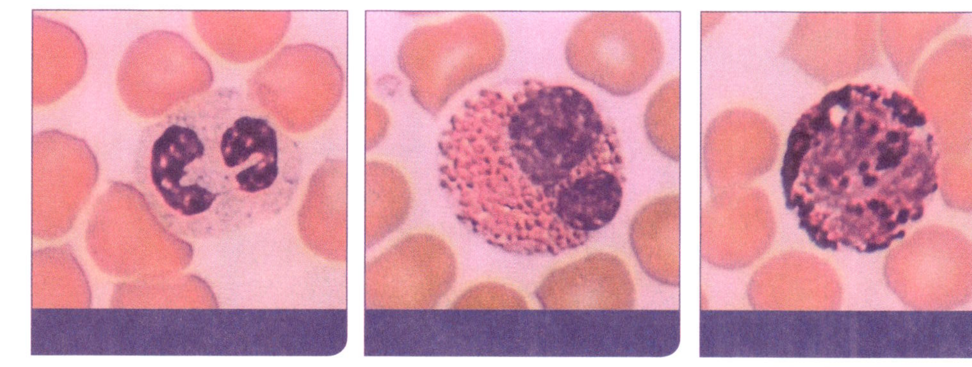

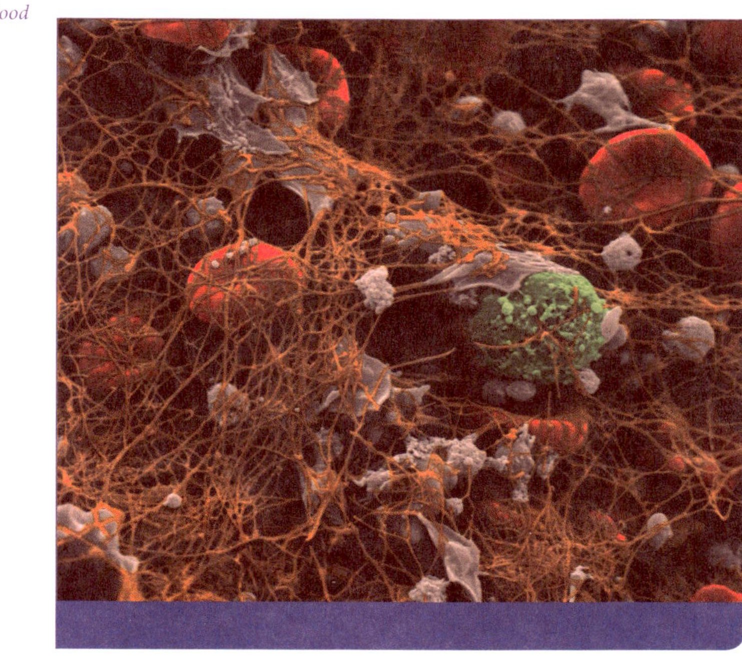

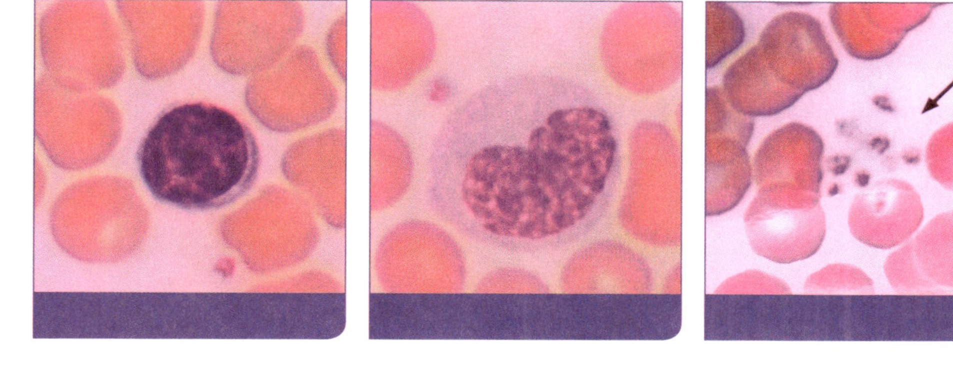

Photomicrograph of blood smear

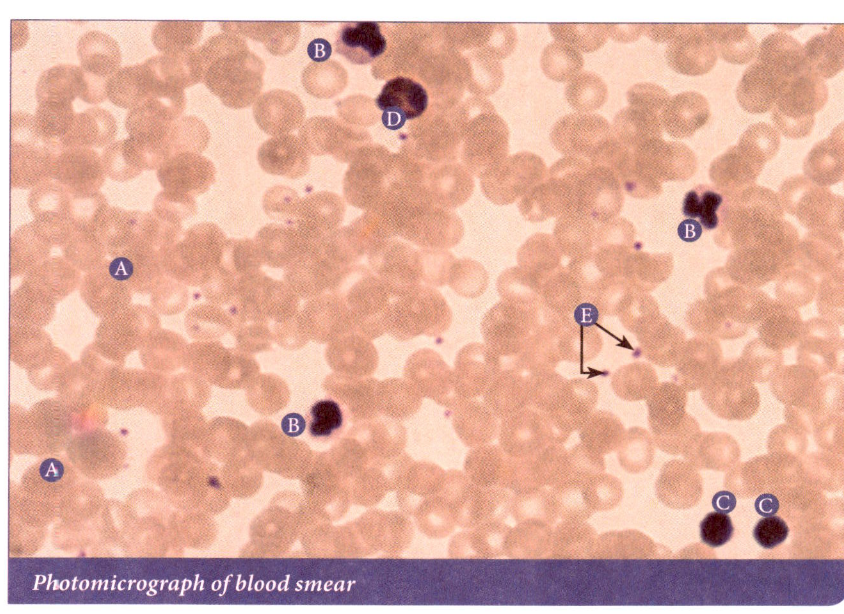

- Ⓐ Erythrocytes (majority of blood cells—99%)
- Ⓑ Neutrophils
- Ⓒ Lymphocytes
- Ⓓ Eosinophil
- Ⓔ Platelets

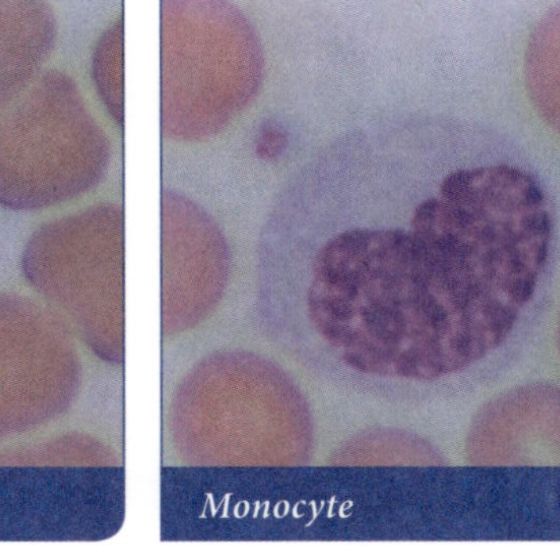

Lymphocyte

25–30% of all leukocytes

Have a variety of types and functions. T cells and B cells are responsible for specific immunity (antibody production in response to non-self antigens. NK cells hunt and kill bacteria, cells with viruses, and cancer cells.

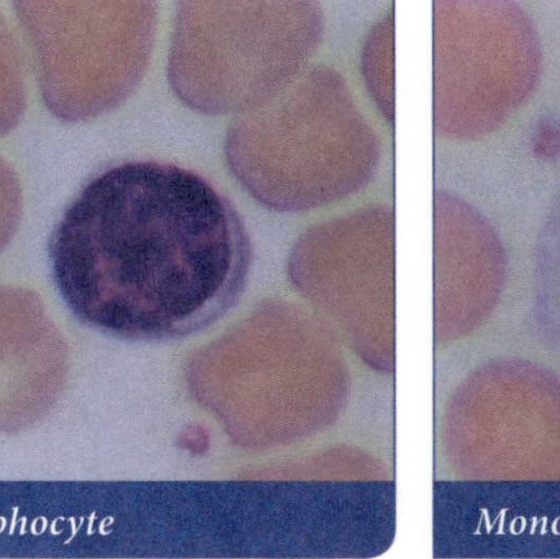

Monocyte

3–8% of all leukocytes

Leave the circulatory system and become wandering or fixed macrophages. Macrophages are phagocytosis machines that reside in connective tissue throughout the body but are no longer classified as leukocytes.

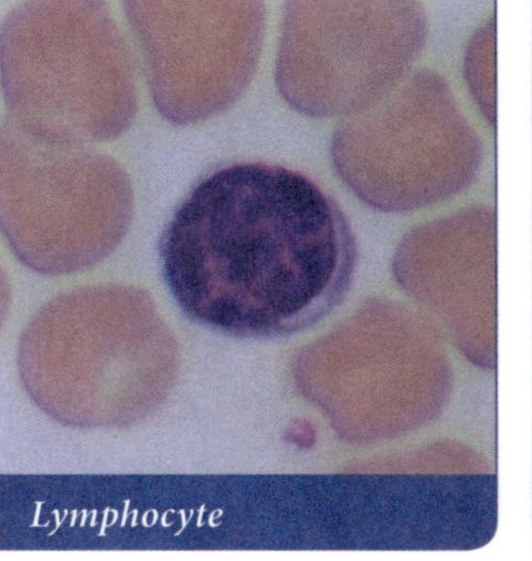

Platelets

Platelets are abundant (250,000,000/µl) but very small (2–3 µm in diameter).

Arise from a huge cell called a megakaryocyte that lives in red bone marrow. Projections extend into sinusoid areas and break off. These fragments of membrane and cytoplasm are platelets.

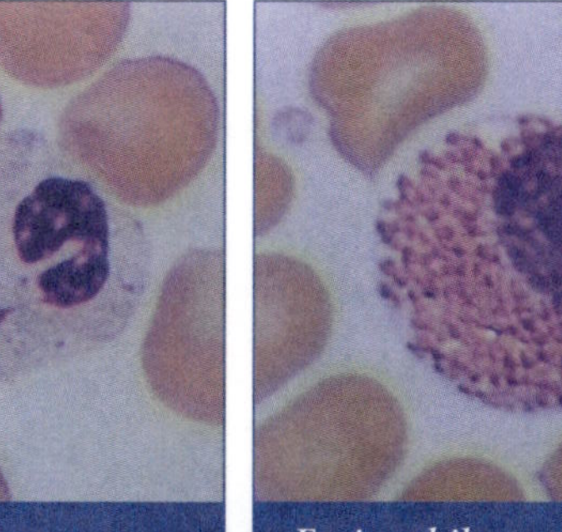

Neutrophil

60–70% of all leukocytes

Bacteria killers. Perform phagocytosis of bacteria and release hydrogen peroxide, hypochlorite, and superoxide into tissue, killing many bacteria and themselves.

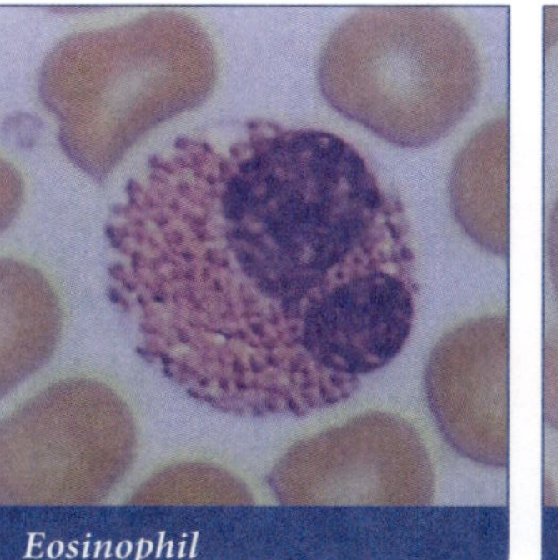

Eosinophil

2–4% of all leukocytes

Produce hydrogen peroxide and superoxide, phagocytize antibody tagged antigen, promote basophil action, and release enzymes that break down histamine.

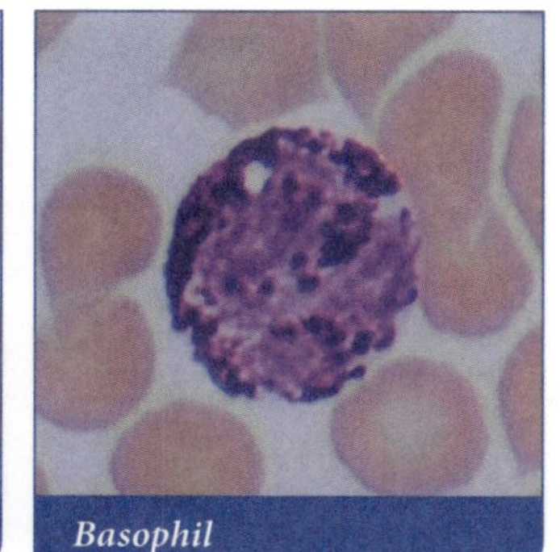

Basophil

0.4–1% of all leukocytes

Secrete histamine (vasodilator) and heparin (anticoagulant) to increase and maintain blood flow and leukotrienes that attract and promote neutrophils and eosinophils.

Coagulation (blood clotting)

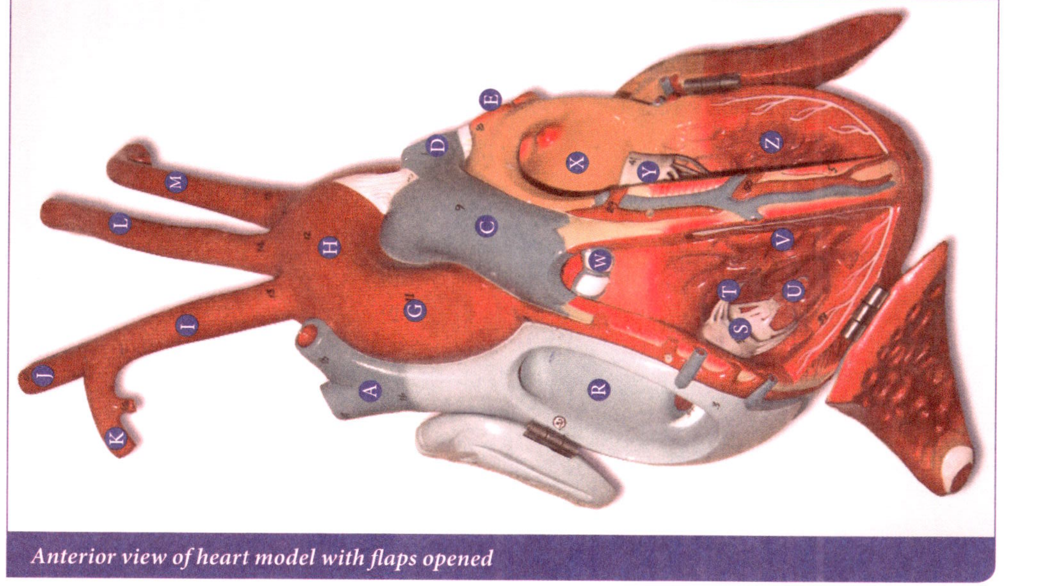

Anterior view of heart model with flaps opened

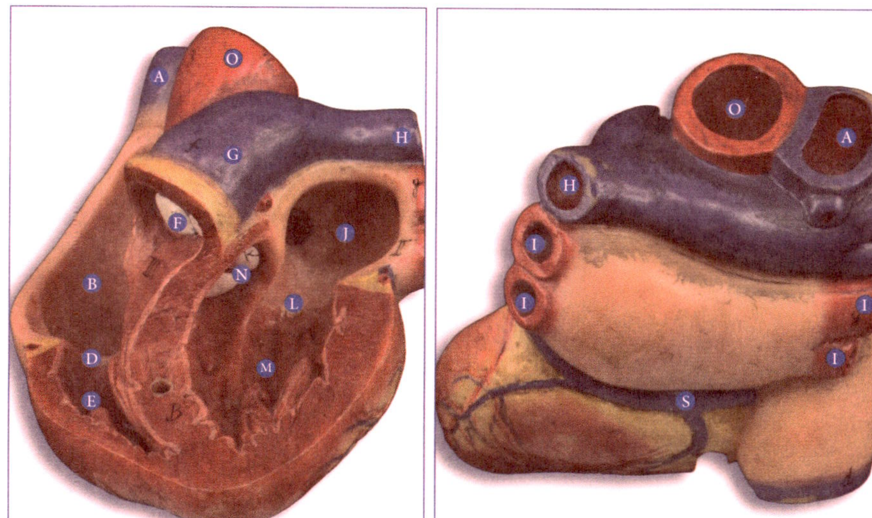

Anterior view of heart model

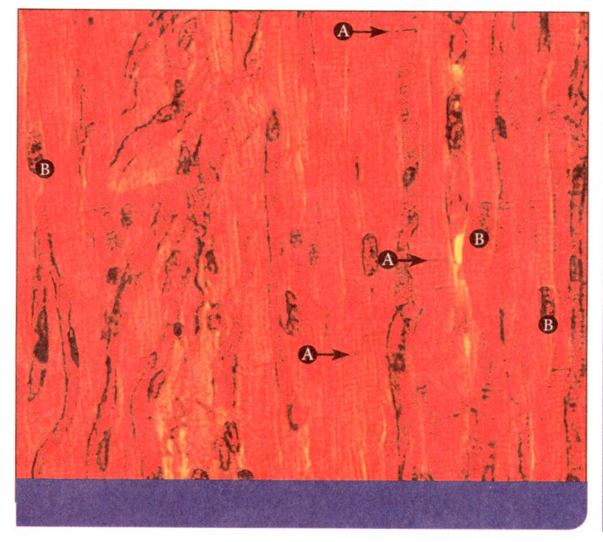

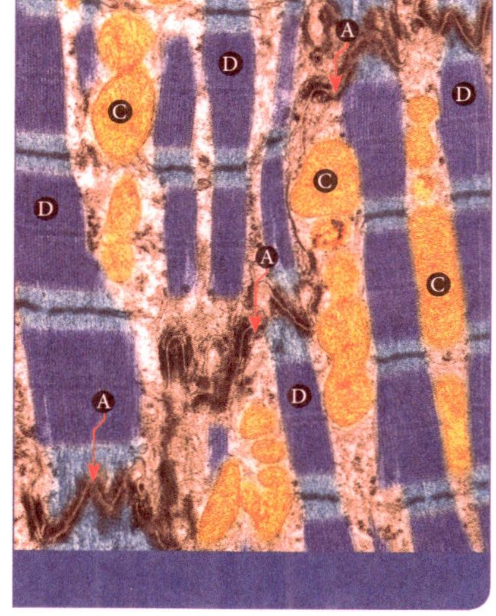

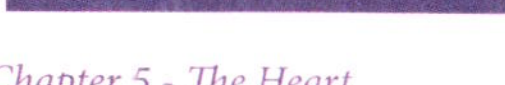

Anterior view of half of the heart model

Posterior view of heart model

A Superior vena cava
B Right auricle
C Pulmonary trunk
D Pulmonary artery
E Pulmonary veins
F Left auricle
G Aorta
H Aortic arch
I Brachiocephalic trunk
J Right common carotid artery
K Right subclavian artery
L Left common carotid artery
M Left subclavian artery
N Anterior interventricular artery
O Great cardiac vein
P Right coronary artery
Q Circumflex artery

A Superior vena cava
C Pulmonary trunk
D Pulmonary artery
E Pulmonary veins
G Aorta
H Aortic arch
I Brachiocephalic trunk
J Right common carotid artery
K Right subclavian artery
L Left Common carotid artery
M Left subclavian artery
R Right atrium
S Tricuspid (right AV) valve
T Chordae tendineae
U Papillary muscle
V Right ventricle
W Pulmonary semilunar valve
X Left atrium
Y Bicuspid (mitral or left AV) valve
Z Left ventricle

A Superior vena cava
B Right atrium
D Tricuspid (right AV) valve
E Right ventricle
F Pulmonary semilunar valve
G Pulmonary trunk
H Pulmonary artery
I Pulmonary veins
J Left atrium
L Bicuspid (mitral or left AV) valve
M Left ventricle
N Aortic semilunar valve
O Aorta
S Coronary sinus

Photomicrograph of cardiac muscle fibers 400×

EM of cardiac muscle fibers 25,100×

A Intercalated disks
B Nuclei
C Mitochondria
D Actin and myosin filaments

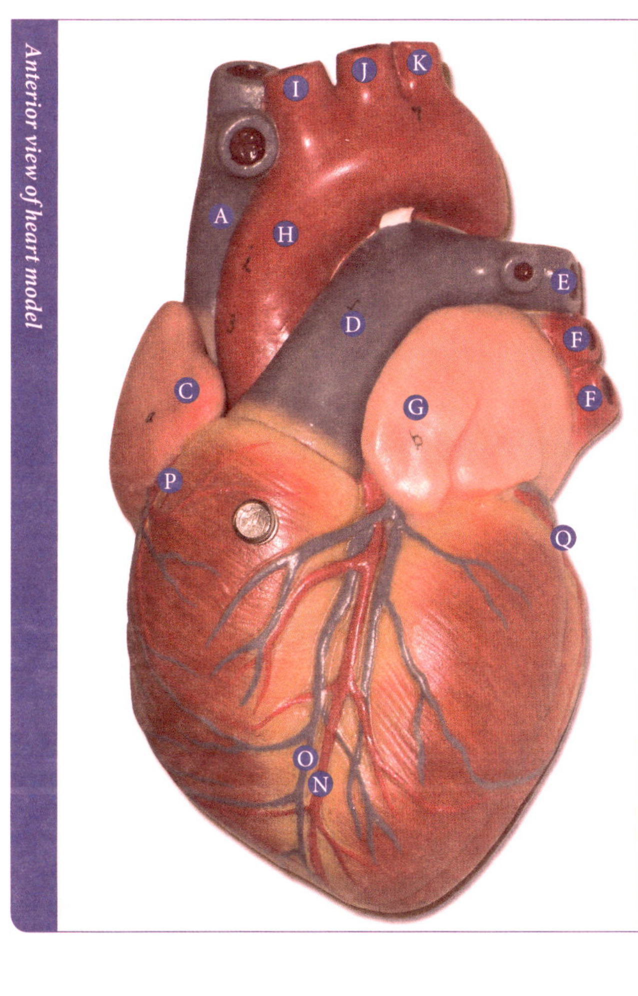

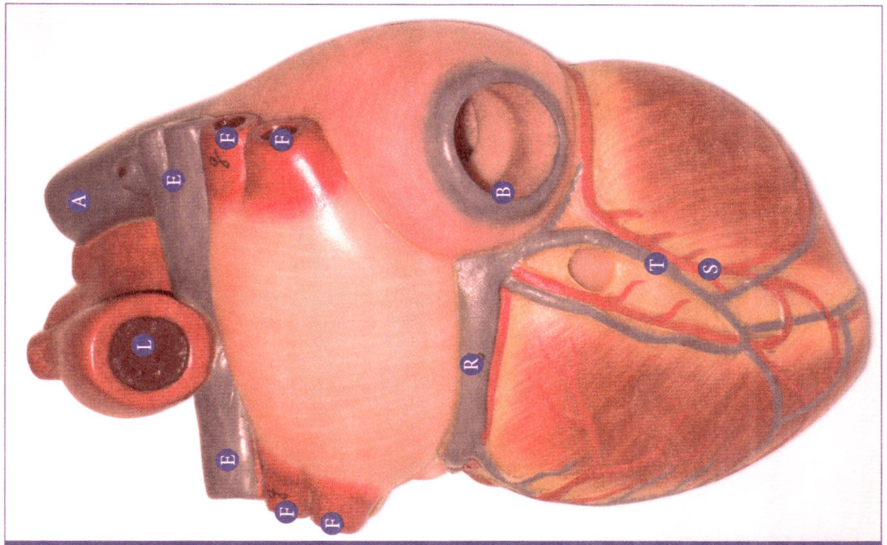

Anterior view of heart model

Chapter 5 - The Heart

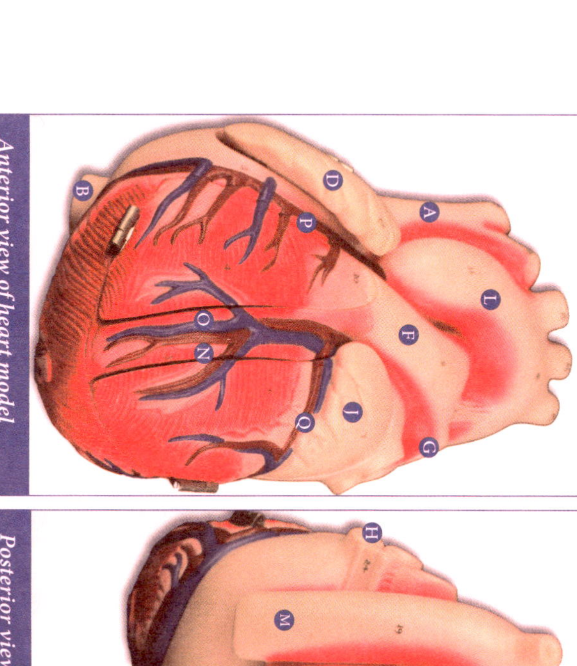

Posterior view of heart model

Chapter 5 - The Heart

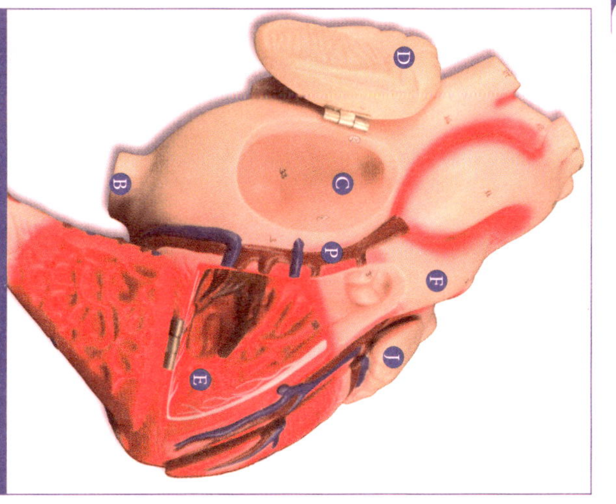

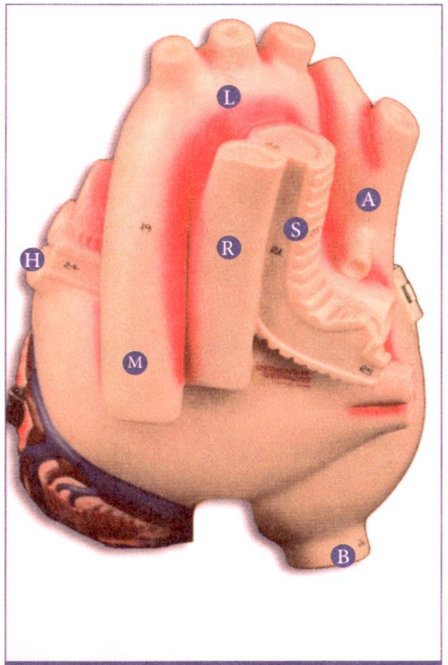

Lateral view of heart model

Chapter 5 - The Heart

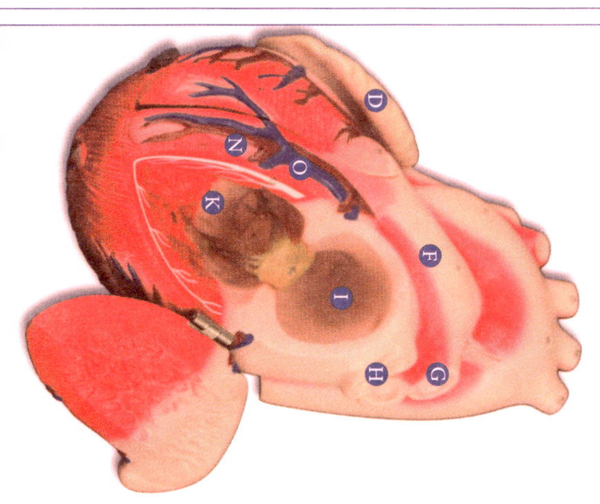

Anterior view of heart model

Chapter 5 - The Heart

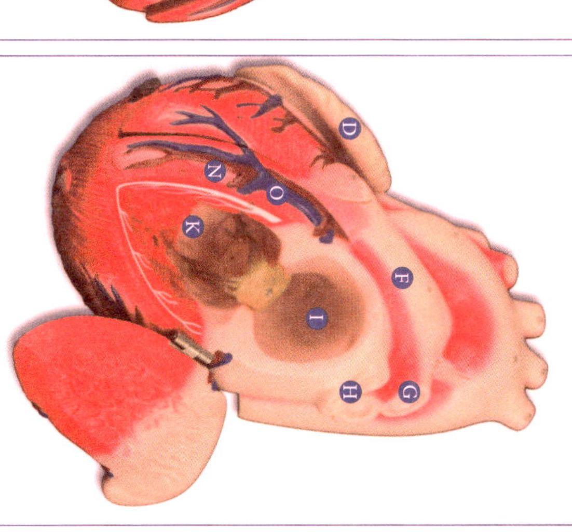

Lateral view of heart model

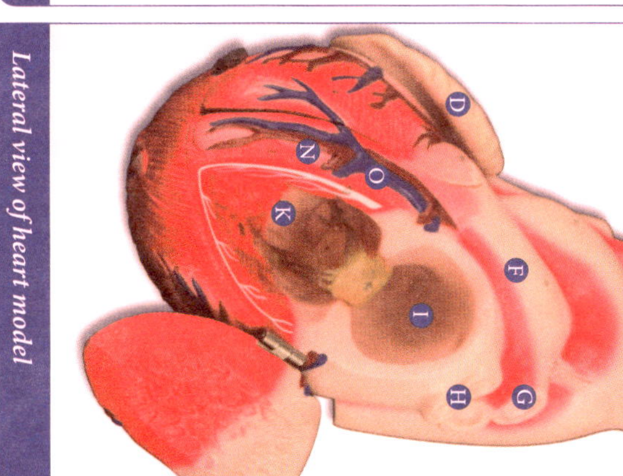

Posterior view of heart model

Ⓐ Superior vena cava
Ⓑ Inferior vena cava
Ⓓ Right auricle
Ⓕ Pulmonary trunk
Ⓖ Pulmonary artery
Ⓗ Pulmonary veins
Ⓙ Left auricle
Ⓛ Aorta
Ⓜ Descending aorta

Ⓝ Anterior interventricular a.
Ⓞ Great cardiac vein
Ⓟ Right coronary artery
Ⓠ Circumflex artery
Ⓡ Esophagus
Ⓢ Trachea

Ⓐ Superior vena cava
Ⓑ Inferior vena cava
Ⓔ Pulmonary artery
Ⓕ Pulmonary veins
Ⓛ Descending aorta

Ⓡ Coronary sinus
Ⓢ Posterior interventricular artery
Ⓣ Posterior interventricular vein

Ⓑ Inferior vena cava
Ⓒ Right atrium
Ⓓ Right auricle
Ⓔ Right ventricle
Ⓕ Pulmonary trunk
Ⓖ Pulmonary artery
Ⓗ Pulmonary veins
Ⓘ Left atrium
Ⓙ Left auricle
Ⓚ Left ventricle
Ⓝ Anterior interventricular a.
Ⓞ Great cardiac vein
Ⓟ Right coronary artery

Ⓐ Superior vena cava
Ⓒ Right auricle
Ⓓ Pulmonary trunk
Ⓔ Pulmonary artery
Ⓕ Pulmonary veins
Ⓖ Left auricle
Ⓗ Aorta
Ⓘ Brachiocephalic trunk
Ⓙ Left common carotid artery
Ⓚ Left subclavian artery

Ⓝ Anterior interventricular artery
Ⓞ Great cardiac vein
Ⓟ Right coronary artery
Ⓠ Circumflex artery

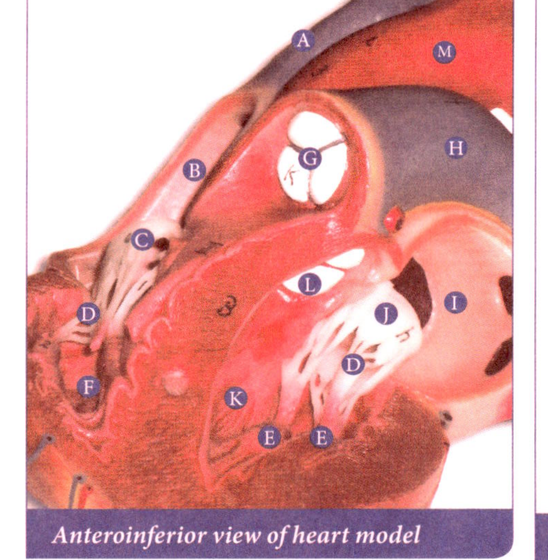

Anteroinferior view of heart model

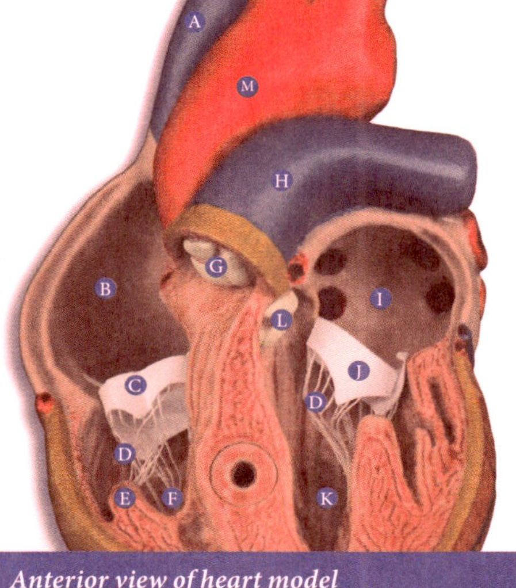

Anterior view of heart model

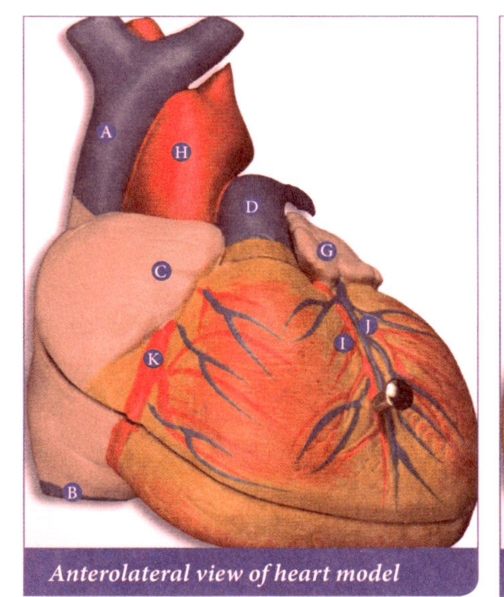

Anterolateral view of heart model

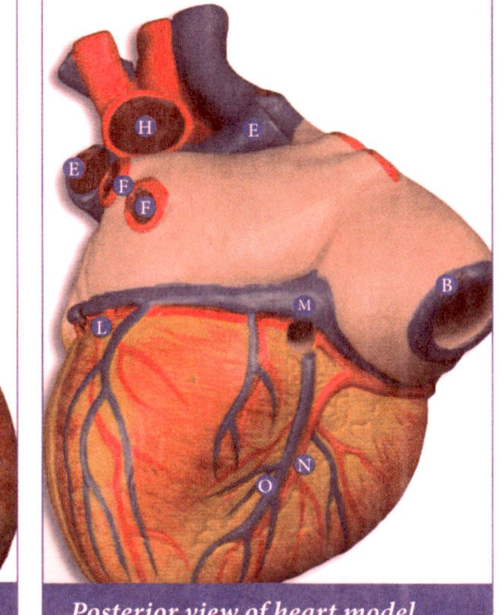

Posterior view of heart model

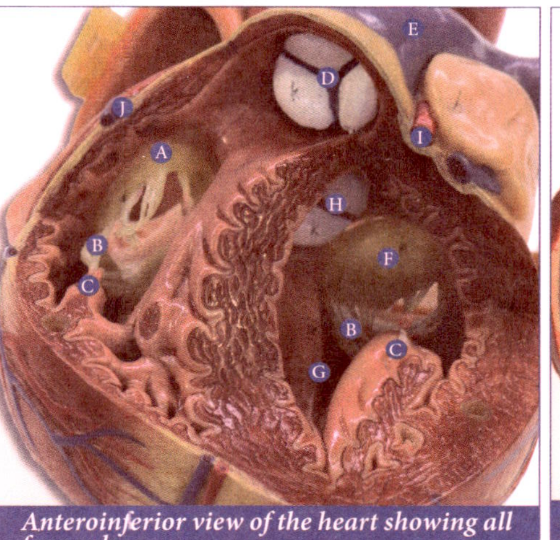

Anteroinferior view of the heart showing all four valves

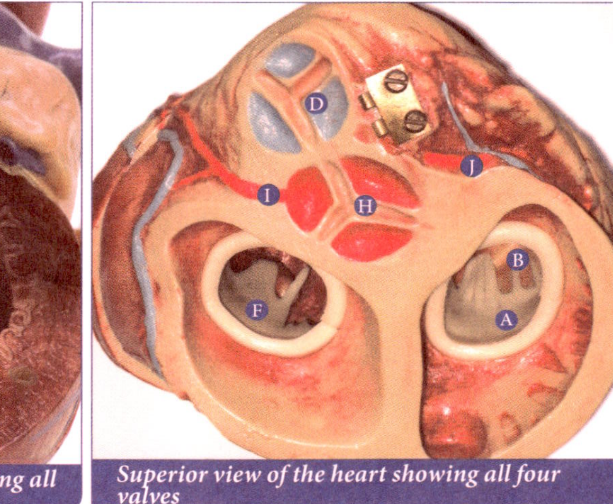

Superior view of the heart showing all four valves

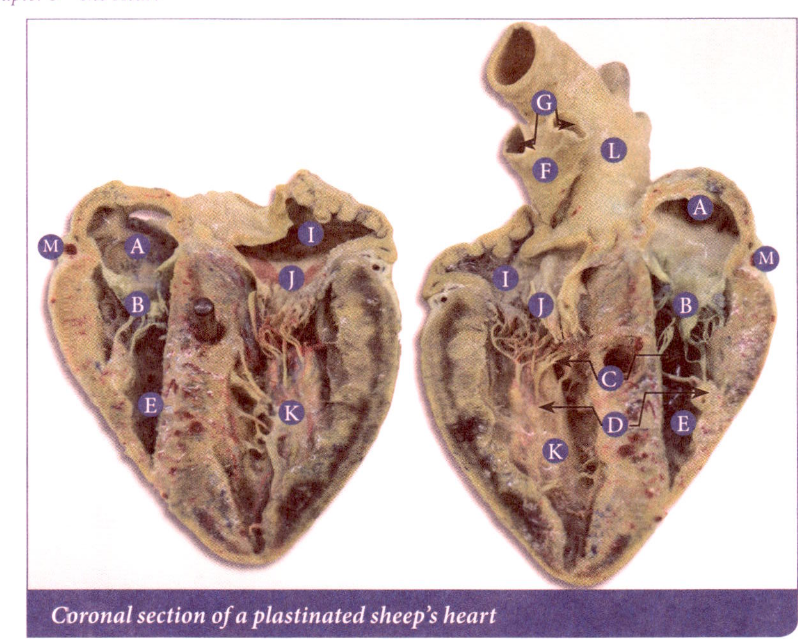

Coronal section of a plastinated sheep's heart

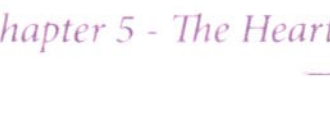
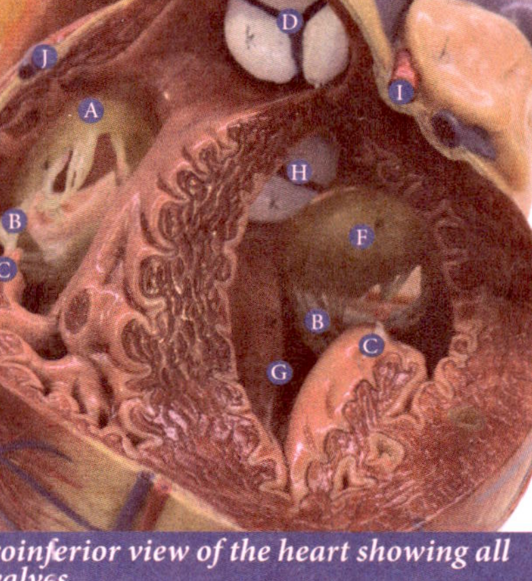

(A) Tricuspid (right AV) valve
(B) Chordae tendineae
(C) Papillary muscle
(D) Pulmonary semilunar valve
(E) Pulmonary trunk
(F) Bicuspid (mitral or left AV) valve
(G) Left ventricle
(H) Aortic semilunar valve
(I) Left Coronary artery
(J) Right Coronary artery

(A) Superior vena cava
(B) Right atrium
(C) Tricuspid (right AV) valve
(D) Chordae tendineae
(E) Papillary muscle
(F) Right ventricle
(G) Pulmonary semilunar valve
(H) Pulmonary trunk
(I) Left atrium
(J) Bicuspid (mitral or left AV) valve
(K) Left ventricle
(L) Aortic semilunar valve
(M) Aorta

(A) Right atrium
(B) Tricuspid (right AV) valve
(C) Chordae tendineae
(D) Papillary muscle
(E) Right ventricle
(F) Pulmonary trunk
(G) Pulmonary arteries
(I) Left atrium
(J) Bicuspid (mitral or left AV) valve
(K) Left ventricle
(L) Aorta
(M) Right coronary artery

(A) Superior vena cava
(B) Inferior vena cava
(C) Right auricle
(D) Pulmonary trunk
(E) Pulmonary artery
(F) Pulmonary veins
(G) Left auricle
(H) Aorta
(I) Anterior interventricular artery
(J) Great cardiac vein
(K) Right coronary artery
(L) Circumflex artery
(M) Coronary sinus
(N) Posterior interventricular artery
(O) Posterior interventricular vein

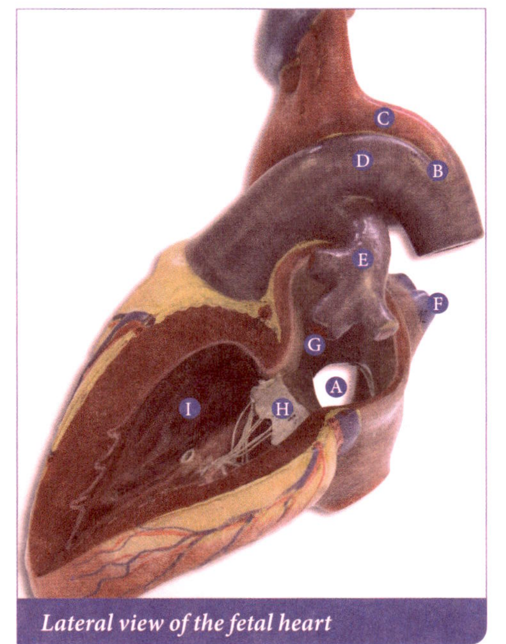

Anterior view of heart's electrical system

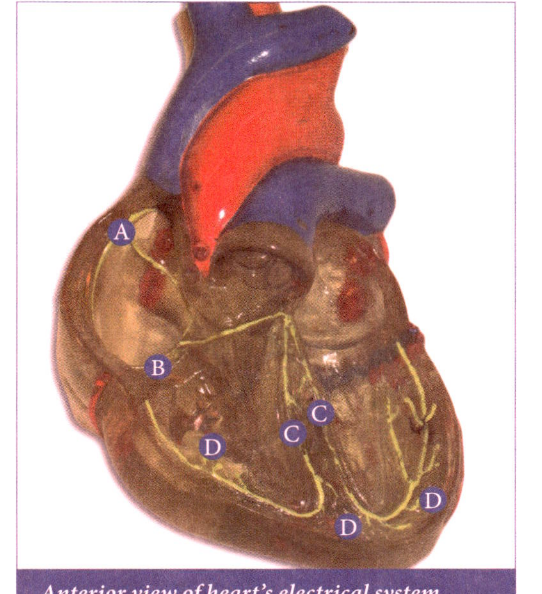

An ECG recording of a heart beat

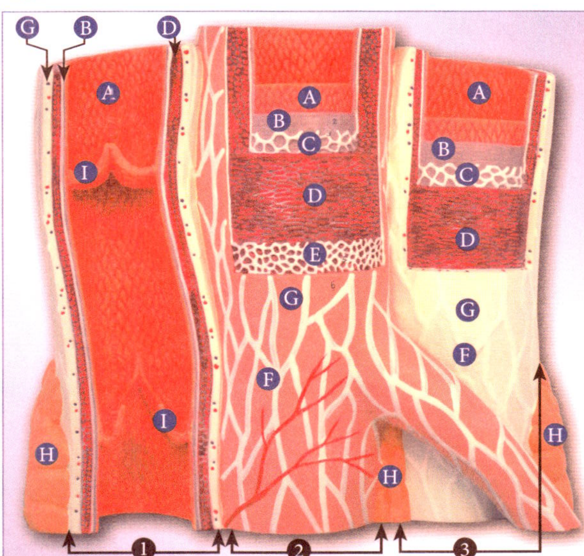

Lateral view of the fetal heart

Model showing a vein, distributing artery, and an arteriole

Model showing a capillary

P P-wave

Q
R } QRS complex
S

T T-wave

A Sinoatrial (SA) node
B Atrioventricular (AV) node
C Bundle branches
D Purkinje fibers

1 Vein
2 Medium size (distributing) artery
3 Arteriole
4 Capillary

A Endothelium
B Basement membrane   } Tunica intima
C Internal elastic lamina
D Tunica media (smooth muscle)
E External elastic membrane
F Collagen fibers
G Tunica externa
H Adipose tissue
I Valves – ensure one directional flow towards heart

A Foramen ovale
B Ductus arteriosus
C Aorta
D Pulmonary trunk
E Pulmonary arteries
F Pulmonary veins
G Left atrium
H Bicuspid (mitral or left AV) valve
I Left ventricle

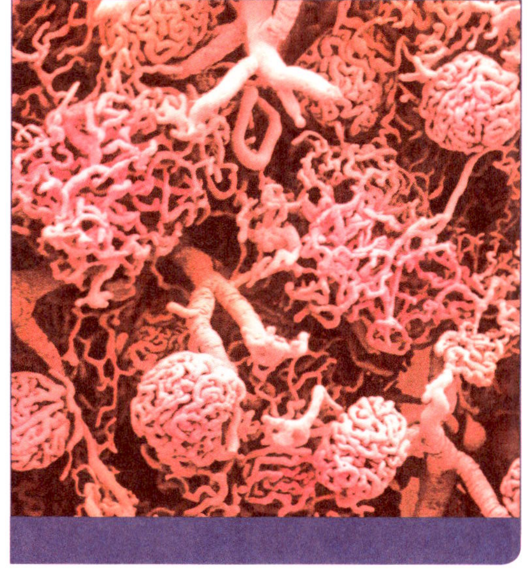

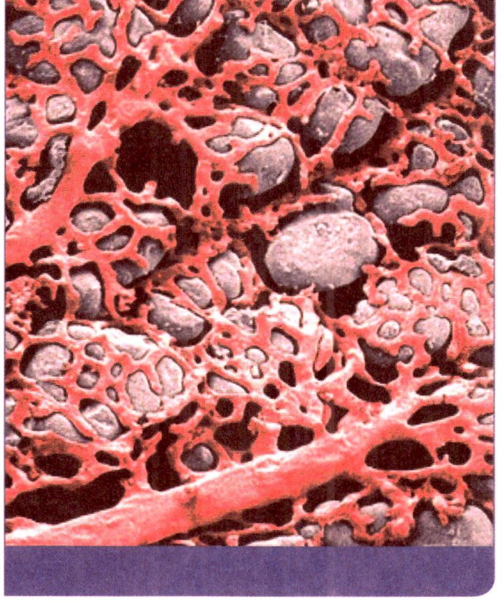

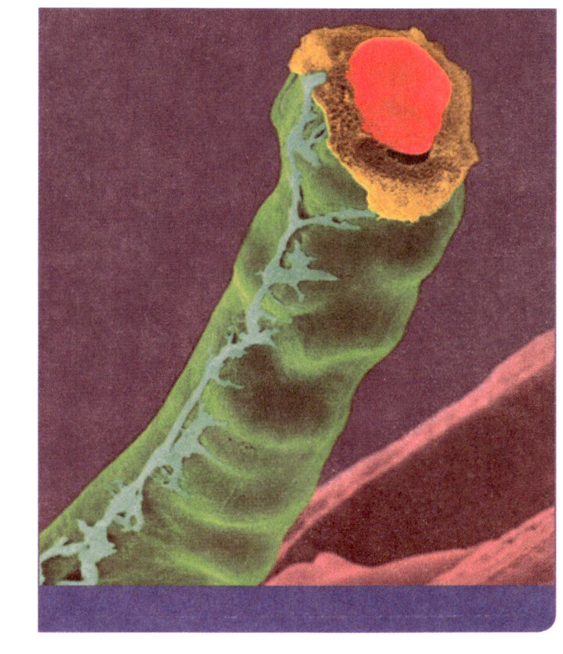

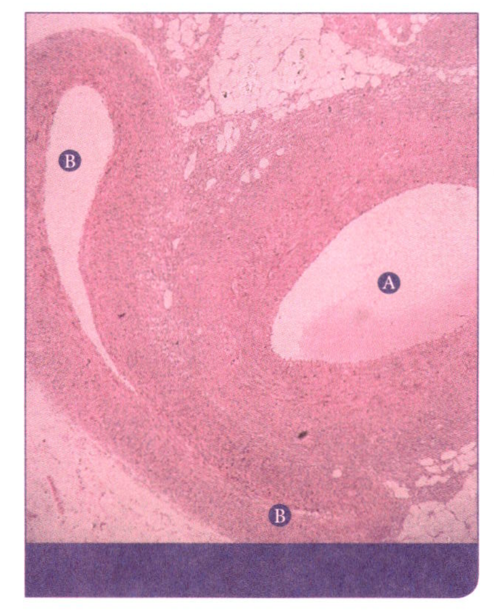

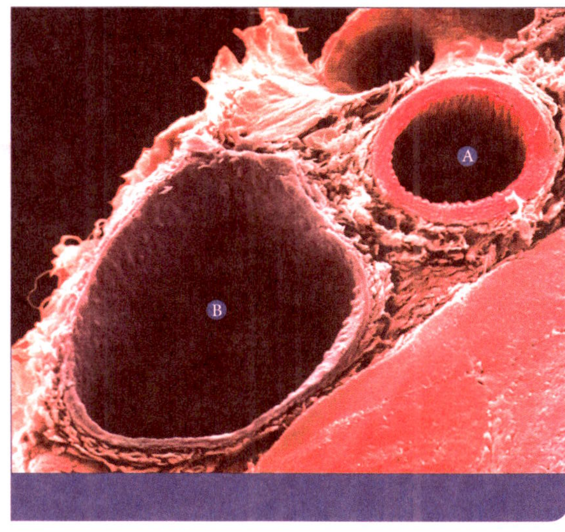

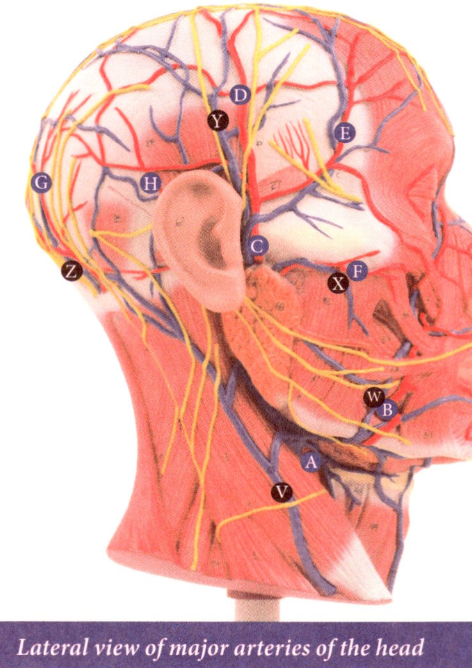

Lateral view of major arteries of the head

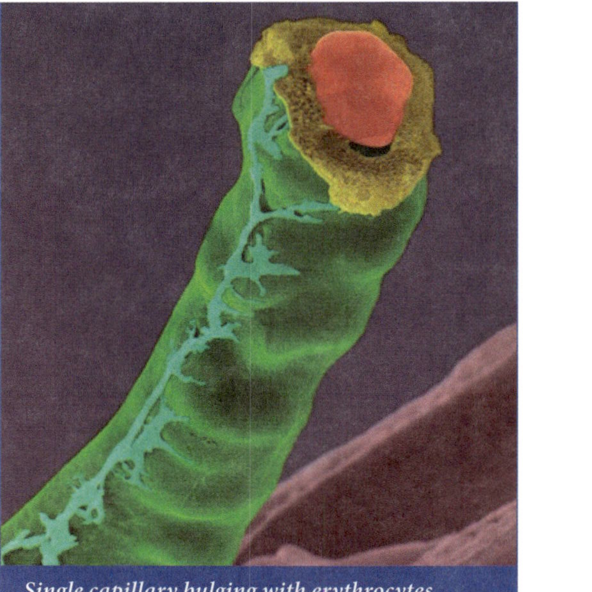

Single capillary bulging with erythrocytes

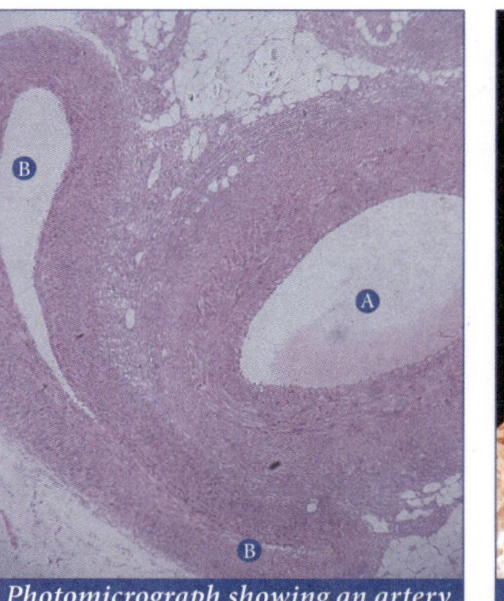

SEM of specialized capillary networks in the kidney

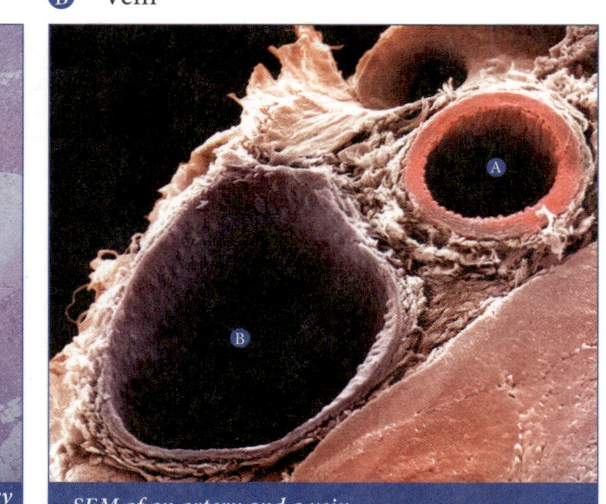

SEM of alveolar capillaries in the lung

A Artery

B Vein

A Common carotid a.

B Facial a.

C Superficial temporal a.

D Parietal branch of superficial temporal a.

E Frontal branch of superficial temporal a.

F Transverse facial a.

G Occipital a.

H Posterior auricular a.

V Jugular v.

W Facial v.

X Transverse facial v.

Y Parietal branch of superficial temporal v.

Z Occipital v.

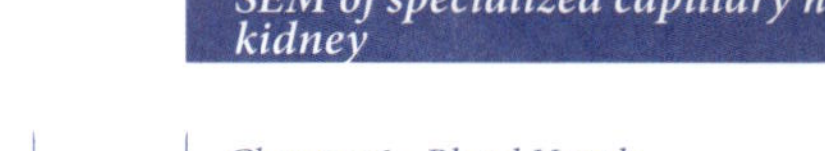

Photomicrograph showing an artery and a vein

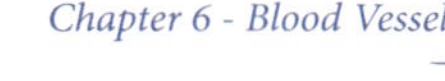

SEM of an artery and a vein

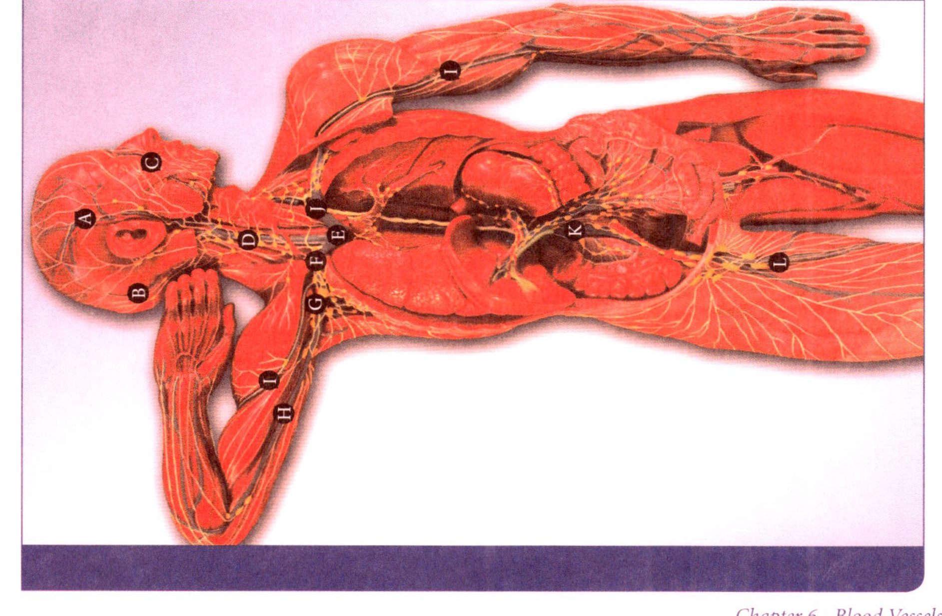

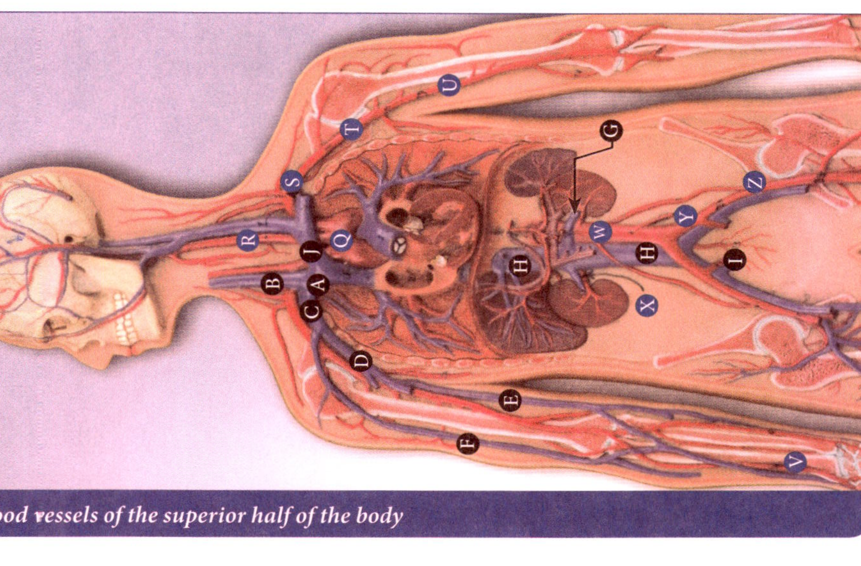

Blood vessels of the superior half of the body

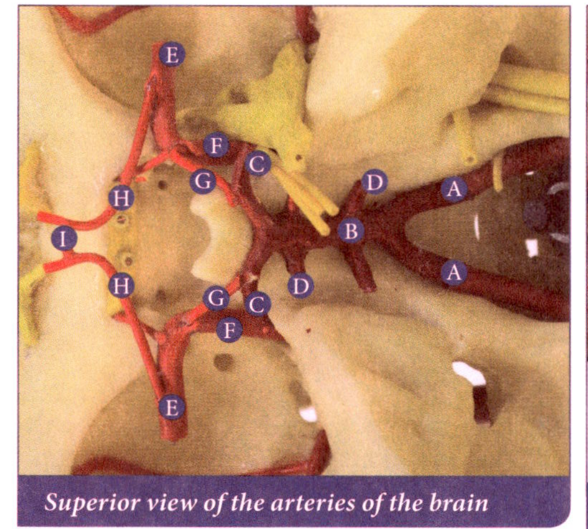

Superior view of the arteries of the brain

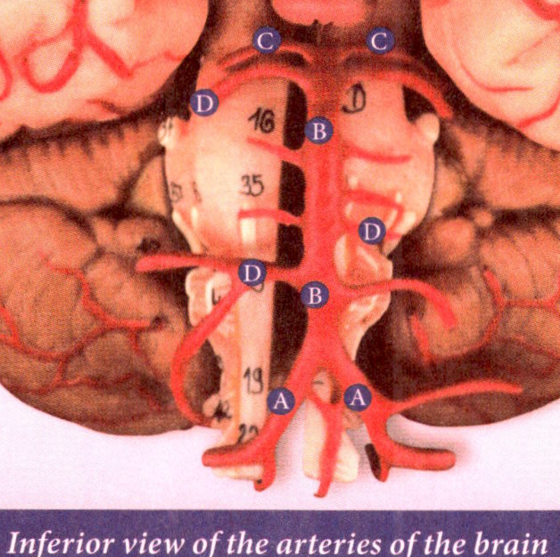

Inferior view of the arteries of the brain

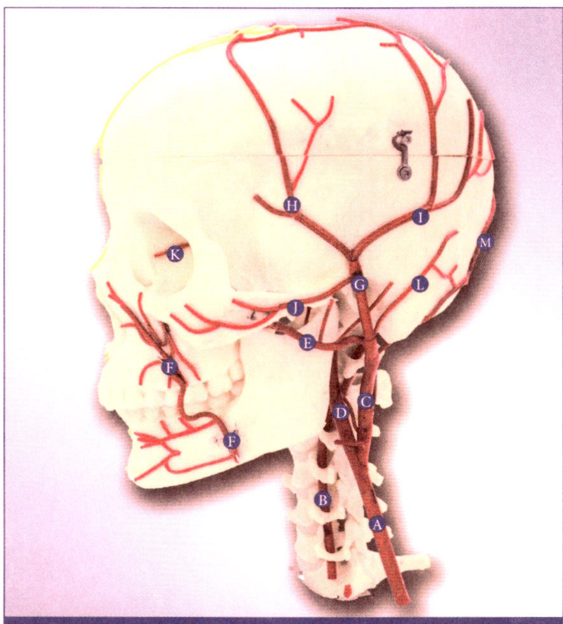

Lateral view of major arteries of the head

VEINS

- Ⓐ r. Brachiocephalic v.
- Ⓑ Internal jugular v.
- Ⓒ Subclavian v.
- Ⓓ Axillary v.
- Ⓔ Basilic v.
- Ⓕ Cephalic v.
- Ⓖ Renal v.
- Ⓗ Inferior vena cava
- Ⓘ Internal iliac v.
- Ⓙ l. Brachiocephalic v.

ARTERIES

- Ⓠ Aorta
- Ⓡ Carotid a.
- Ⓢ Subclavian a.
- Ⓣ Axillary a.
- Ⓤ Brachial a.
- Ⓥ Radial a.
- Ⓦ Descending aorta
- Ⓧ Renal a.
- Ⓨ Common iliac a.
- Ⓩ External iliac a.

- Ⓐ Superficial temporal v.
- Ⓑ Occipital v.
- Ⓒ Facial v.
- Ⓓ Internal jugular v.
- Ⓔ r. Brachiocephalic v.
- Ⓕ Subclavian v.
- Ⓖ Axillary v.
- Ⓗ Basilic v.
- Ⓘ Cephalic v.
- Ⓙ l. Brachiocephalic v.
- Ⓚ Common iliac v.
- Ⓛ Great saphenous v.

- Ⓐ Common carotid a.
- Ⓑ Vertebral a.
- Ⓒ External carotid a.
- Ⓓ Internal carotid a.
- Ⓔ Maxillary a.
- Ⓕ Facial a.
- Ⓖ Superficial temporal a.

- Ⓗ Frontal branch of superficial temporal a.
- Ⓘ Parietal branch of superficial temporal a.
- Ⓙ Transverse facial a.
- Ⓚ Opthalmic a.
- Ⓛ Posterior auricular a.
- Ⓜ Occipital a.

- Ⓐ Vertebral a.
- Ⓑ Basilar a.
- Ⓒ Posterior cerebral a.
- Ⓓ Cerebellar a. (several are labeled)
- Ⓔ Middle cerebral a.
- Ⓕ Internal carotid a.
- Ⓖ Posterior communicating a. ←┐
- Ⓗ Anterior cerebral a.   ├ Cerebral arterial circle (circle of Willis)
- Ⓘ Anterior communicating a. ←┘

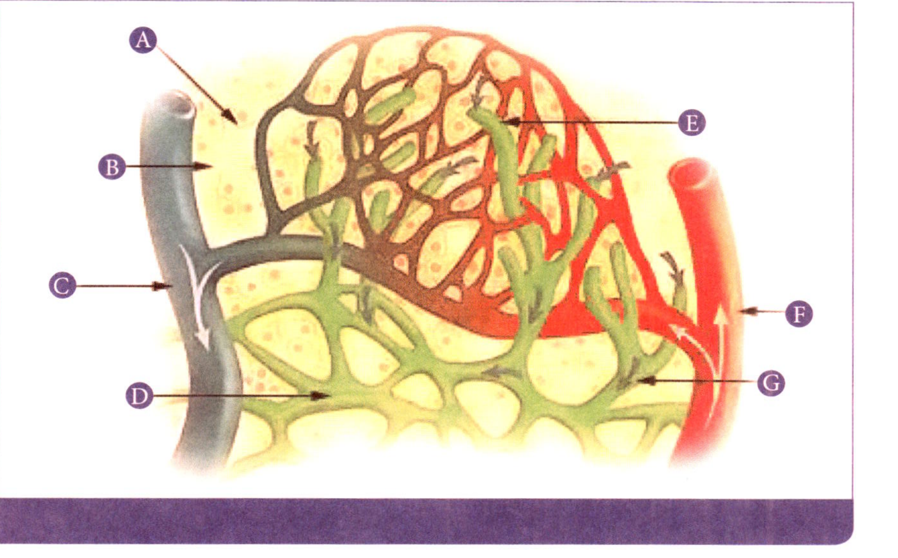

A
B
C
D
E
F
G

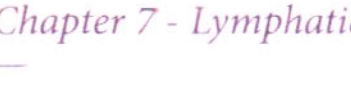

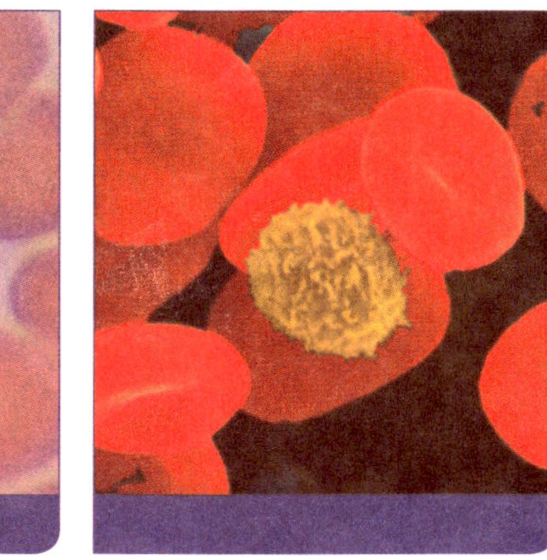

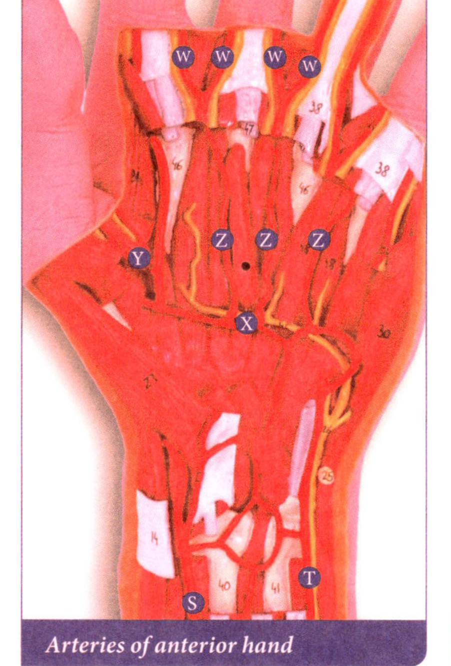

W W W W
Y
Z Z Z
X
S
T
Arteries of anterior hand

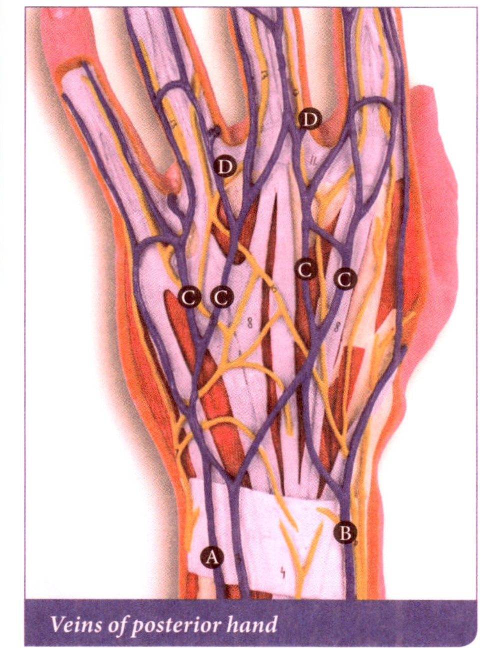

D
D
C C C
A
B
Veins of posterior hand

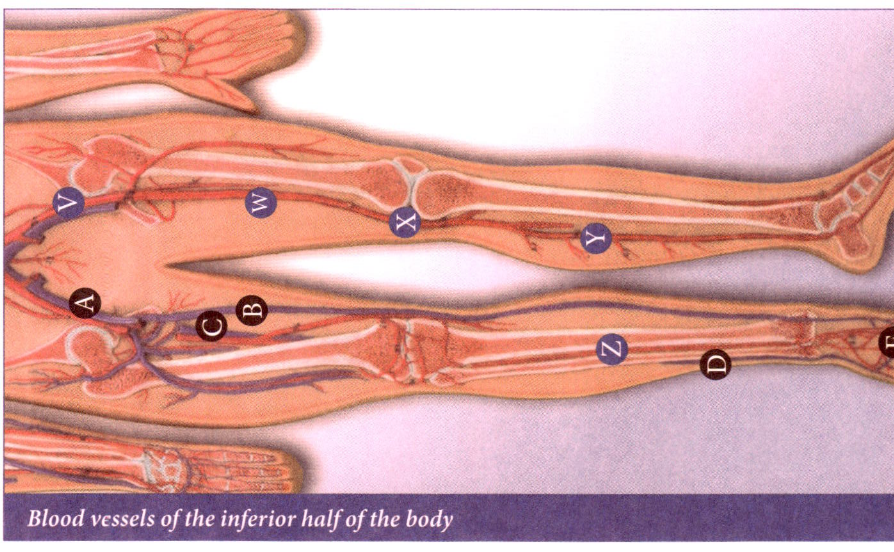

V
W
X
Y
A
C
B
Z
D
E
Blood vessels of the inferior half of the body

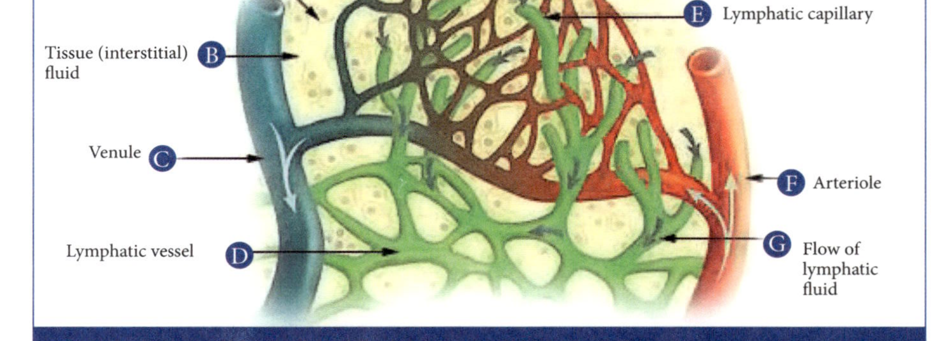

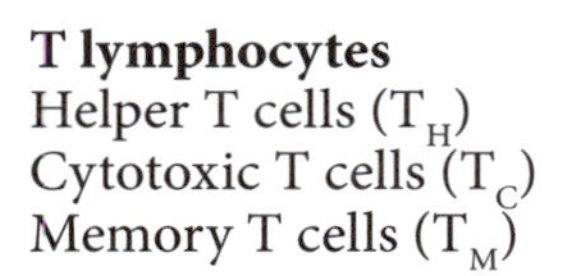

NK (natural killer)

B lymphocytes
Plasma cells
Memory B cells (B_M)

T lymphocytes
Helper T cells (T_H)
Cytotoxic T cells (T_C)
Memory T cells (T_M)

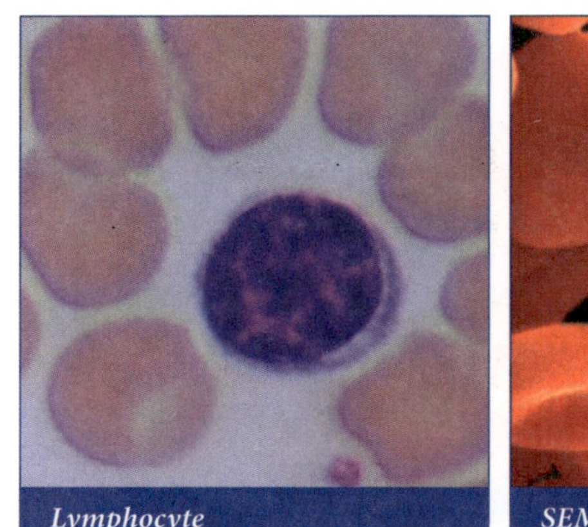

Chapter 7 - Lymphatic

Chapter 7 - Lymphatic

Chapter 6 - Blood Vessels

Chapter 6 - Blood Vessels

VEINS

Ⓐ External iliac v.
Ⓑ Great saphenous v.
Ⓒ Femoral v.
Ⓓ Small saphenous v.
Ⓔ Dorsal venous arch

ARTERIES

Ⓥ External iliac a.
Ⓦ Femoral a.
Ⓧ Popliteal a.
Ⓨ Posterior tibial a.
Ⓩ Anterior tibial a.

VEINS

Ⓐ Basilic v.
Ⓑ Cephalic v.
Ⓒ Dorsal metacarpal v.
Ⓓ Intercapitular v.

ARTERIES

Ⓢ Radial a.
Ⓣ Ulnar a.
Ⓦ Proper palmar digital a.
Ⓧ Deep palmar arch
Ⓨ Principal pollicis a.
Ⓩ Palmar metacarpal a.

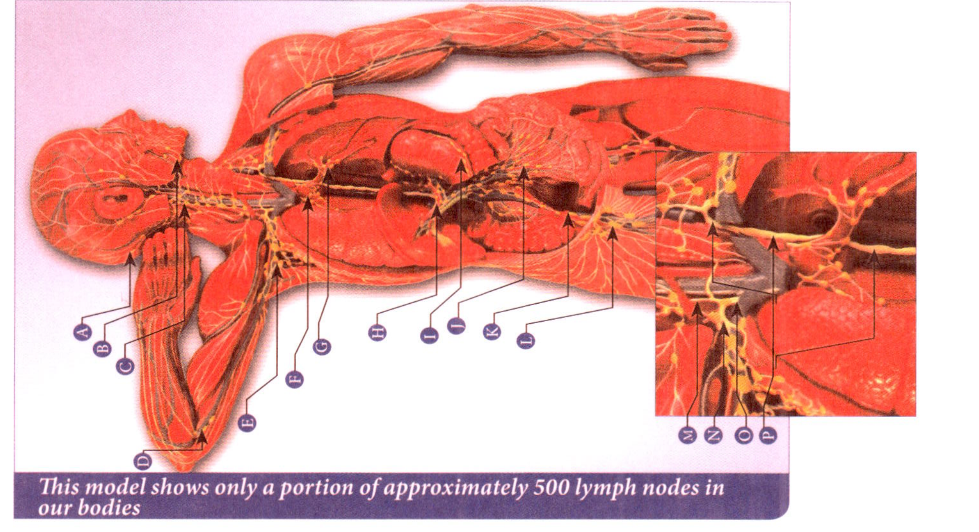

This model shows only a portion of approximately 500 lymph nodes in our bodies

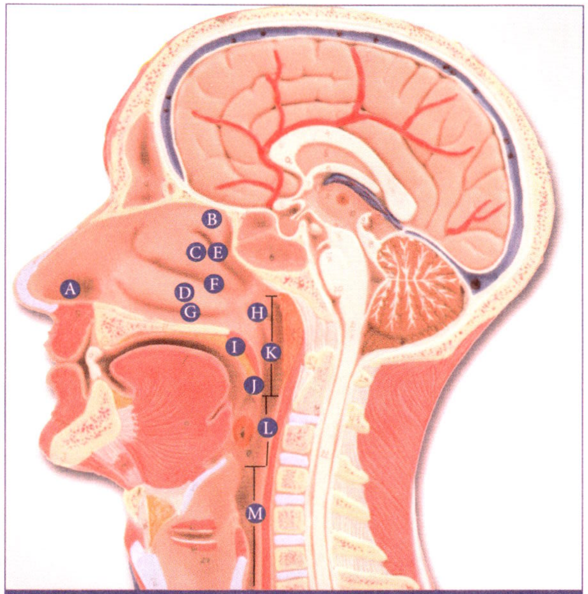

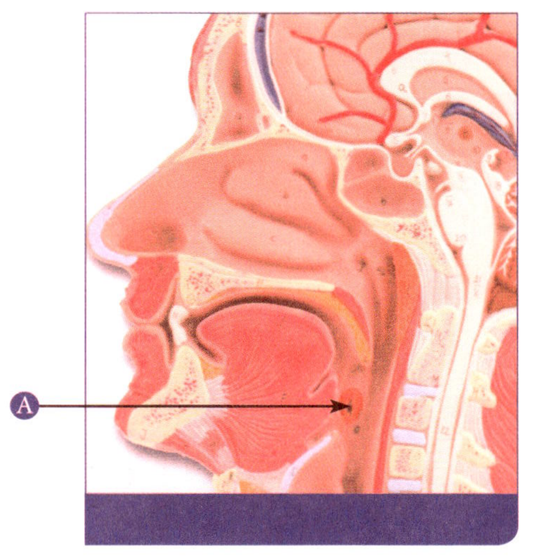

Midsagittal section of the head showing the upper respiratory tract

Chapter 7- Lymphatic

Chapter 8 - Respiration

Chapter 7- Lymphatic

Chapter 7 - Lymphatic

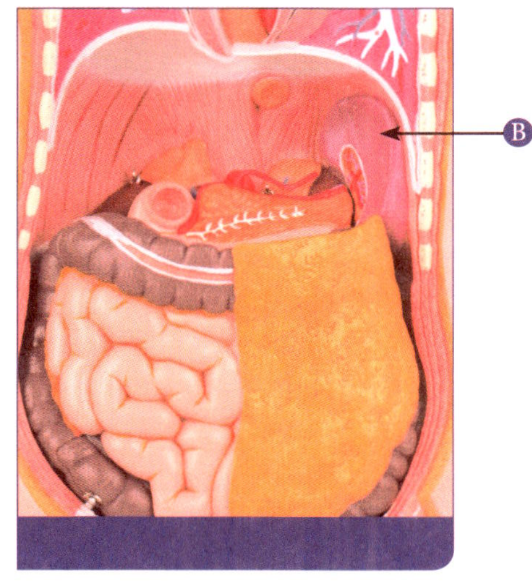

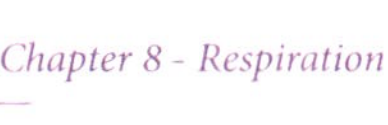

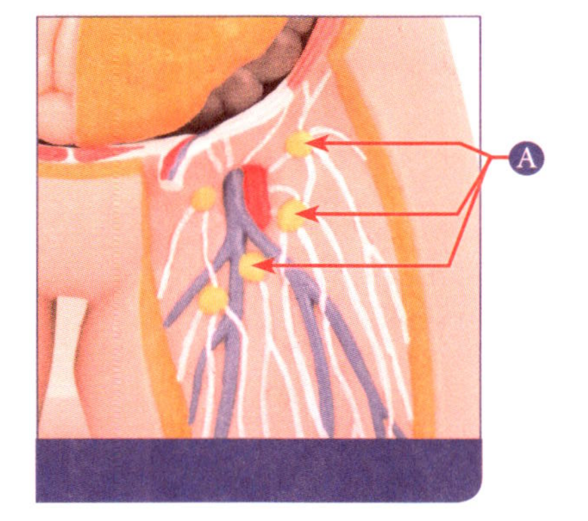

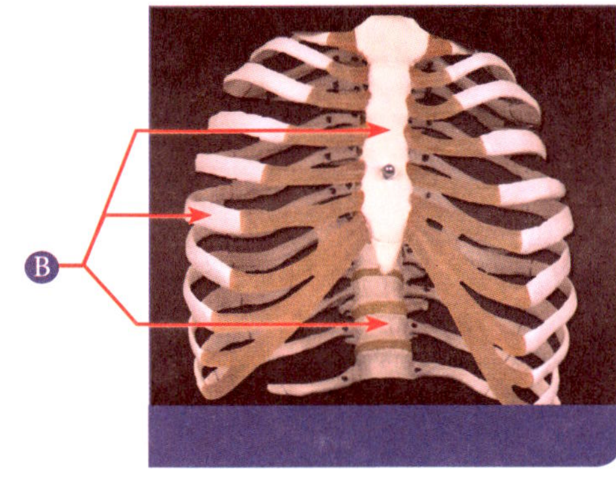

A Vestibule
B Superior conchae
C Middle conchae
D Inferior conchae
E Superior meatus
F Middle meatus
G Inferior meatus

H Auditory tube
I Soft palate
J Uvula
K Nasopharynx
L Oropharynx
M Laryngopharynx

LYMPH NODES

A Occipital
B Submandibular
C Deep cervical
D Cubital
E Axillary
F Tracheobronchial
G Bronchial
H Hepatic

I Gastric
J Mesenteric
K Common iliac
L Inguinal

LYMPHATIC VESSELS

M Jugular trunk
N Subclavian trunk
O Right lymphatic duct
P Thoracic duct

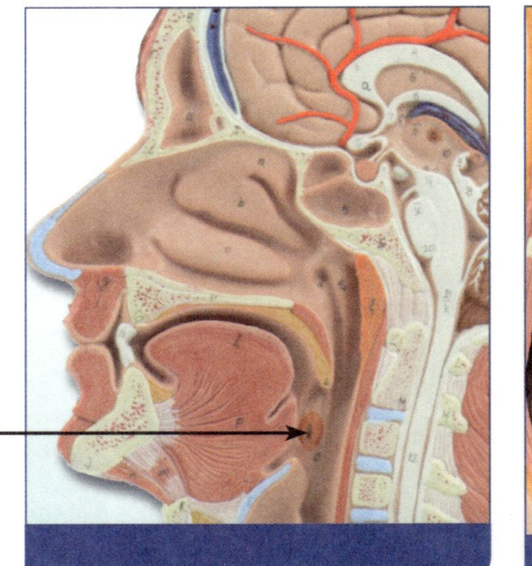

Lymph nodes

Red bone marrow

A Palatine tonsils

B The spleen

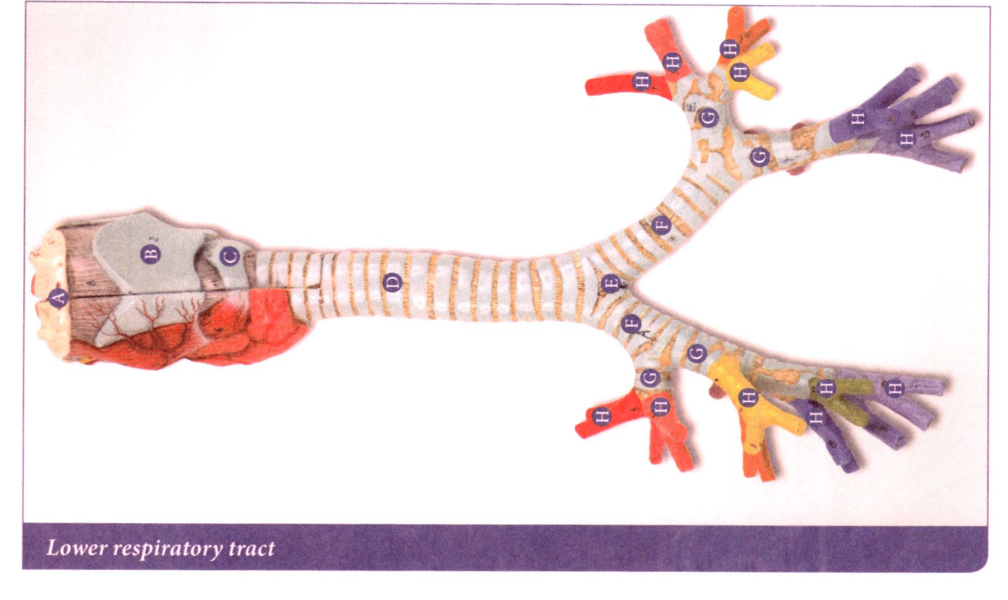

Lower respiratory tract

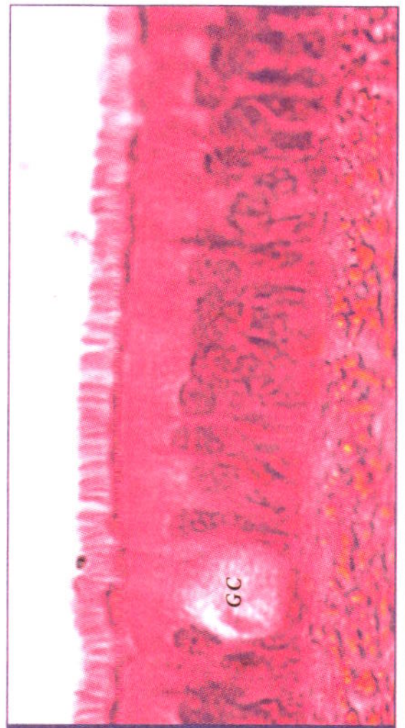

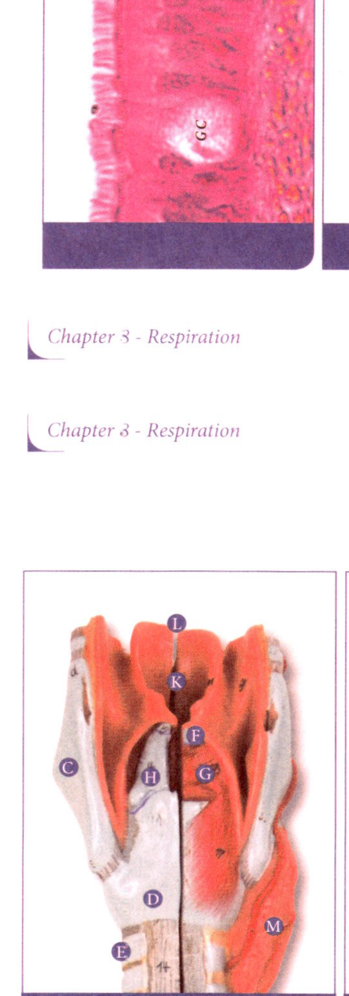

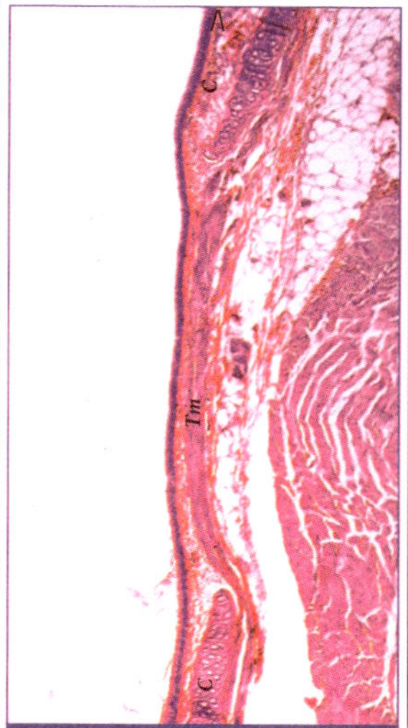

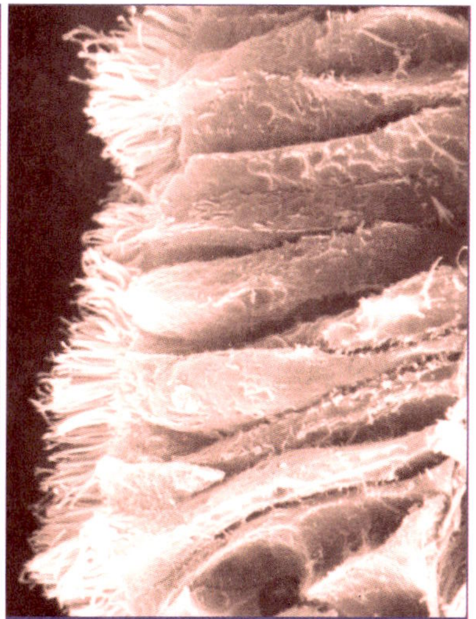

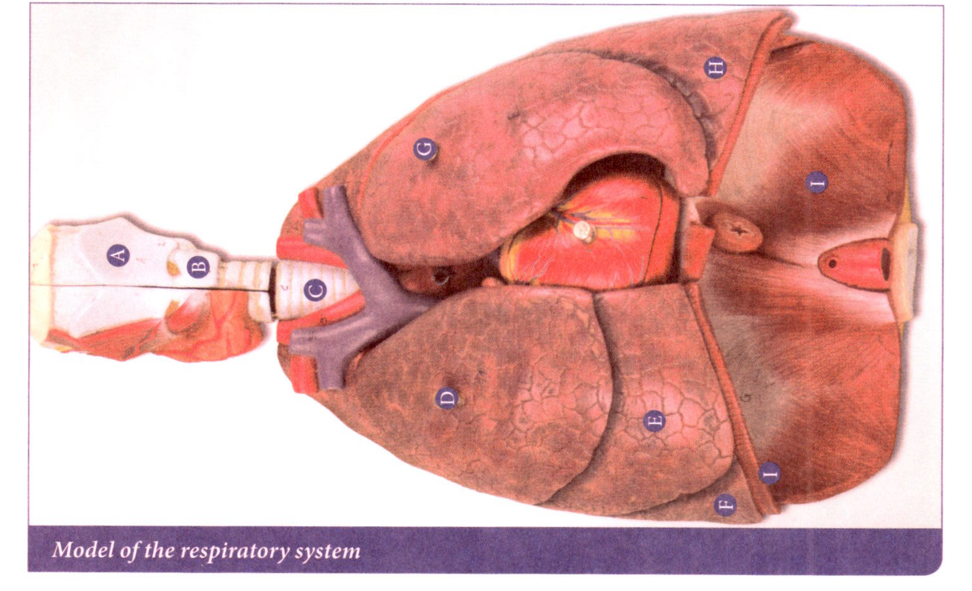

Model of the respiratory system

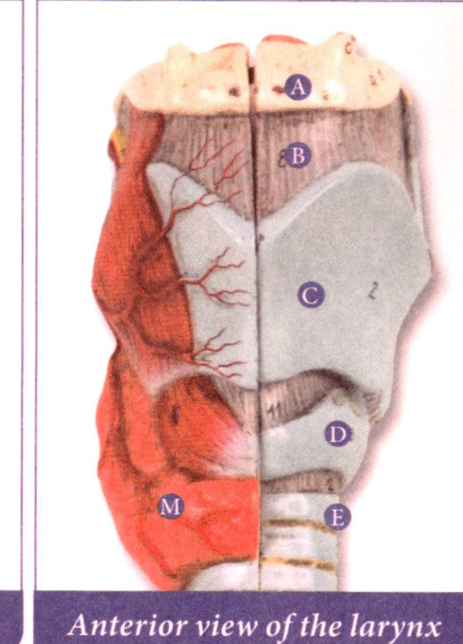

Lateral view of the larynx

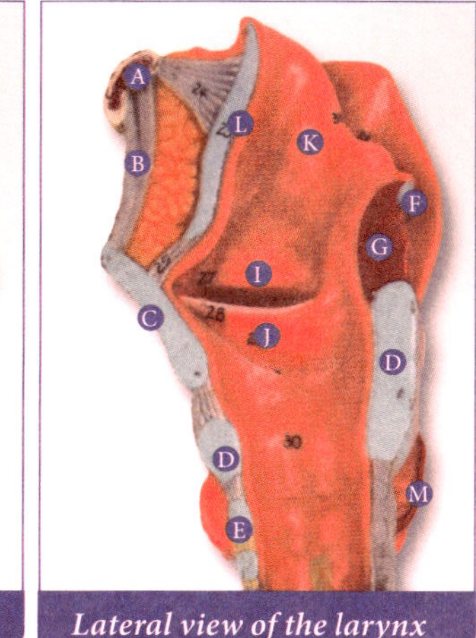

Anterior view of the larynx

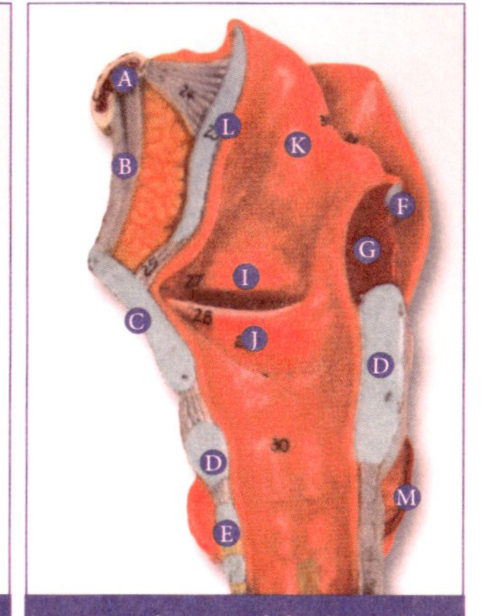

Lateral view of the larynx

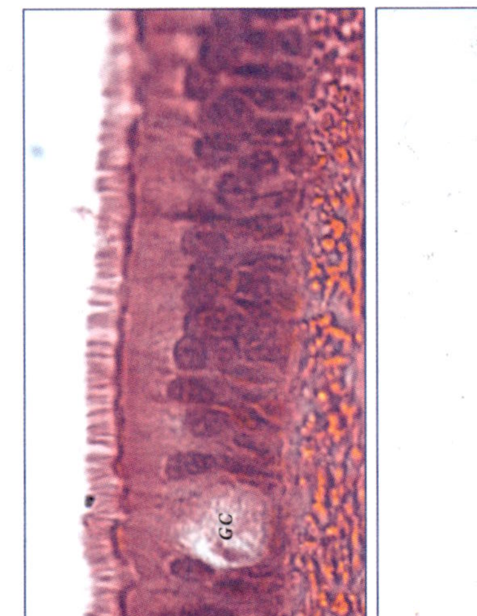

Repiratory epithelium (pseudostratified ciliated columnar epithelium and goblet cell, GC) 400×

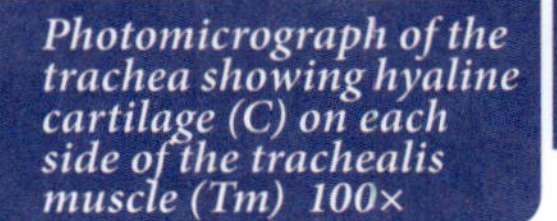

Photomicrograph of the trachea showing hyaline cartilage (C) on each side of the trachealis muscle (Tm) 100×

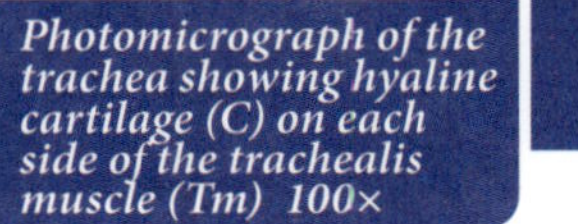

SEM of repiratory epithelium (pseudostratified columnar epithelium with cilia) 2,650×

Ⓐ Hyoid bone
Ⓑ Thyroid cartilage
Ⓒ Cricoid cartilage
Ⓓ Tracheal cartilage
Ⓔ Carina
Ⓕ Main (primary) bronchus
Ⓖ Lobar (secondary) bronchi
Ⓗ Segmental (tertiary) bronchi

Ⓐ Hyoid bone
Ⓑ Thyrohyoid ligament
Ⓒ Thyroid cartilage
Ⓓ Cricoid cartilage
Ⓔ Tracheal cartilage
Ⓕ Corniculate cartilage
Ⓖ Arytenoid muscle

Ⓗ Arytenoid cartilage
Ⓘ Vestibular fold (false vocal cord)
Ⓙ Vocal cord
Ⓚ Glottis (opening into the trachea)
Ⓛ Epiglottis
Ⓜ Thyroid gland

Ⓐ Thyroid cartilage
Ⓑ Cricoid cartilage
Ⓒ Trachea
Ⓓ Superior lobe of right lung
Ⓔ Middle lobe of right lung
Ⓕ Inferior lobe of right lung
Ⓖ Superior lobe of left lung
Ⓗ Inferior lobe of left lung
Ⓘ Diaphragm

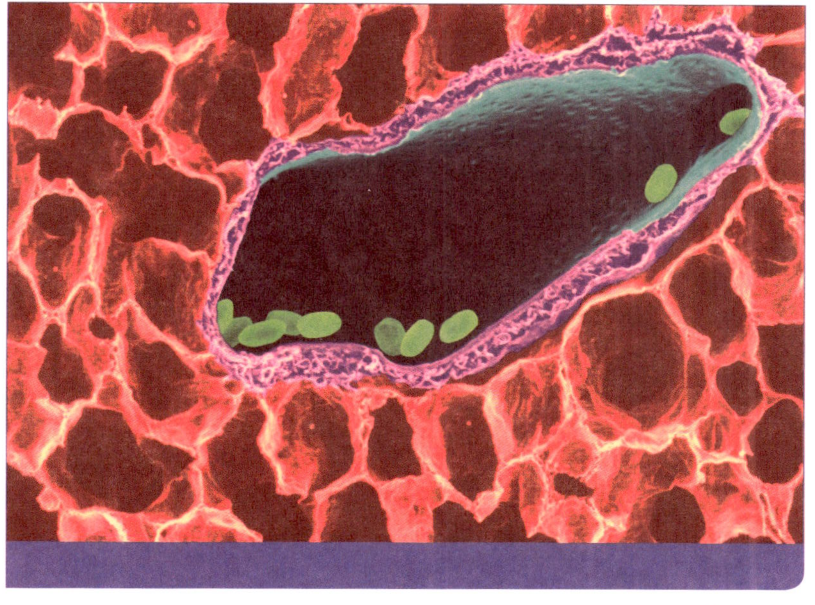

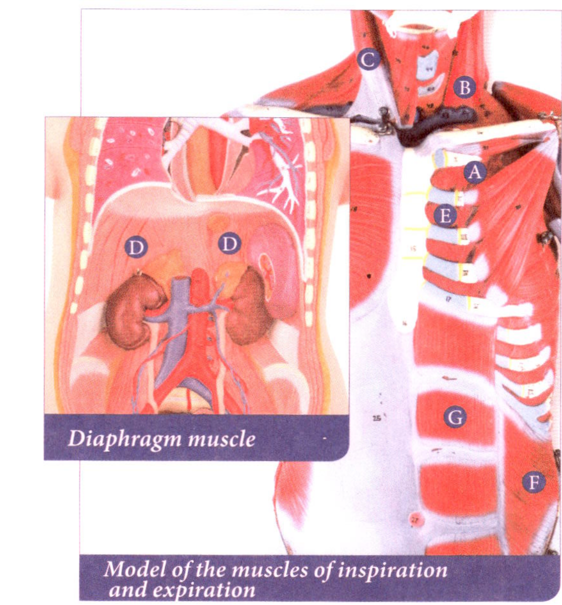
Diaphragm muscle
Model of the muscles of inspiration and expiration

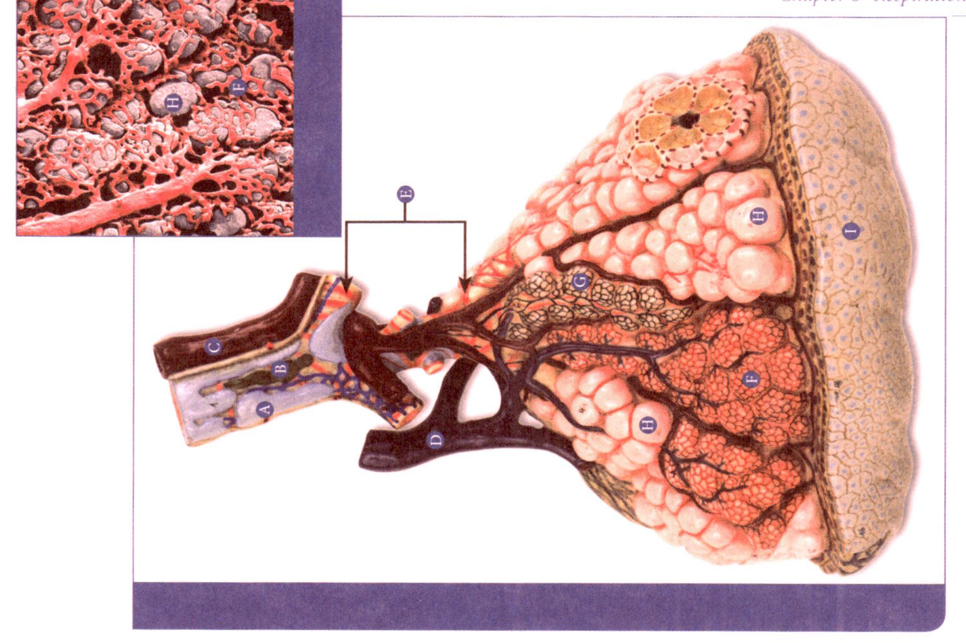

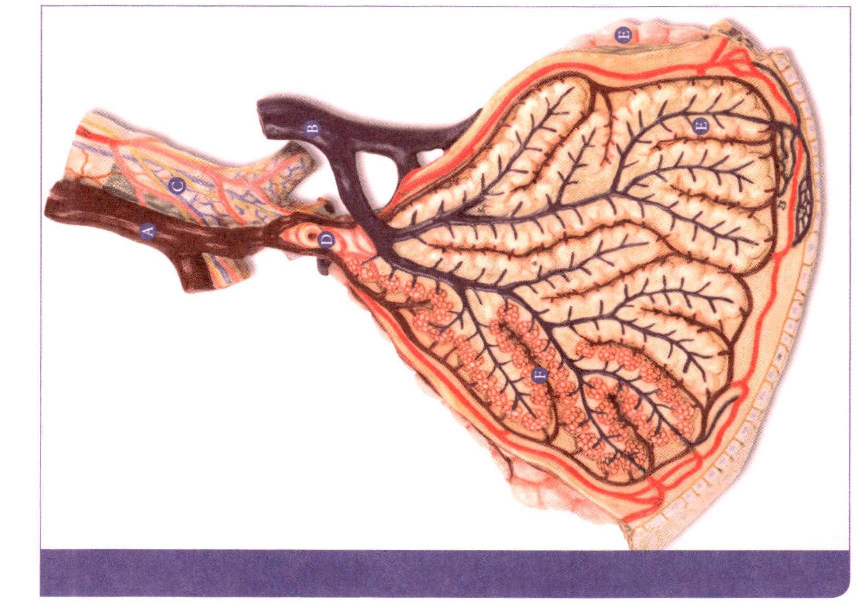

MUSCLES OF INSPIRATION

- Ⓐ External intercostals
- Ⓑ Scalenes
- Ⓒ Sternocleidomastoid
- Ⓓ Diaphragm

MUSCLES OF EXPIRATION

- Ⓔ Internal intercostals
- Ⓕ External abdominal obliques
- Ⓖ Rectus abdominis

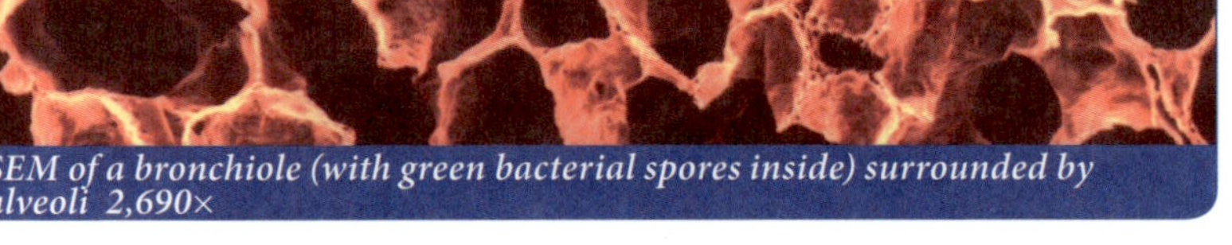

SEM of a bronchiole (with green bacterial spores inside) surrounded by alveoli 2,690×

- Ⓐ Pulmonary venule
- Ⓑ Pulmonary arteriole
- Ⓒ Terminal bronchial
- Ⓓ Respiratory bronchiole
- Ⓔ Alveolar sacs (alveoli)
- Ⓕ Alveolar capillaries

- Ⓐ Cartilagenous plates
- Ⓑ Lymphatic vessel
- Ⓒ Pulmonary venule
- Ⓓ Pulmonary arteriole
- Ⓔ Smooth muscle cells
- Ⓕ Alveolar capillaries
- Ⓖ Alveolar lymphatic capillaries
- Ⓗ Alveolar sacs
- Ⓘ Visceral pleura

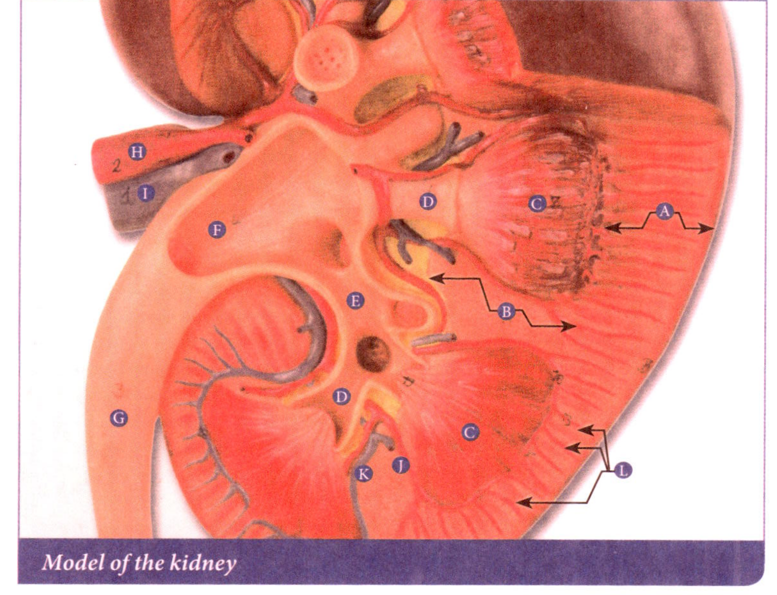

Model of the kidney

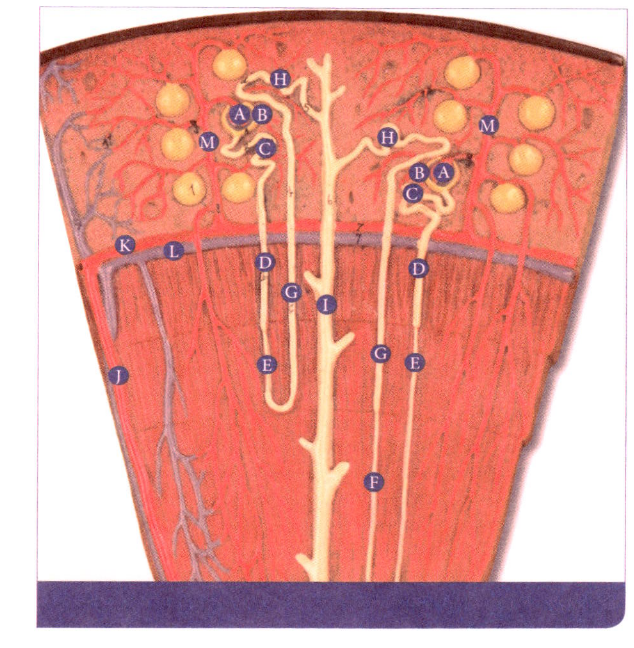

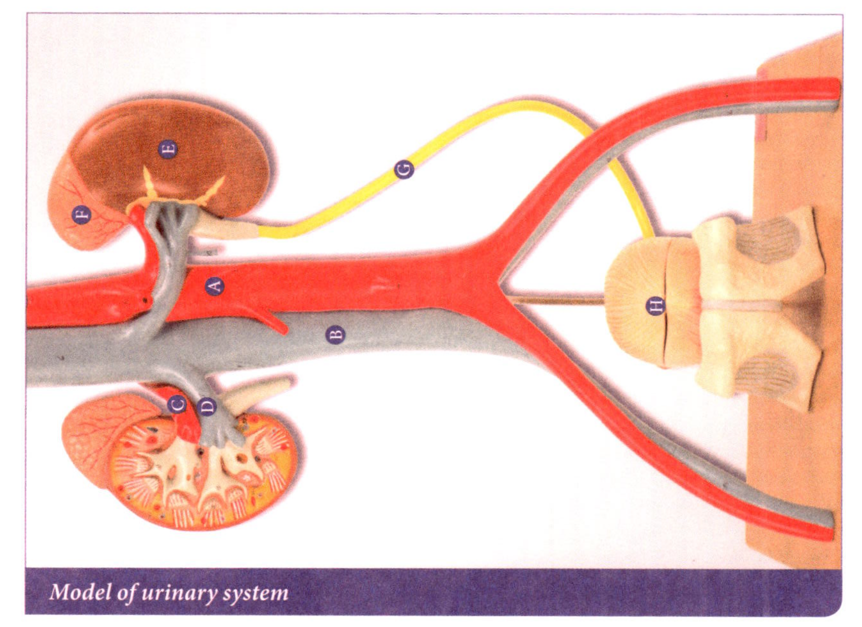

Model of urinary system

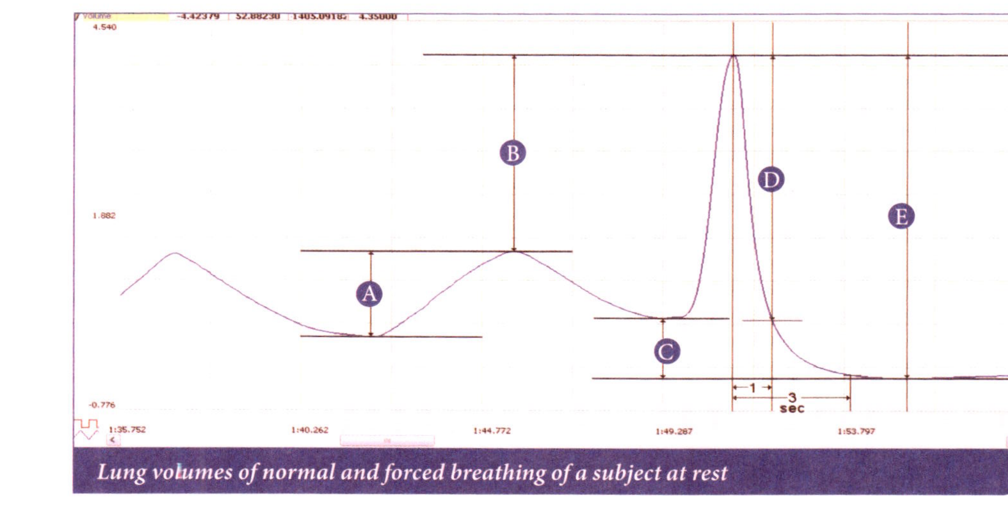

Lung volumes of normal and forced breathing of a subject at rest

Model showing two nephrons (A–H), a collecting duct (I), and blood vessels (J–M)

(A) Glomerulus
(B) Glomerular (Bowman's) capsule
(C) Proximal convoluted tubule
(D) Descending limb (thick segment)
(E) Descending limb (thin segment)
(F) Ascending limb (thin segment)
(G) Ascending limb (thick segment)

Nephron loop (loop of Henle) — D, E, F, G

(H) Distal convoluted tubule
(I) Collecting duct
(J) Interlobar artery
(K) Arcuate artery
(L) Arcuate vein
(M) Interlobular artery

(A) Renal cortex
(B) Renal medulla
(C) Renal pyramid
(D) Minor calyx
(E) Major calyx
(F) Renal pelvis
(G) Ureter
(H) Renal artery
(I) Renal vein
(J) Interlobar artery
(K) Interlobar vein
(L) Interlobular arteries

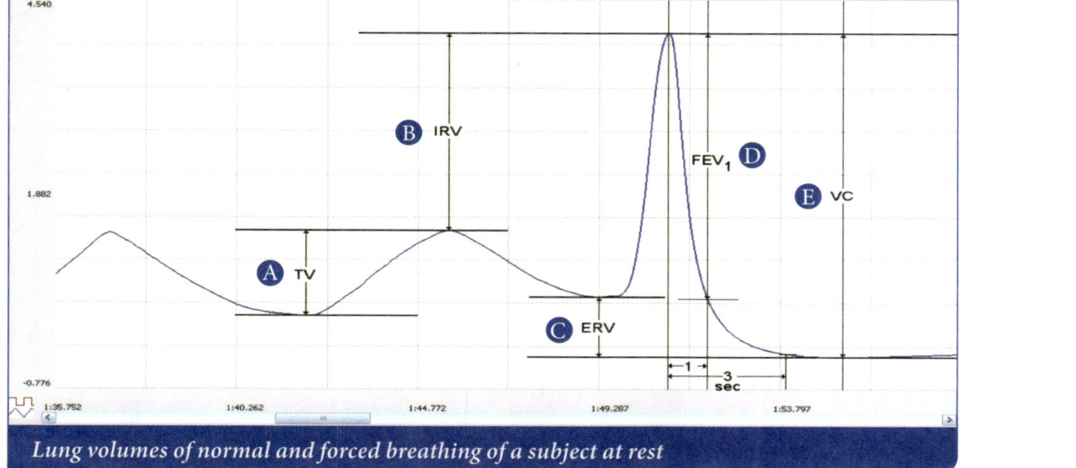

Lung volumes of normal and forced breathing of a subject at rest

(A) Descending aorta
(B) Inferior vena cava
(C) Renal artery
(D) Renal vein
(E) Kidney
(F) Adrenal gland
(G) Ureter
(H) Urinary bladder

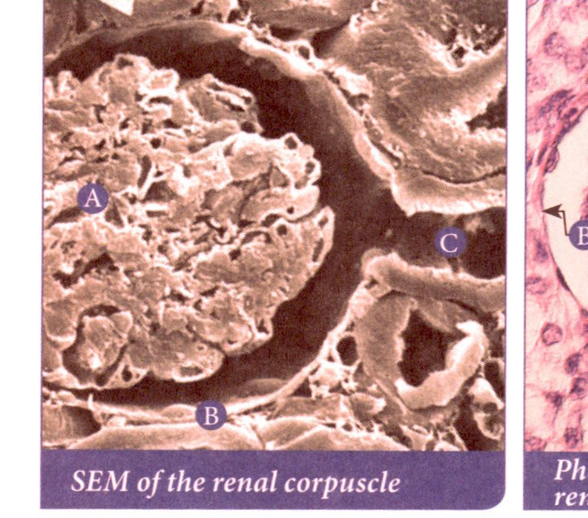

SEM of the renal corpuscle

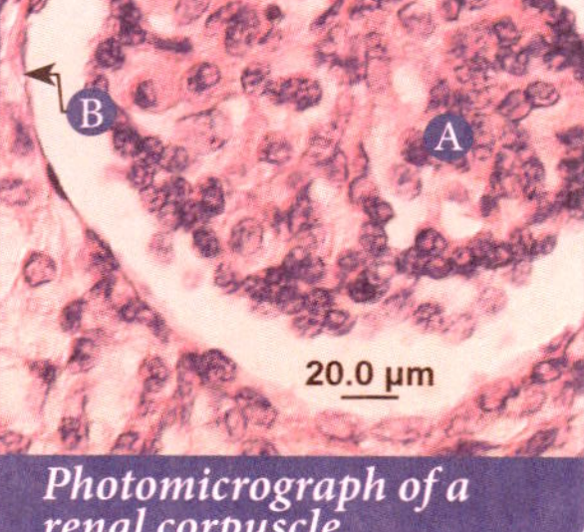

Photomicrograph of a renal corpuscle

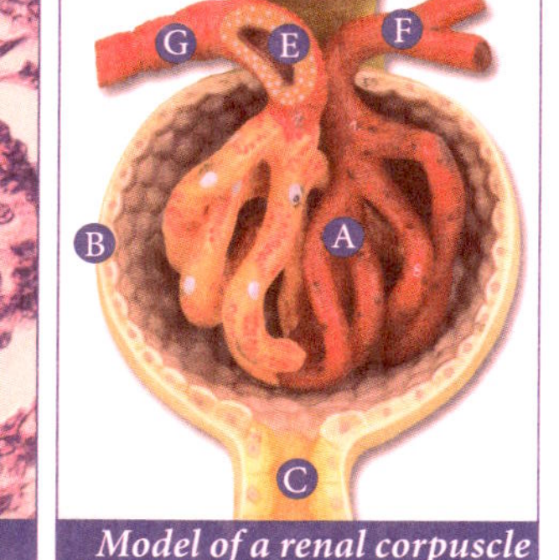

Model of a renal corpuscle and associated structures

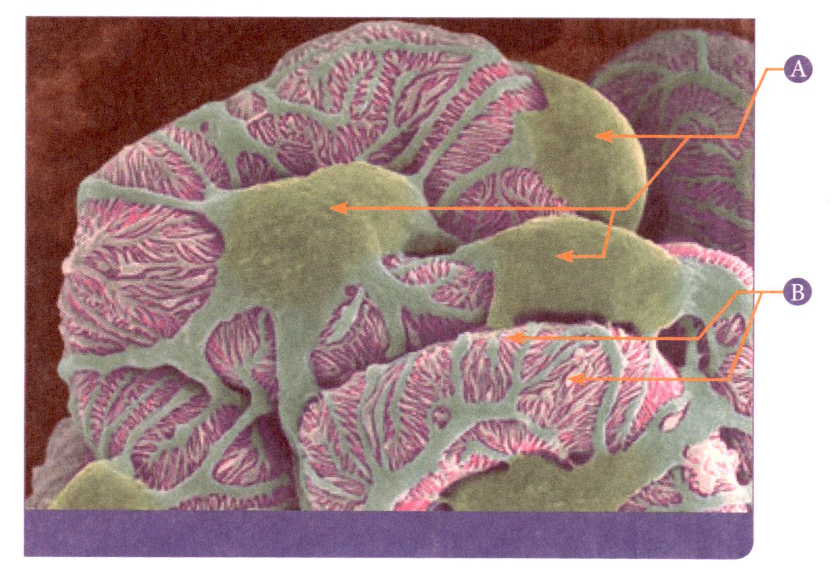

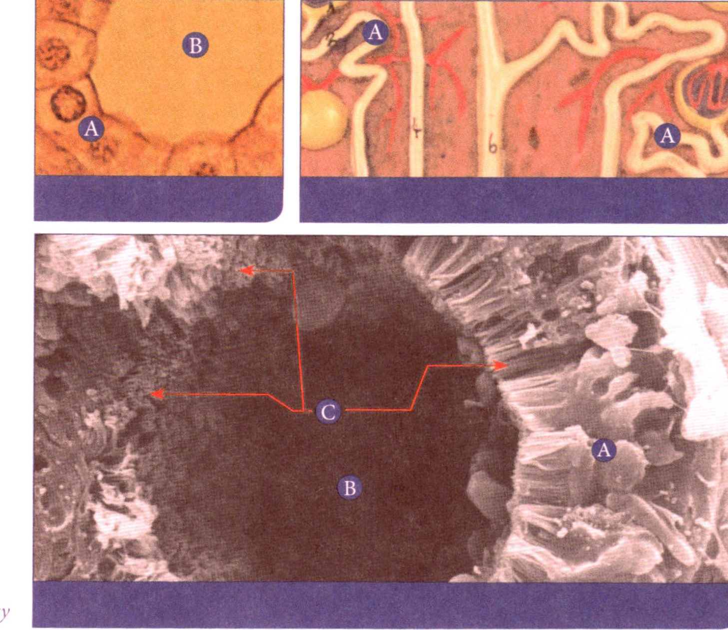

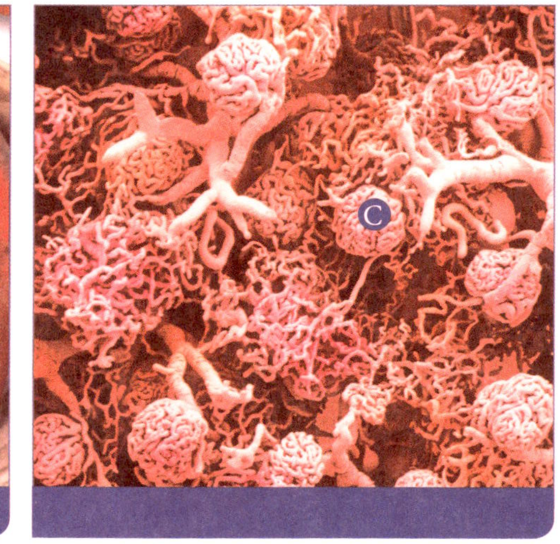

- **A** Proximal convoluted (twisted) tubule (PCT)
- **B** Lumen
- **C** Microvilli

- **A** Glomerulus — Renal corpuscle
- **B** Glomerular (Bowman's) capsule — corpuscle
- **C** Proximal convoluted tubule
- **D** Macula densa of the end of the nephron loop
- **E** Juxtaglomerular cells of the afferent arteriole
- **F** Efferent arteriole
- **G** Afferent arteriole

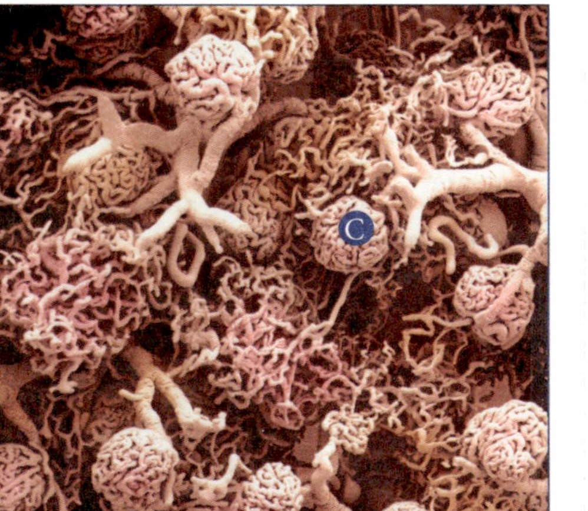

Model of the glomerulus covered with podocytes

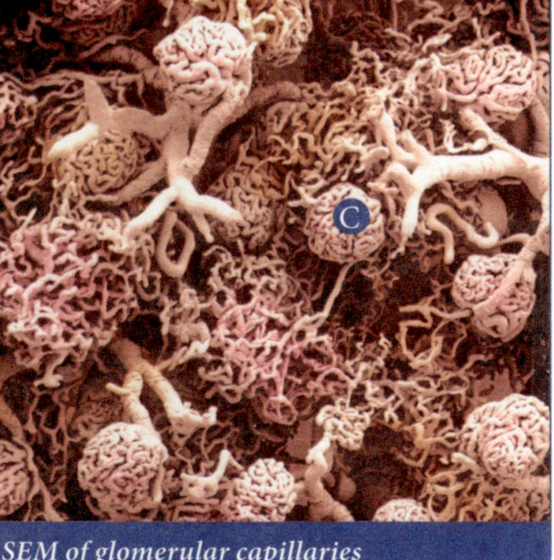

SEM of glomerular capillaries

A Podocyte nuclei **B** Filtration slits **C** Glomerulus

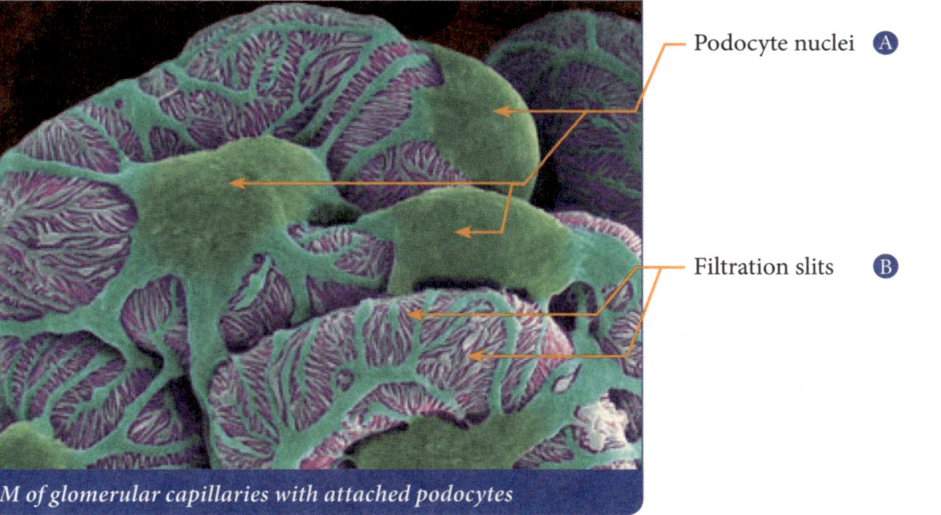

SEM of glomerular capillaries with attached podocytes

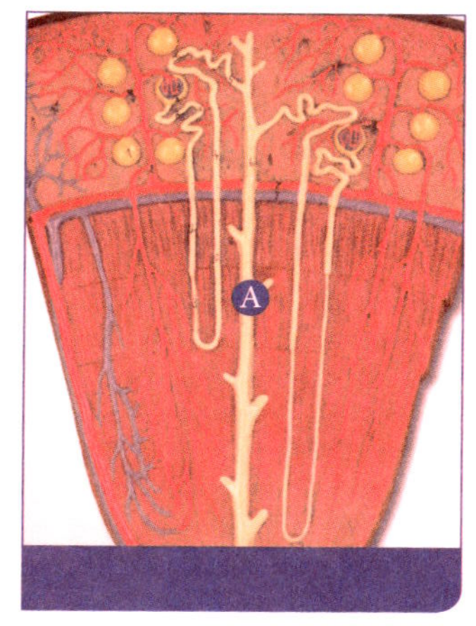
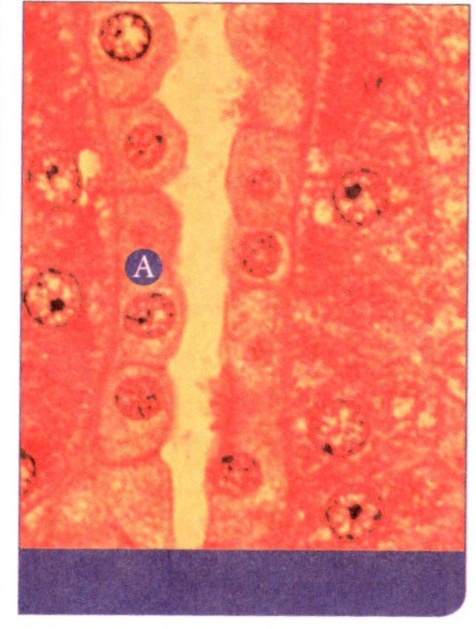
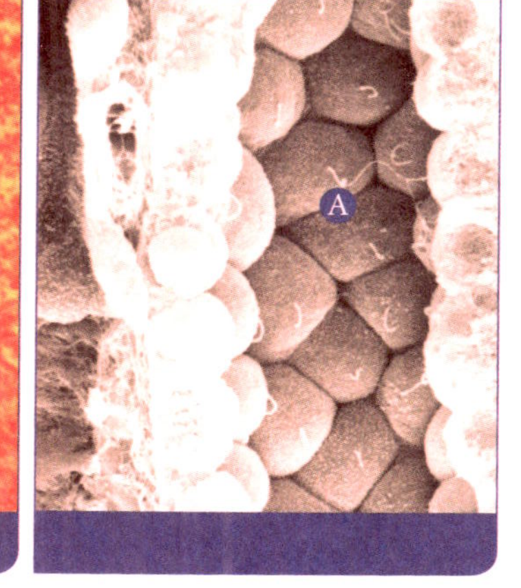

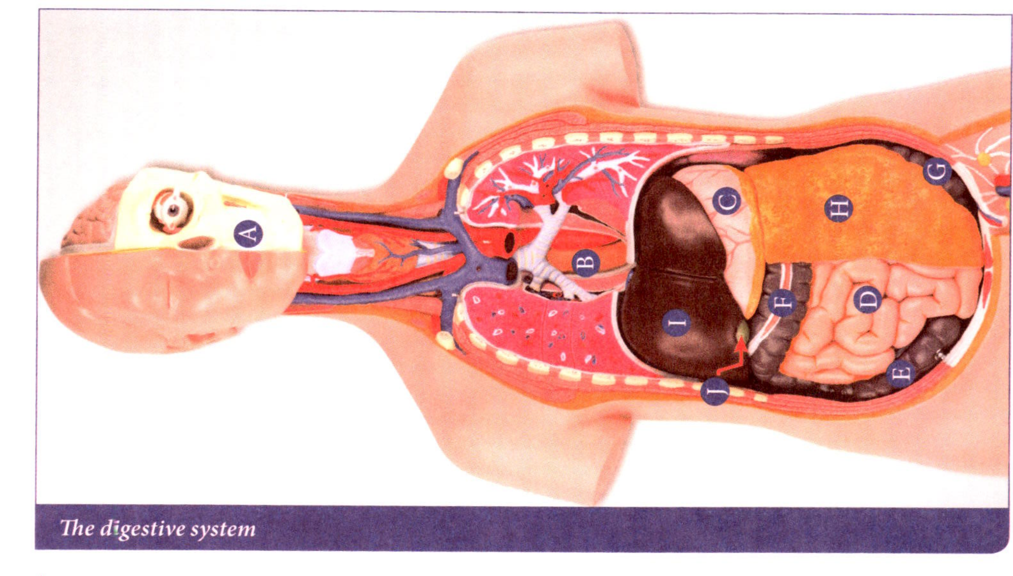
The digestive system

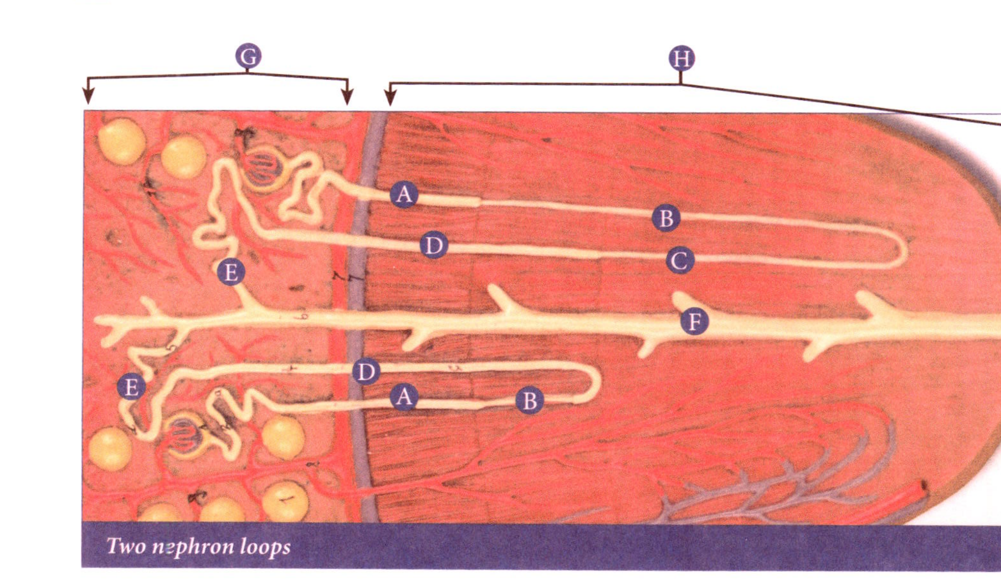

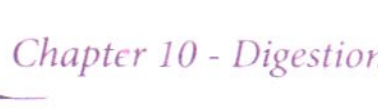
Two nephron loops

Ⓐ Teeth

Ⓑ Esophagus

Ⓒ Stomach

Ⓓ Small intestines

Ⓔ Ascending colon

Ⓕ Transverse colon

Ⓖ Descending colon

Ⓗ Greater omentum (can enlarge to a "beer belly")

Ⓘ Liver

Ⓙ Gallbladder

Ⓐ Collecting duct

Ⓐ Descending limb (thick segment)

Ⓑ Descending limb (thin segment)

Ⓒ Ascending limb (thin segment)

Ⓓ Ascending limb (thick segment)

Nephron loop (loop of Henle)

Ⓔ Distal convoluted tubule

Ⓕ Collecting duct

Ⓖ Cortex

Ⓗ Medulla

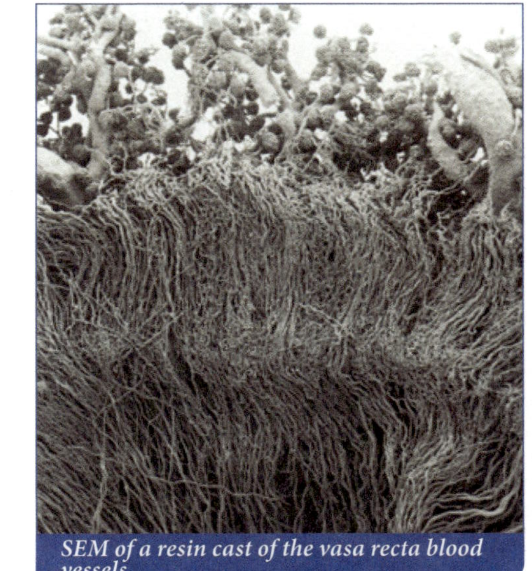

SEM of a resin cast of the vasa recta blood vessels

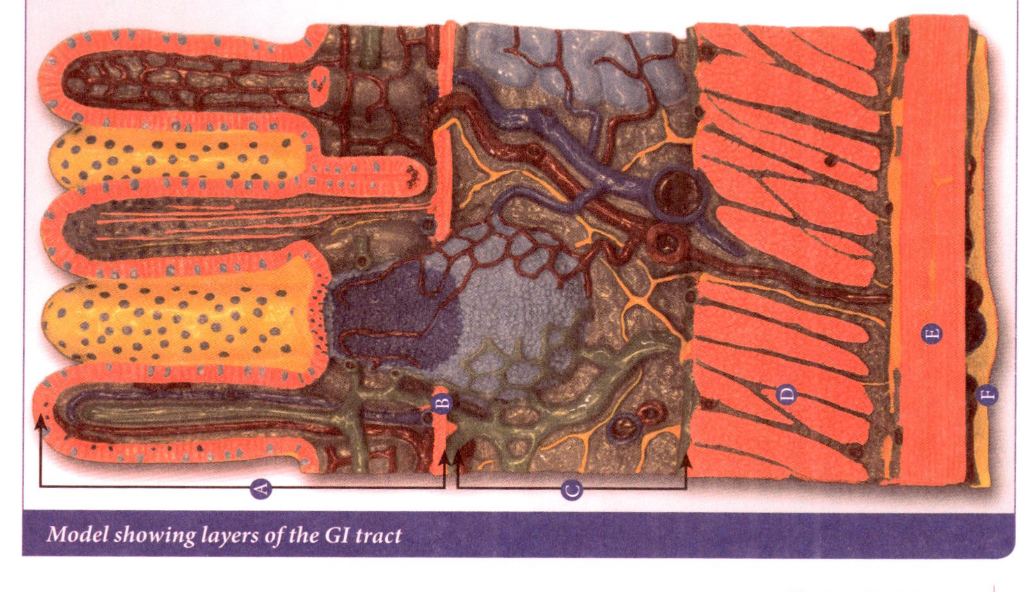

Model showing layers of the GI tract

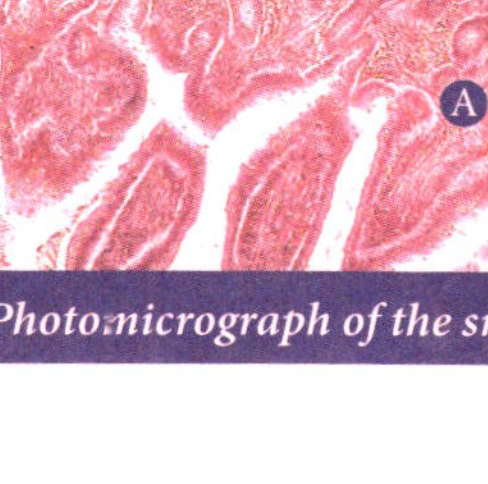

Photomicrograph of the small intestine

The digestive tract layers

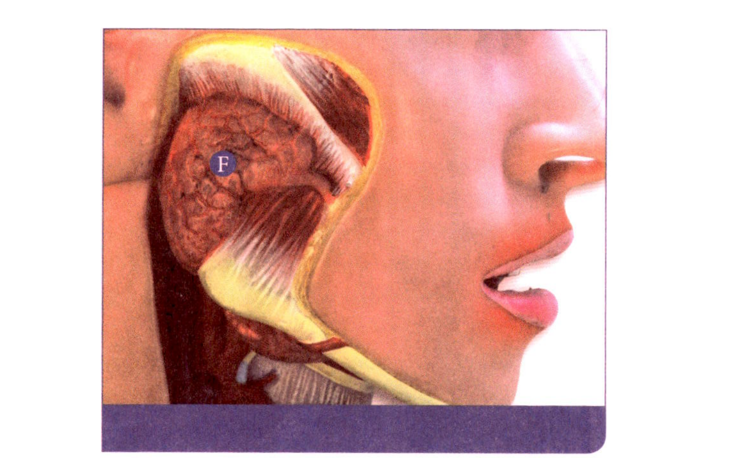

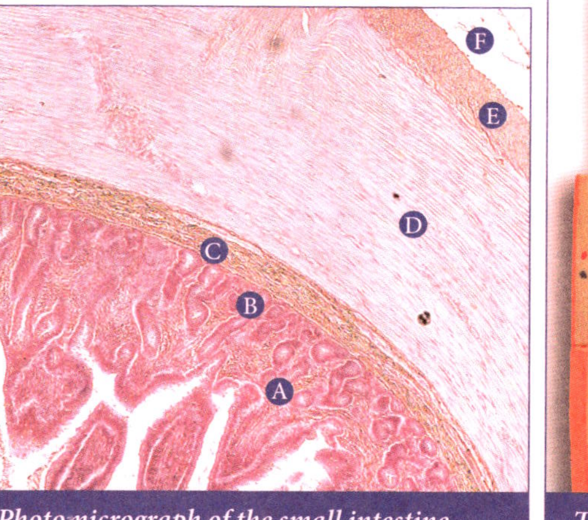

Model of the stages of tooth development

Models of five types of teeth

(A) Mucosa

(B) Muscularis mucosae

(C) Submucosa

(D) Circular muscularis externa

(E) Longitudinal muscularis externa

(F) Serosa

(A) Mucosa

(B) Muscularis mucosae

(C) Submucosa

(D) Circular muscularis externa

(E) Longitudinal muscularis externa

(F) Serosa

(A) Incisor(s) – 8, 4 top and 4 bottom center of smile

(B) Canine – 4, 2 top and 2 bottom, made for ripping flesh

(C) Premolar(s) – 8, 4 top and 4 bottom

(D) Molar(s) – 8, 4 top and 4 bottom

(E) Wisdom teeth (3rd molars) – 4, 2 top and 2 bottom, sometimes do
not descend or have to be pulled to make room

(F) Parotid salivary gland – produces the majority of saliva, which
helps moisten a bolus through the esophagus

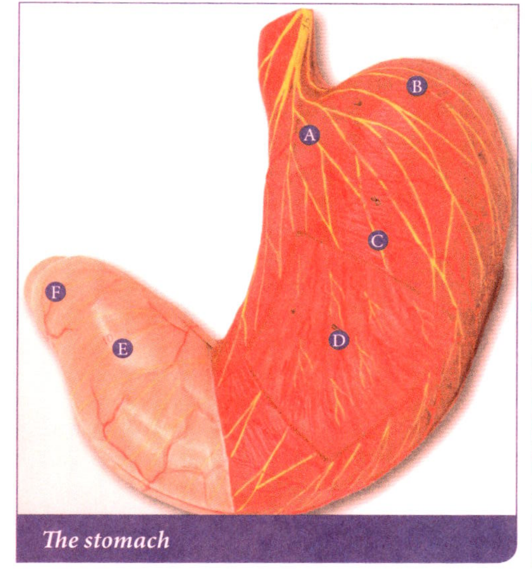

The stomach

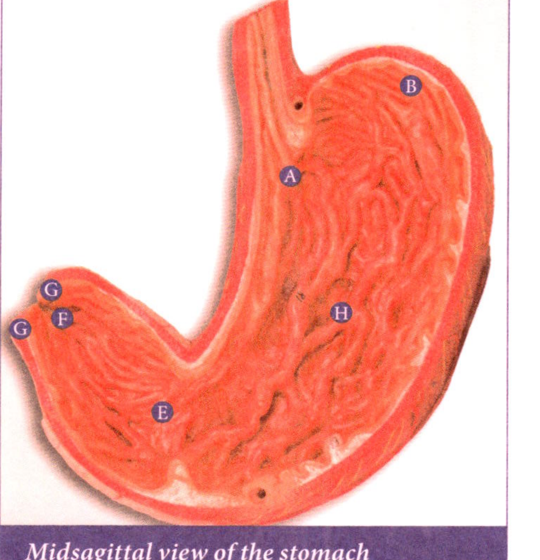

Midsagittal view of the stomach

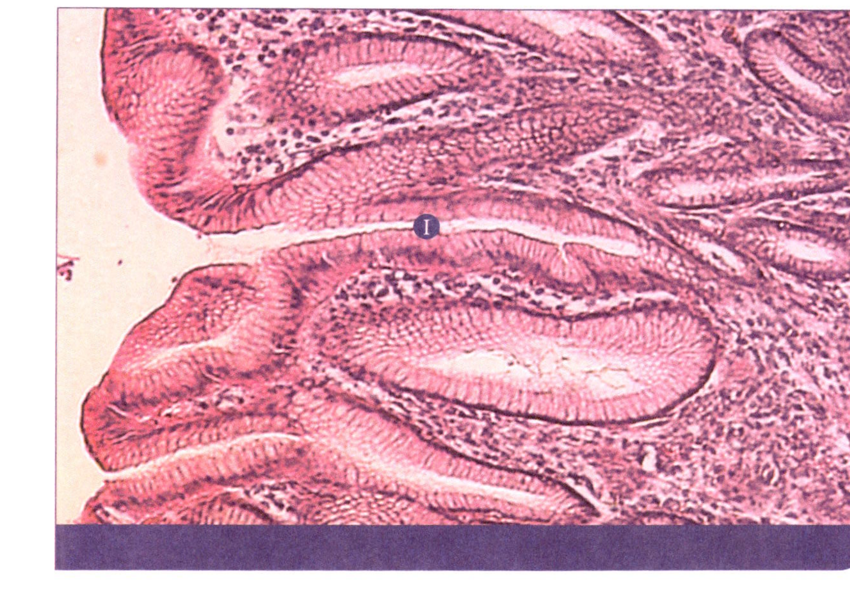

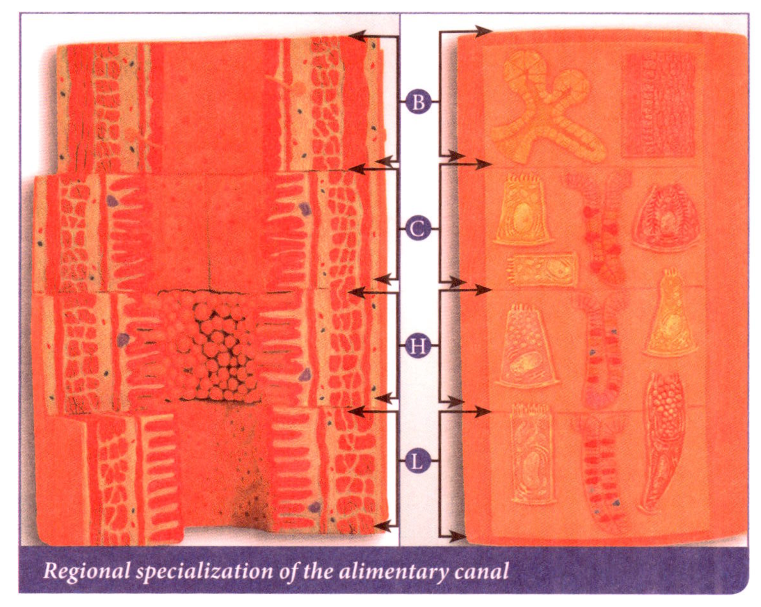

Regional specialization of the alimentary canal

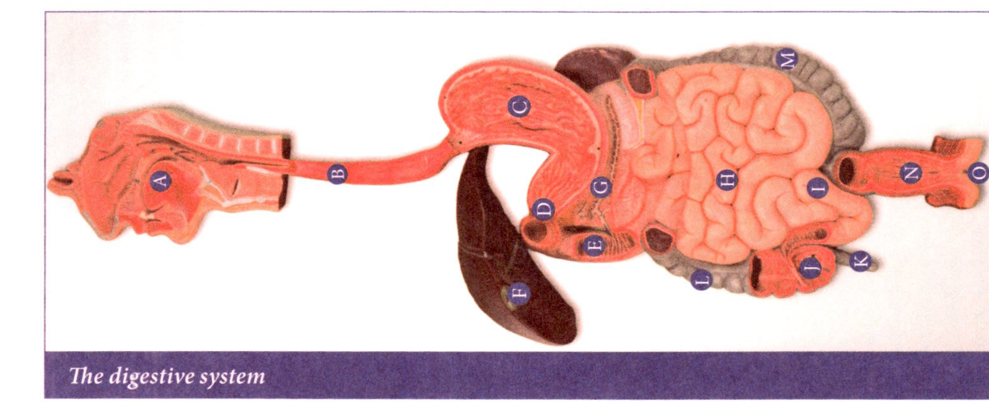

The digestive system

I Gastric pit

A Cardiac region
B Fundic region
C Circular muscle layer
D Oblique muscle layer
E Antrum
F Pylorus
G Pyloric sphincter
H Gastric rugae

A Tongue
B Esophagus
C Stomach
D Pyloric sphincter
E Duodenum
F Gall bladder
G Pancreas
H Jejunum
I Ileum
J Cecum
K Appendix
L Ascending colon
M Descending colon
N Rectum
O Anus

B Esophagus
C Stomach
H Jejunum
L Ascending colon

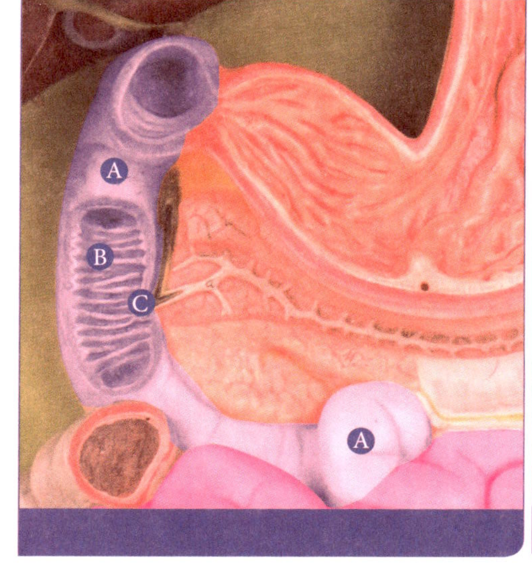

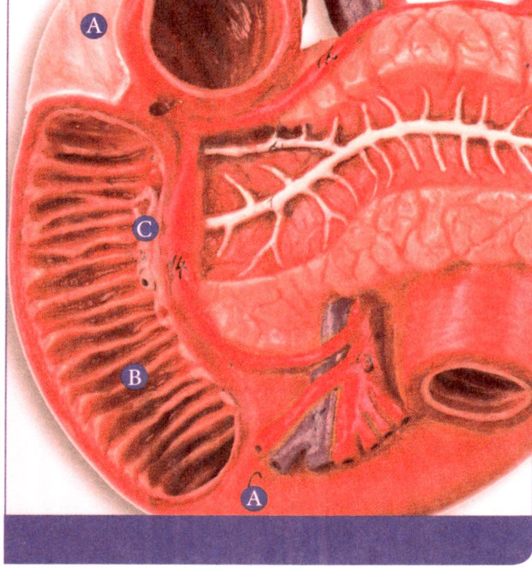

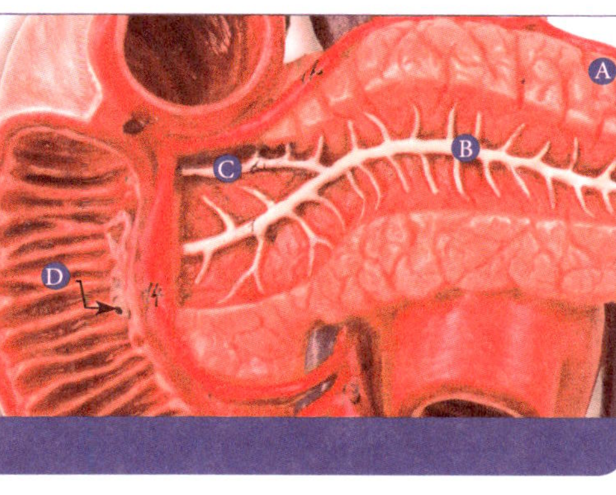

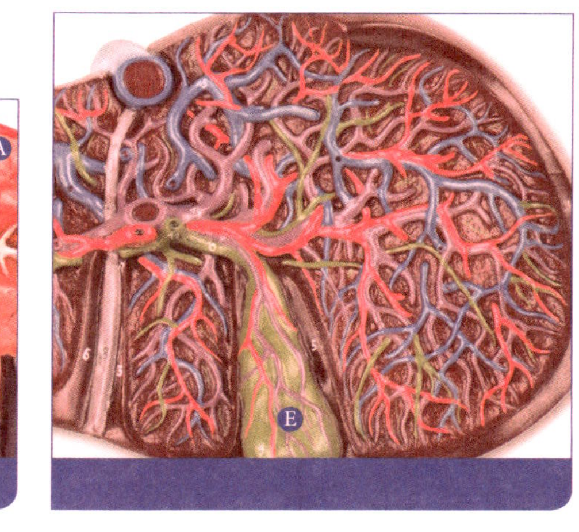

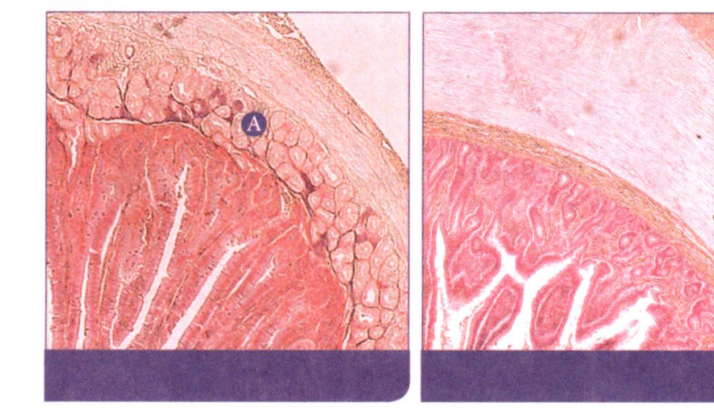

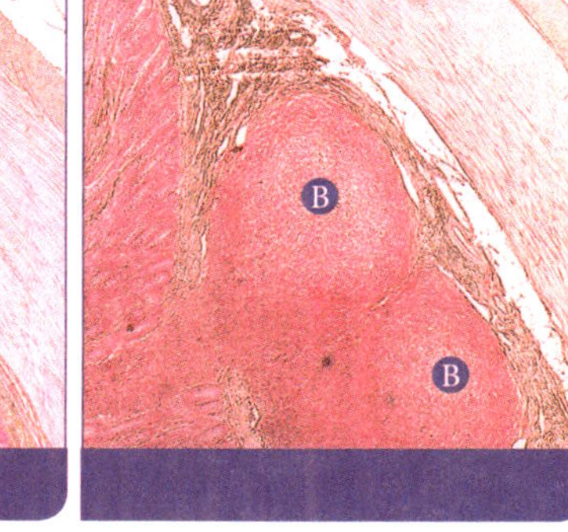

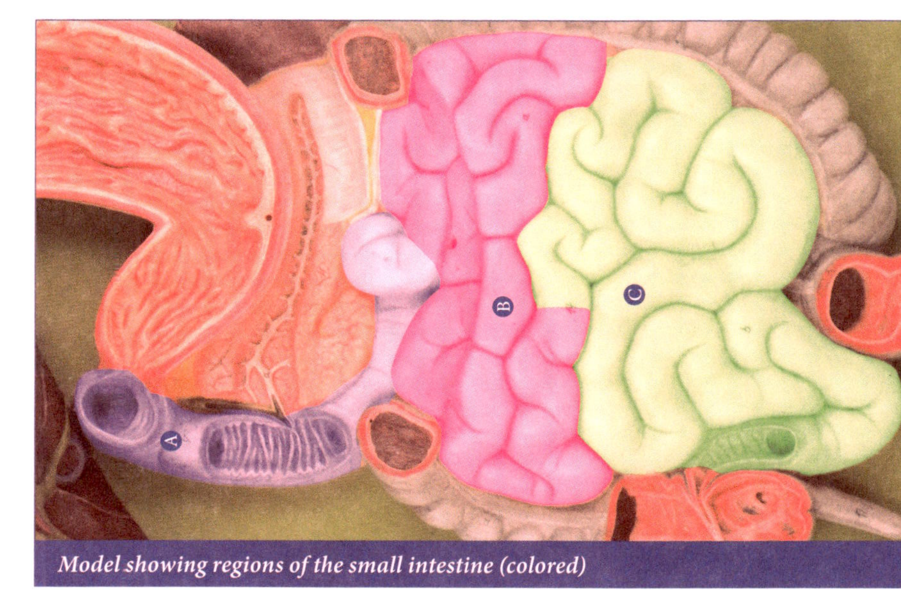

Model showing regions of the small intestine (colored)

A Pancreas
B Pancreatic duct
C Accessory pancreatic duct
D Major duodenal papillae
E Gallbladder

A Duodenum
B Plicae circulares
C Major duodenal papillae

A Duodenum
B Jejunum
C Ileum

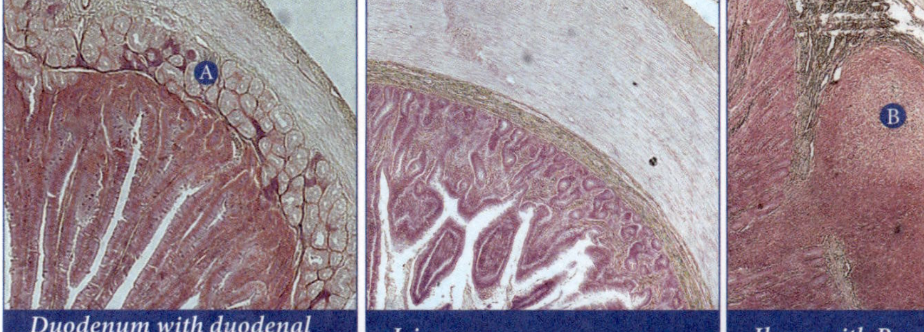

Duodenum with duodenal glands (A)

Jejunum

Ileum with Peyer's patches (B)

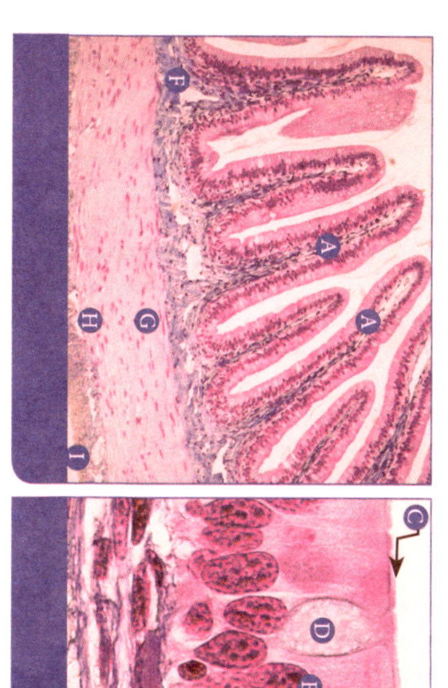

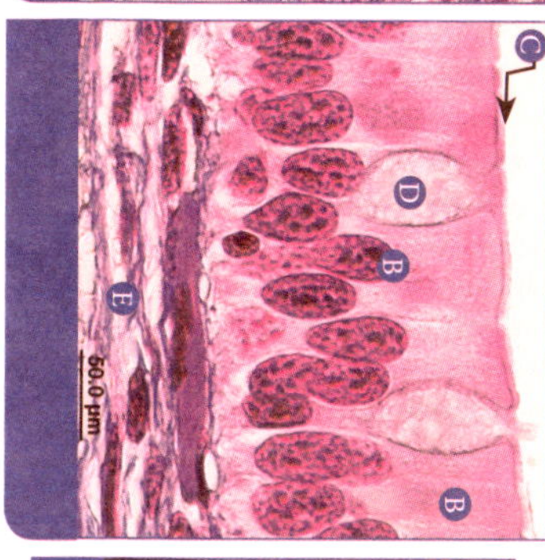

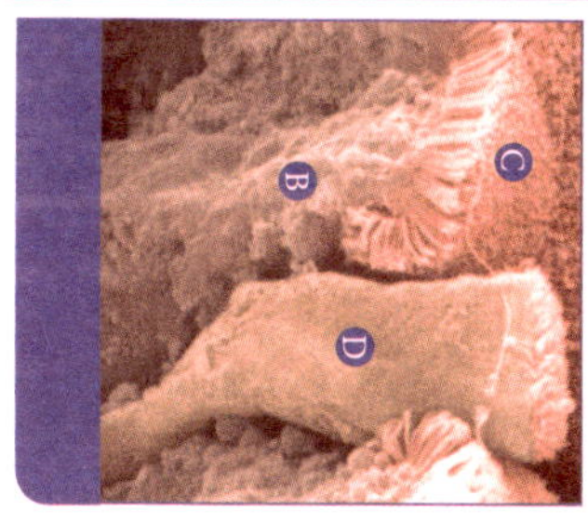

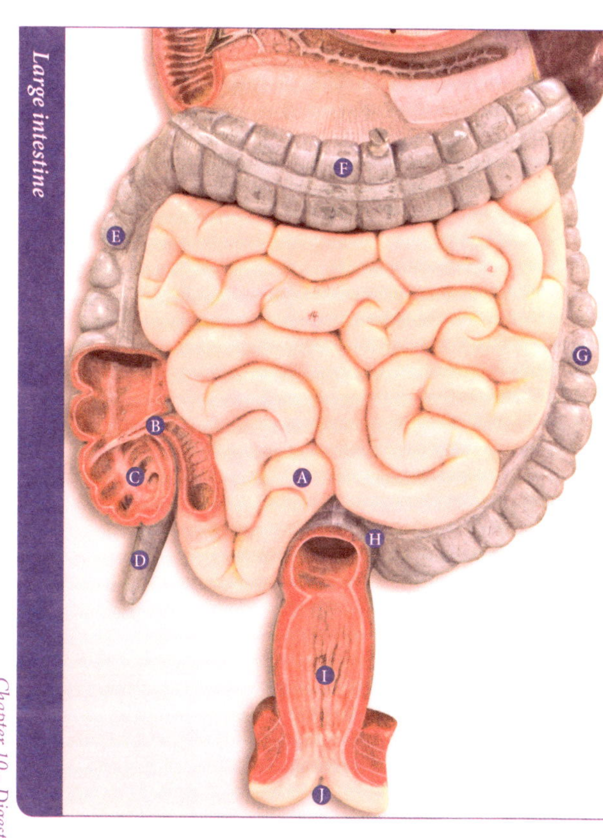

Large intestine

Chapter 10 - Digestion

Chapter 10 - Digestion

Chapter 11 - Reproduction

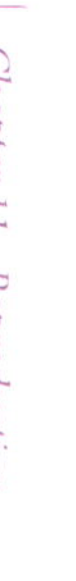

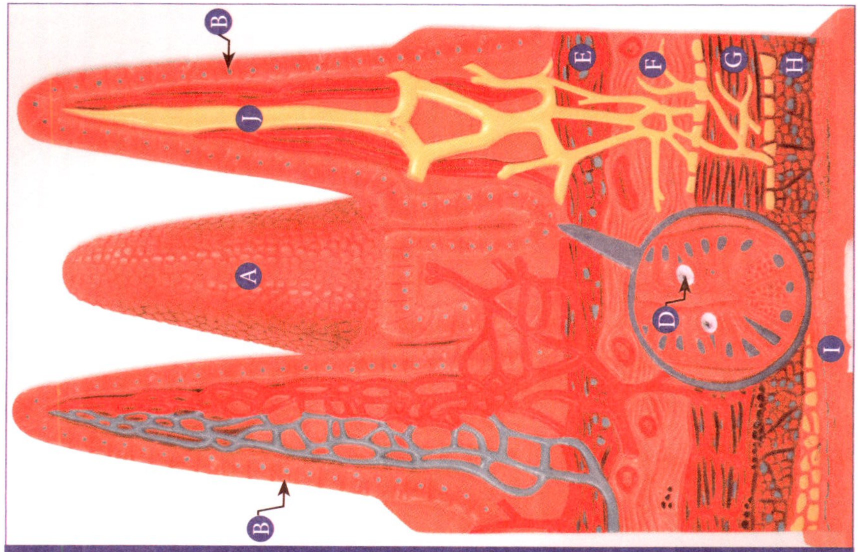

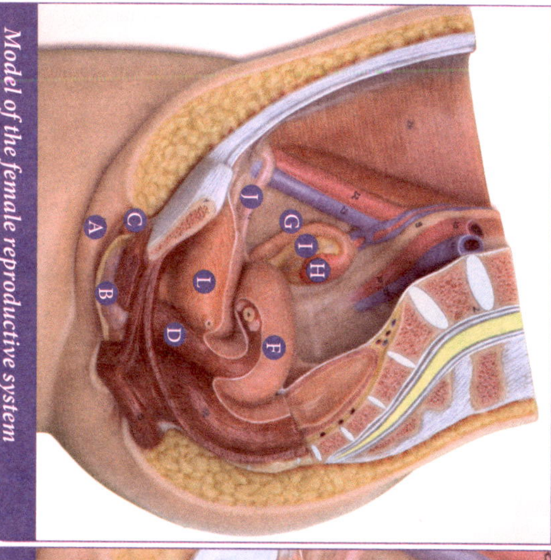

Model of the female reproductive system

Midsagittal view of the female reproductive system

Chapter 10 - Digestion

Model of a section of the small intestine

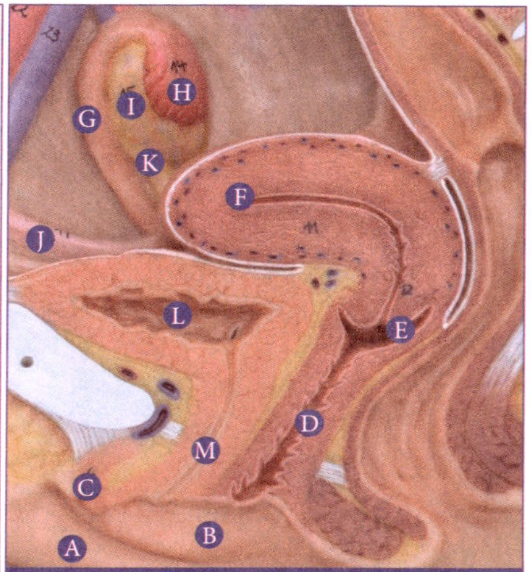

(A) Labium majus
(B) Labium minus
(C) Clitoris
(D) Vagina
(E) Cervix
(F) Uterus
(G) Uterine (Fallopian) tube
(H) Fimbrae of the infundibulum
(I) Ovary
(J) Round ligament
(K) Ovarian ligament
(L) Urinary bladder
(M) Urethra

(A) Ileum
(B) Ileocecal valve
(C) Cecum
(D) Appendix
(E) Ascending colon
(F) Transverse colon
(G) Descending colon
(H) Sigmoid colon
(I) Rectum
(J) Anus

(A) Villus
(B) Simple columnar epithelial cells
(D) Goblet cell
(E) Muscularis mucosae
(F) Submucosa
(G) Circular muscularis externa
(H) Longitudinal muscularis externa
(I) Serosa
(J) Central lacteal

Features of the small intestine

(A) Villi
(B) Simple columnar epithelial cells
(C) Microvilli
(D) Goblet cell
(E) Muscularis mucosae
(F) Submucosa
(G) Circular muscularis externa
(H) Longitudinal muscularis externa
(I) Serosa

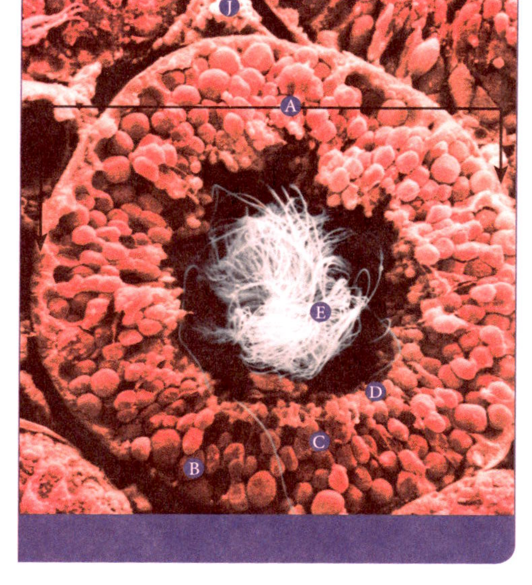

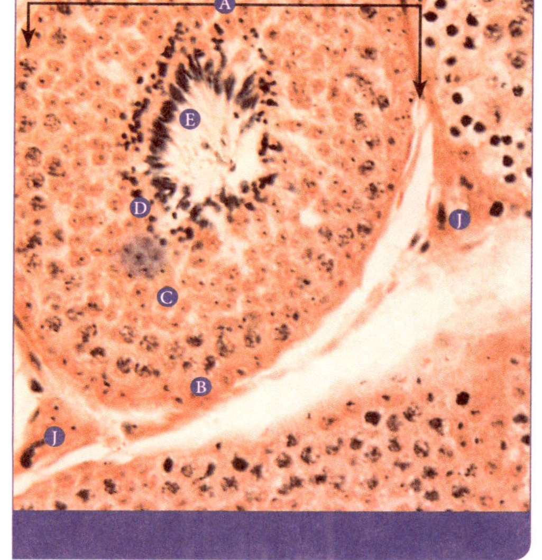

Chapter 11 - Reproduction

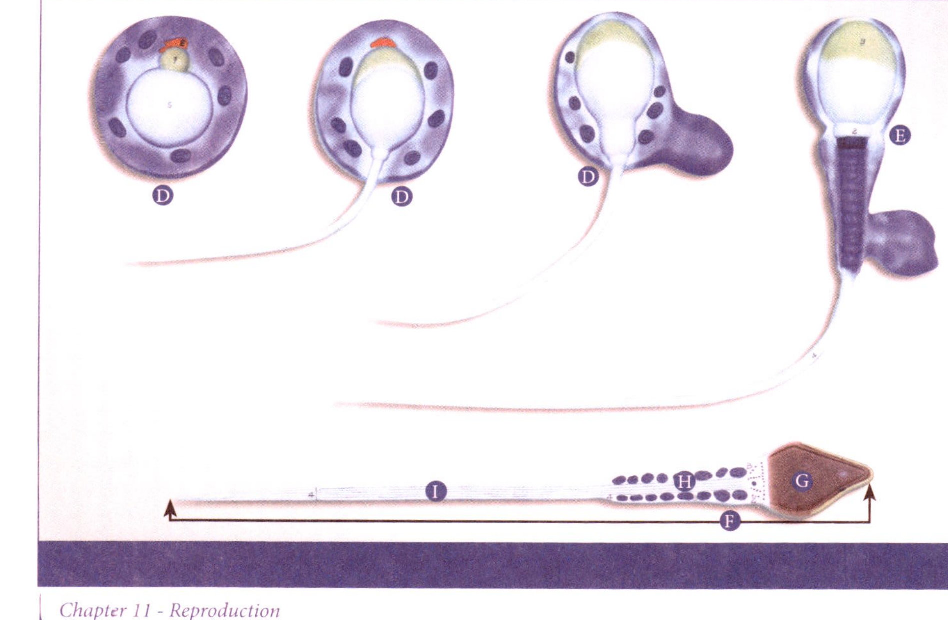

Chapter 11 - Reproduction

Chapter 11 - Reproduction

Chapter 11 - Reproduction

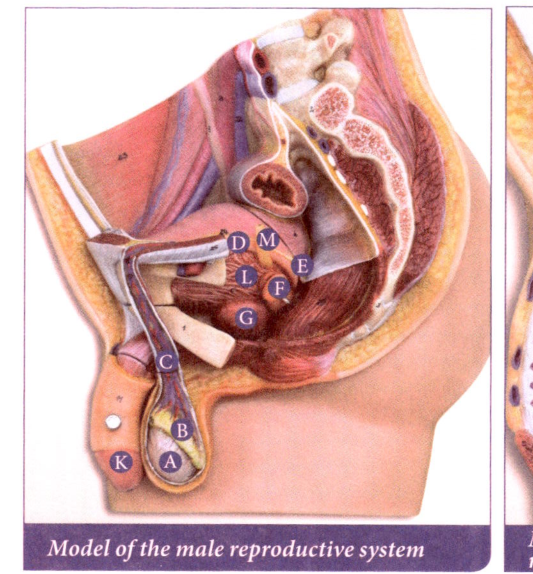

Model of the male reproductive system

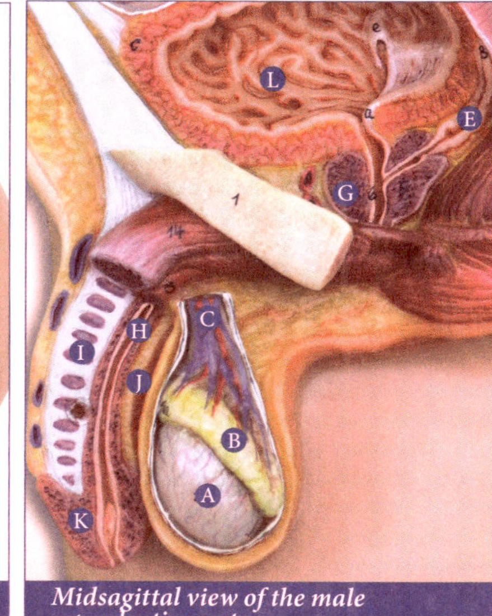

Midsagittal view of the male reproductive system

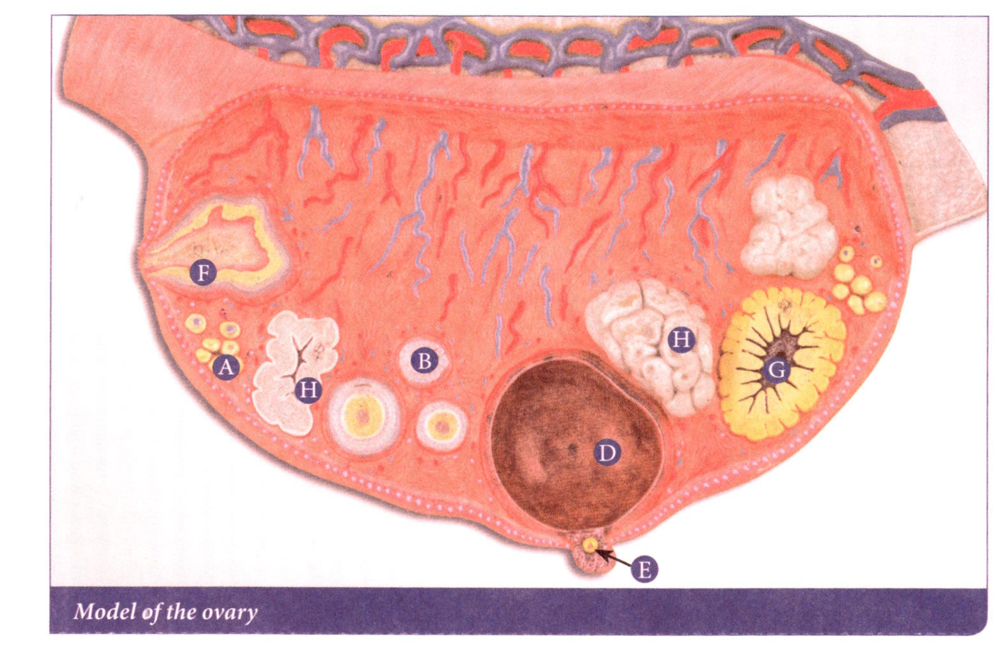

Model of the ovary

D Spermatids

E Immature spermatozoa

F Spermatozoon

G Head

H Midpiece

I Tail

A Seminiferous tubule

B Spermatogonia

C Spermatocytes

D Spermatids

E Immature spermatozoa

J Interstitial (Leydig) cells

A Primordial follicles

B Primary follicle

D Rupturing follicle releasing its oocyte (egg)

E Oocyte

F Post ovulatory follicle

G Corpus luteum

H Corpus albicans

A Testis

B Epididymis

C Pampiniform plexus

D Ductus deferens

E Ampulla of ductus deferens

F Seminal vesicle

G Prostate gland

H Urethra

I Corpus cavernosum

J Corpus spongiosum

K Glans of penis

L Urinary bladder

M Ureter

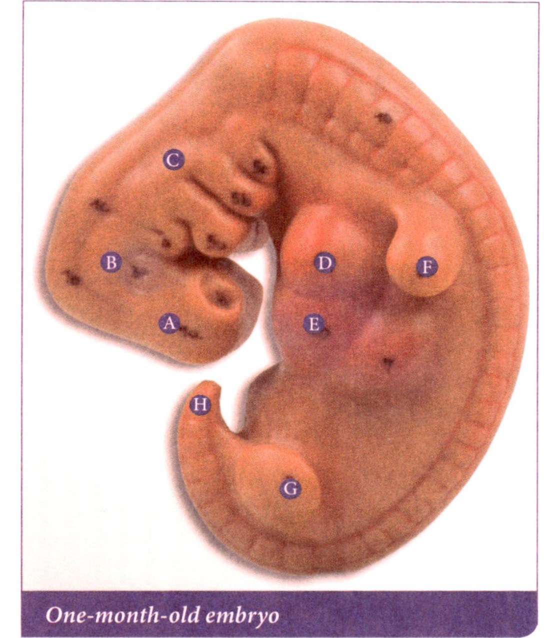

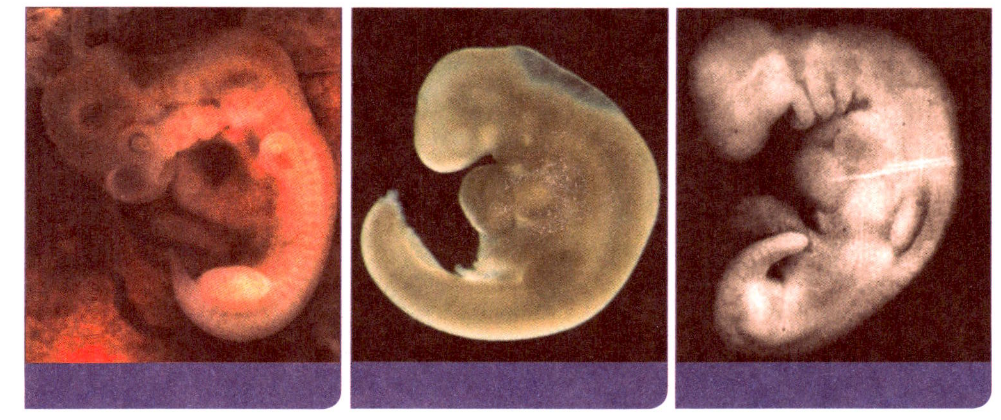

Chapter 12 - Development

Chapter 12 - Development

Chapter 12 - Development

Chapter 12 - Development

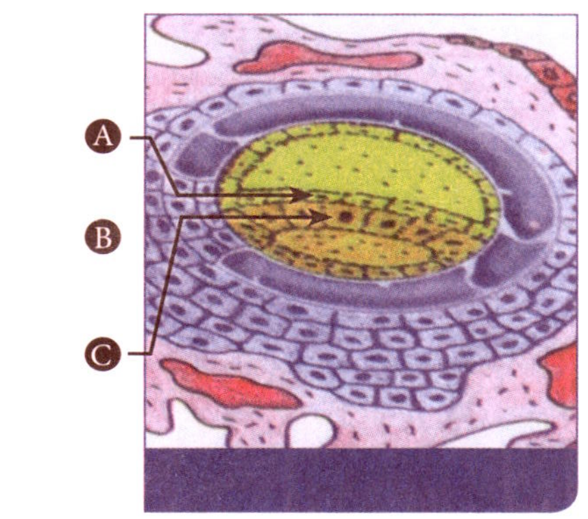

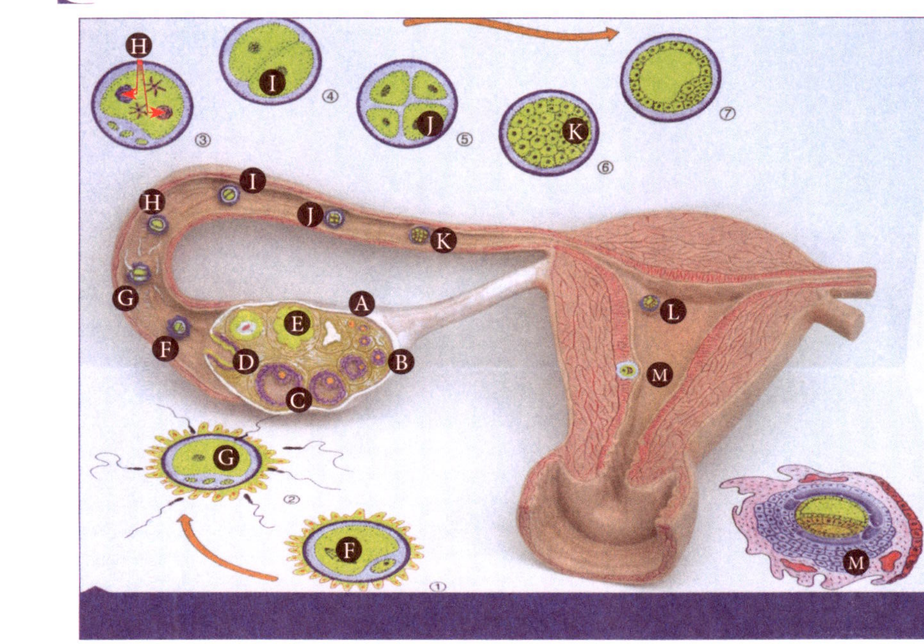

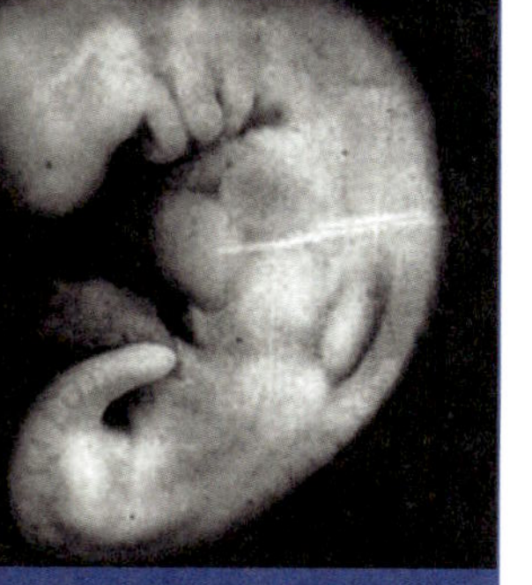
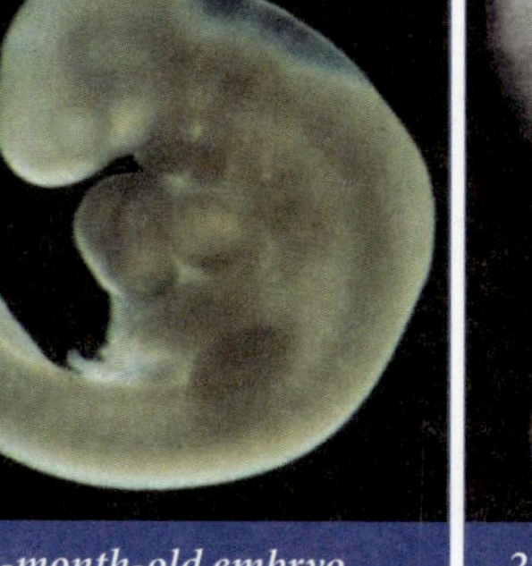
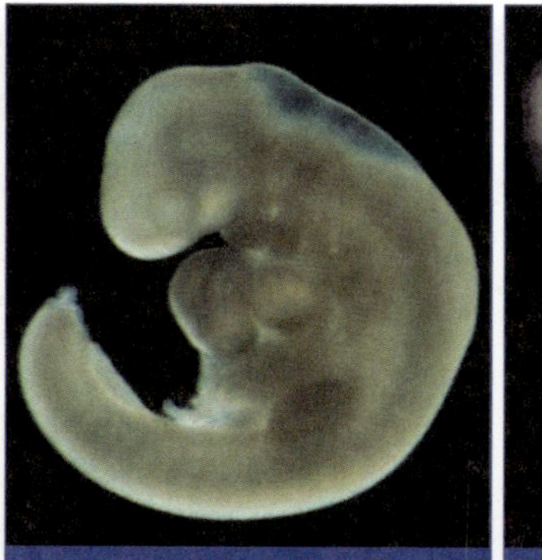

One-month-old embryo

One-month-old embryo

28-day-old embryo

(A) Head
(B) Primitive lens
(C) Pharyngeal arches
(D) Heart bulge

(E) Liver bulge
(F) Arm bud
(G) Leg bud
(H) Tail

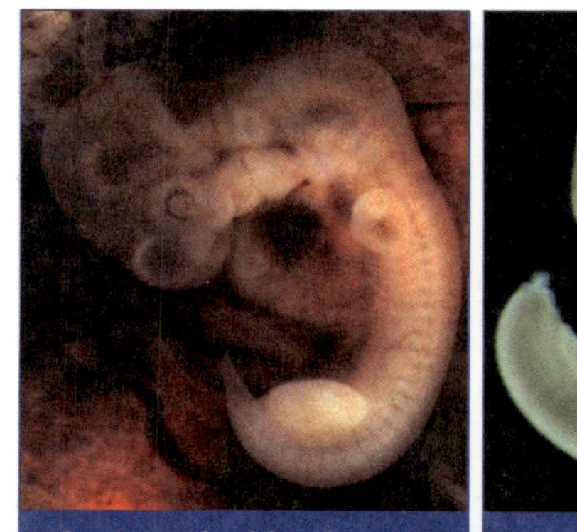

(A) Primordial follicles
(B) Primary follicle
(C) Secondary follicles
(D) Rupturing follicle
(E) Corpus luteum
(F) Oocyte
(G) Fertilization (union of sperm and egg)

(H) Sperm and egg nuclei swell and become pronuclei (arrows) then join, forming the zygote
(I) 2-cell stage
(J) 4-cell stage
(K) Morula
(L) Blastocyst
(M) Implanted embryo (embryoblast)

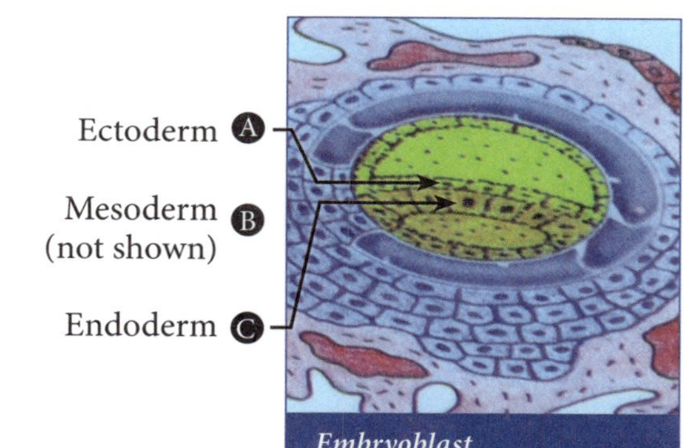

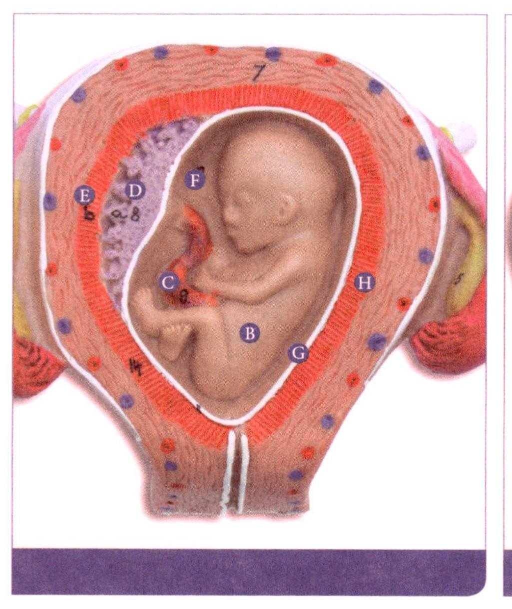
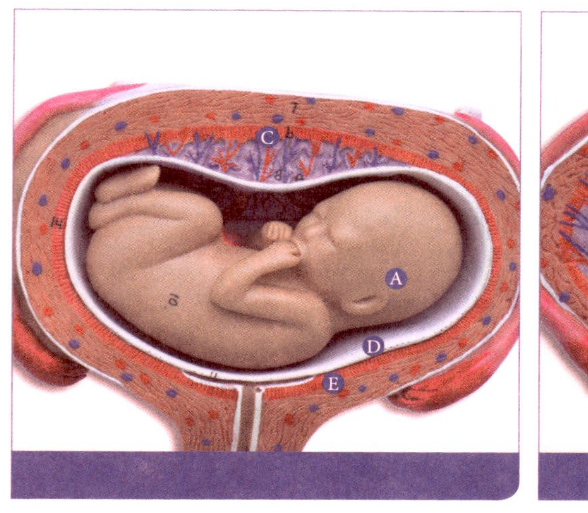

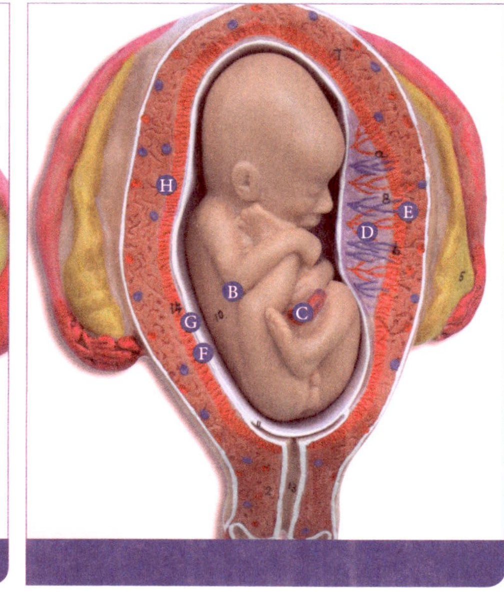
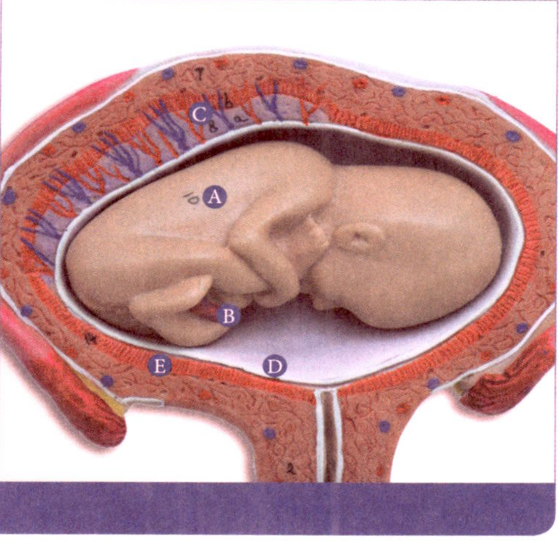

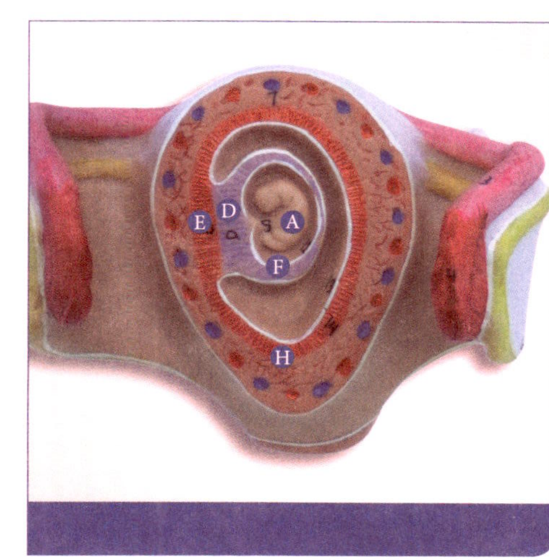
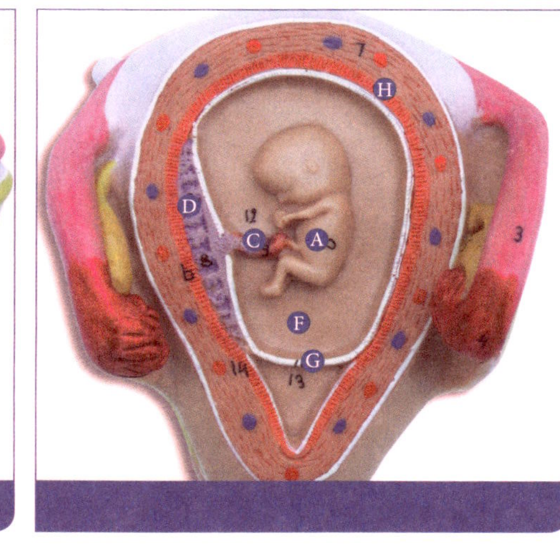
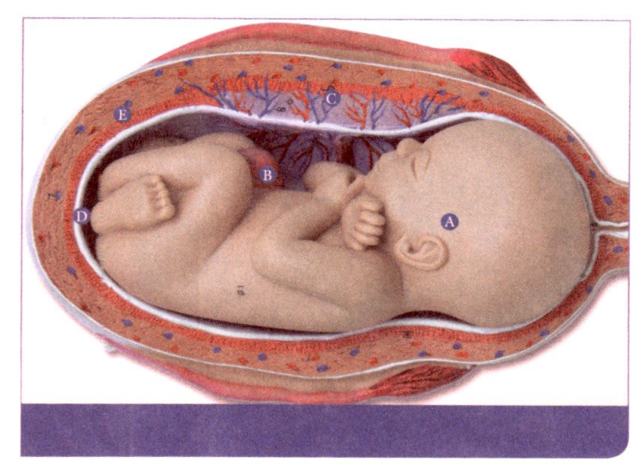
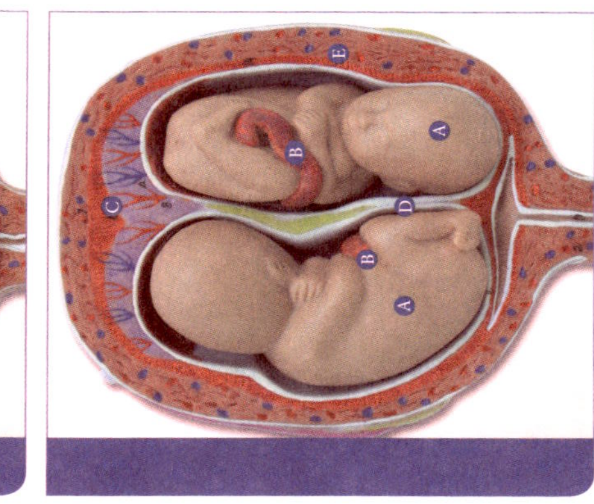

(A) Fetus
(B) Umbilical cord
(C) Placenta (chorion and endometrium)
(D) Amnion and chorion
(E) Decidua parietalis of endometrium

(A) Fetus
(B) Umbilical cord
(C) Placenta (chorion and endometrium)
(D) Amnion and chorion
(E) Decidua parietalis of endometrium

(A) Embryo
(C) Umbilical cord
(D) Child's portion of the placenta (chorion)
(E) Mother's portion of the placenta (endometrium)
(F) Amnion
(G) Chorion
(H) Decidua parietalis of endometrium

(B) Fetus
(C) Umbilical cord
(D) Child's portion of the placenta (chorion)
(E) Mother's portion of the placenta (endometrium)
(F) Amnion
(G) Chorion
(H) Decidua parietalis of endometrium